Proceedings of the

Dalgarno Celebratory Symposium

Contributions to Atomic, Molecular, and Optical Physics, Astrophysics, and Atmospheric Physics

Proceedings of the

Dalgarno Celebratory Symposium

Contributions to Atomic, Molecular, and Optical Physics, Astrophysics, and Atmospheric Physics

Cambridge, Massachusetts 10 – 12 September 2008

Editors

James F Babb

Kate Kirby

Hossein Sadeghpour

ITAMP, Harvard-Smithsonian Center for Astrophysics, USA

ICP

Imperial College Press

Published by

Imperial College Press
57 Shelton Street
Covent Garden
London WC2H 9HE

Distributed by

World Scientific Publishing Co. Pte. Ltd.
5 Toh Tuck Link, Singapore 596224
USA office: 27 Warren Street, Suite 401-402, Hackensack, NJ 07601
UK office: 57 Shelton Street, Covent Garden, London WC2H 9HE

British Library Cataloguing-in-Publication Data
A catalogue record for this book is available from the British Library.

PROCEEDINGS OF THE DALGARNO CELEBRATORY SYMPOSIUM
Contributions to Atomic, Molecular, and Optical Physics, Astrophysics, and Atmospheric Physics

ISBN-13 978-1-84816-469-7
ISBN-10 1-84816-469-6

Printed by FuIsland Offset Printing (S) Pte Ltd, Singapore

PREFACE

On September 10, 2008, more than 125 friends, colleagues, and former students and postdocs gathered in Cambridge to join Professor Alex Dalgarno in celebrating his 80th birthday (which actually occurred on January 5, 2008). A symposium* highlighting Dalgarno's many scientific contributions was sponsored by the Institute for Theoretical Atomic, Molecular and Optical Physics (ITAMP) and held in Radcliffe Yard, September 10–12, and it included several receptions, a conference banquet, and closed with an evening concert in Cronkhite Courtyard.

Alex began his scientific career with a publication in the Proceedings of the Royal Society of London (1952). He has made seminal contributions in theoretical atomic and molecular physics, physics of planetary atmospheres and comets, molecular astrophysics, and astrochemistry. He has published nearly 750 peer-reviewed papers, conference proceedings and book chapters, including 7 papers in 2008. His average annual publication rate currently stands at 13. His most cited paper (*The Theory of Scattering by a Rigid Rotator*) published in 1960 with Arthurs has been referenced more than 935 times. He has five papers each with more than 500 citations. His h-index at 87 surely places him near the top of anyone's ranking, should he decide to leave Harvard for academic positions elsewhere.

Alex is clearly unique and his contributions have been wide-ranging, insightful, and influential.

The organizers of the Symposium—the editors of this book—intended the Dalgarno Celebratory Symposium to feature a range of topics representative of Alex's on-going scientific legacy and to be focused on his contributions over the last 20–25 years. The scientific program was divided into seven sessions.

The session on Calculations of Atomic and Molecular Properties was chaired by Ray Flannery. It reflected Alex's long-held belief that "Astrophysics is almost entirely applied Atomic, Molecular, and Optical Physics", and that to understand atomic and molecular reactions in astrophysical environments, one must first obtain an accurate picture of such processes through detailed and insightful calculations. Gordon Drake, in his review,

*http://www.cfa.harvard.edu/itamp/DalgarnoSymposium.html

describes his work with Alex and later at the University of Windsor on the development of highly-accurate pseudospectral theory for the dipole response of atoms and molecules. Bernard Zygelman outlines the extension of Alex's successful spin exchange theory, based solely on the elastic approximation, to the collision of two spin-polarized hydrogen atoms, and its relation to the population of the hydrogen hyperfine levels, important in understanding emission or absorption of radiation in the 21 cm line. Michael Jamieson describes the current implementations of numerical techniques for the celebrated sum-over-all-state Dalgarno-Lewis method. Xi Chu supplies even more evidence for Alex's dictum, "the key to successful theoretical work is always to have more than one way of doing the same thing." She shows how to use the workhorse of theoretical chemistry—density functional theory (DFT)—to calculate the optical response of atoms and molecules.

The session on Astrochemistry was chaired by Bill Klemperer. The topic which Alex famously posed as a question in 1986, "Is Astrochemistry Useful?" was addressed by the four speakers in this session, each of whom offered an answer to that question. John Black pointed out that even in the Early Universe epoch of recombination, chemistry played a significant role, a point further emphasized by Phillip Stancil during his presentation of the H^- chemistry. Ewine van Dishoeck noted the myriad molecules which have thus far been detected and are archived on the web†. The review by van Dishoeck of the historical context and of Alex's seminal contributions to this interdisciplinary field is illuminating. Stephen Lepp writes about his experience working with Alex on the discovery of the molecular ions H_3^+ and HeH^+ in the spectra of the supernova remnant 1987A.

The session on the Interstellar Medium and Molecular Astrophysics resonated with the session on Astrochemistry. It was chaired by Pat Thaddeus. Tom Hartquist, in his review touches upon Alex's vast fundamental knowledge fueling his long term vision to construct detailed, robust models which can be applied to astrophysical environments. Reactions involving atomic deuterium allow for placing constraints on deuterium fractional ionization in dark molecular clouds, which in turn influences the role magnetic fields play in star formation in the dark clouds. The review describes how Alex recognized that HD was a much more effective coolant—due to its permanent dipole moment and its larger mass—than H_2, and the cooling rate increased with decreasing temperature. Amiel Sternberg describes Alex's insight in incorporating detailed molecular interactions (bond formation,

---

†http://www.astrochemistry.org

bond destruction, and bond rearrangement) into quantitative interstellar cloud models. David Neufeld discusses the range of fine structure emissions from molecular ions, neutral molecule vibrational transitions, and rotational transitions of metal hydrides, all of which result from shock wave propagation in interstellar gases, and observation of interstellar hydrides using the Infrared Spectrograph on board the Spitzer Space Telescope. Cesare Cecchi-Pestellini—who unfortunately was not able to attend the Symposium—postulates that local turbulence thermal spikes may be responsible for the increased levels of $H_3{}^+$ and CH^+ in the diffuse interstellar medium.

The session on Atomic and Molecular Processes in the Solar System and Planetary Atmospheres, chaired by Jane Fox, was yet another example of the enormous breadth of Alex's research interests. Tom Cravens chronicles the many important contributions that Alex made to understanding the ionization, airglow, recombination and chemistry of planetary and cometary atmospheres. One important atomic process which regulates the radiation and heating in ionospheric plasmas is charge exchange between neutral species of interstellar or atmospheric origin and fast solar wind ions. This process, accompanied by emission of EUV or soft x-ray photons, is responsible for the x-ray radiation detected in cometary and planetary atmospheres and is described by Vasili Kharchenko in this volume. Peng Zhang describes his work with Alex on the thermalization path taken by hot atoms.

The session on Ultracold Collisions/Long-Range Interactions was chaired by Françoise Masnou. This session reflected Alex's more recent interest in cold collisions, but many of the collisional processes in ultracold physics have relied on Alex's seminal work four decades ago on the van der Waals and dispersive interactions between atoms, and between atoms and molecules. Using these ideas, important parameters for cold collisions, such as the s-wave scattering length, can be obtained from precise measurements of molecular transitions in ultracold regimes. Such processes are described in Randy Hulet's review of the field of photoassociation of ultracold atoms. Robin Côté's description of the formation of ultracold molecules by photoassociating two atoms at very large distances relies heavily on understanding the long-range atomic interactions and scattering information. Roman Krems writes about the development of numerical molecular basis representations for collisions involving molecules at cold temperatures and in the presence of external electrostatic fields. Very soon after the first atoms were cooled and trapped Alex began exploring the topic of cold molecular collisions and relaxation. He also has pioneered the study of reactions at ultracold temperatures, thus establishing the field of ultracold chemistry which Balakrishnan describes in detail.

Alex was available not only to his graduate students, postdocs or more senior fellows, but also to undergraduate students. Several of those who carried out undergraduate research projects with Alex and are now in graduate programs or in faculty positions described their work and their connections with Alex. The session was chaired by James Babb. In his contribution he discusses Alex's papers on the theory of the hyperfine structure of H_2^+ and some related studies carried out by several undergraduates working with Alex. Jean Turner, who studied under Alex as a junior and senior at Harvard (and now is a Professor of Astronomy at UCLA), gives an appreciation and discusses the frontiers of extragalactic astronomy. Ron Pepino was a visitor from the University of Connecticut and worked with Alex and Vasili Kharchenko for a few formative years starting with a summer visit. Now a graduate student in physics at University of Colorado-Boulder, he introduces the nascent field of 'atomtronics.' Jake Taylor started a project on the electronic structure of H_2^+ while a freshman at Harvard under Alex's guidance. After completing the project, and graduating, he went on to do his Ph.D. at Harvard with Mikhail Lukin. Presently a NIST Fellow, Jake discusses the theory and practice of magnetometry using diamond nitrogen vacancy centers and aspects of simulating quantum mechanical systems using optical lattices.

Alex was instrumental in bringing the Institute for Theoretical Atomic, Molecular, and Optical Physics to the Harvard-Smithsonian as an NSF-funded center. In his absence, ITAMP would not be where it is today. Many of the key players of the early and current days of ITAMP (Richard Pratt and Barry Schneider from the NSF and Kate Kirby from the CfA) offer historical perspectives on ITAMP and Alex's exceptional leadership in establishing this recognized institute which continues to thrive more than 20 years later.

Alex Dalgarno remains a towering figure in atomic and molecular physics and in astrophysics. The editors of this special edition focusing on his scientific legacy consider it an honor to be tasked with this historical endeavor and are grateful to Alex for his continued support and intellectual energy.

The organizers of this symposium are grateful for the support, including financial support, offered by the Harvard College Observatory and the Astronomy Department, in particular to Ms. Maureen Connors and Prof. Charles Alcock. We also thank Marion Shore for editing the concert program and several of the contributed papers. The local organizer, and the person most responsible for ensuring the success of this symposium was Lisa Bastille, ITAMP Administrative Coordinator. The editors of this volume are indebted to her for her amazing organizational and diplomatic skills in coordinating the conference events and planning a memorable banquet at the Harvard Faculty Club.

James F. Babb
Kate Kirby
Hossein Sadeghpour

Cambridge, MA
August 2009

CONTENTS

Session I

Calculations of Atomic and Molecular Properties

First row, kneeling: Graeme Lister, David Schultz, Vasili Kharchenko, Hossein Sadeghpour. **First row, standing:** John Black, Ionel Simbotin, Evelyne Roueff, Kate Kirby, James Babb, Lisa Bastille, Michael Jamieson, Thomas Rescigno, Bernard Zygelman, Françoise Masnou-Seeuws. **Second row:** Tom Cravens, William Stwalley, Stephen Lepp, Paul Rimmer, Alfred Msezane, Donghui Quan, Amiel Sternberg, Neal Lane, Franco Gianturco, Alex Dalgarno, Deborah Watson, Ray Flannery, Jane Fox, Bidhan Saha, Peter Beiersdorfer, Richard Pratt, Barry Schneider, Ewine van Dishoeck, Bill McCurdy, Verne Jacobs, Stefano Bovino. **Third row:** Ronald Pepino, John Raymond, Derrick Crothers, Jacob Taylor, William McConkey, Jean Turner, Phillip Stancil, Jaan Lepson, Tom Hartquist, Steven Manson, Cecil Laughlin, Shih-I Chu, Guo-Xin Chen, Turgay Uzer, Robert Kurucz, Kelly Chance, Eric Heller, Walter Johnson, Daniel Vrinceanu, Roman Krems. **Fourth row:** Enrico Bodo, unknown, Charles Weatherford, Balakrishnan Naduvalath, Timur Tscherbul, Philippe Pellegrini, Robert Forrey, Gordon Drake, Brian Burrows, Maurice Cohen, Brendan McLaughlin, Tom Gorczyca, Marko Gacesa, Hyun-Kyung Chung, Xi Chu, Peng Zhang.

THE TRANSITION FROM MATHEMATICIAN TO ASTROPHYSICIST

M. R. FLANNERY

School of Physics, Georgia Institute of Technology,
Atlanta, GA 30332-0430, USA
E-mail: r.flannery@physics.gatech.edu
www.physics.gatech.edu/people/faculty/rflannery.html

Various landmarks in the evolution of Alexander Dalgarno from a gifted mathematician to becoming the acknowledged Father of Molecular Astrophysics are noted. His researches in basic atomic and molecular physics, aeronomy (the study of the upper atmosphere) and astrophysics are highlighted.

Keywords: atomic and molecular physics, aeronomy, astrophysics, Dalgarno-Lewis method, associative detachment, rotational excitation.

1. Some Distinct Landmarks

As this "Dalgarno Celebratory Symposium" in honor of Alex Dalgarno's 80th birthday continues, I would like to welcome you all to this morning's session "*Calculation of Atomic and Molecular Properties*". This title is particularly well suited to Alex's philosophy because, in a recent reminiscence[1] of his career, he mentions that, "... It is often said, by theorists, that physics is embodied in its equations, but I think it is to be found in the solutions to the equations". And so, emboldened/accelerated by this realization, Alex embarked on making landmark advances in all of the following subjects:

(1) Atomic and Molecular Structure (perturbation variational and expansion methods)
(2) Interactions (polarization, dispersion, model, pseudo and long-range potentials) and
(3) Collisions (near-resonant electronic transfer, excitation and charge transfer radiative transitions, rotational and vibrational excitation in molecules, spin-exchange).

to be covered today by this title. Those advances began with hard work in the 1950's and still continue today,[2] particularly in relation to ultracold gases.[3]

1.1. *The Dalgarno-Lewis Method*

I first met Alex when I entered Queen's University, Belfast as an undergraduate in 1958. In my second year, Alex taught me Classical Mechanics out of Goldstein's book[4] and, in my third (honours) year, David Bates taught me Quantum Mechanics and Scattering Theory, using his notes later published.[5] Even by that time–from when he accepted Bates's invitation to become a faculty member at Queen's in 1951– Alex had already made many valuable contributions to the subjects I have enumerated above in Sect. 1. One that I particularly liked was the Dalgarno-Lewis method[6] developed for "The exact calculation of long-range forces between atoms by perturbation theory". It provides an elegant method which enabled exact results to be obtained for various orders in perturbation theory, and which replaced the infinite sums that arise in each order by the solution of differential equations. Its popularity became further enhanced by its detailed inclusion in the book "*Quantum Mechanics*" by Leonard Schiff[7] and it is still being used today.

1.2. *Associative Detachment*

Also during that year (1958), Alex pointed out that it was the process of associative detachment

$$H^- + H \rightarrow H_2 + e^- \tag{1}$$

that controlled the ionization balance in the solar atmosphere and he gave a verbal estimate of its relatively rapid rate ($3 \times 10^{-10} cm^3 s^{-1}$), later quoted by Pagel.[8] His suggestion led to a new picture of the solar atmosphere. Alex commented that associative detachment (1) is a source of H_2. Others later recognized that (1) is the major source of H_2 in the early universe. Alex's recognition that associative detachment would ordinarily be a fast reaction[9] also led to a new picture of the D-region of the terrestrial ionosphere. After Alex's suggestion, the associative detachment reaction has been included in all discussions of weakly ionized plasmas, laboratory as well as astrophysical.

1.3. *The Arthurs-Dalgarno Method*

Molecular hydrogen, H_2, is the most abundant molecule in the Universe and, because of Alex's studies of its properties, H_2 has become a potent diagnostic probe. Recognizing the need from his work on diffuse interstellar clouds, Alex then embarked on extensive investigations of the radiative and collisional properties of H_2. They involved the formulation of quantum-mechanical scattering theory for rotational excitation by Arthurs and Dalgarno[10] and the development of numerical procedures to solve the resulting close- coupling equations. Allison and Dalgarno[11] reported the first solutions of close-coupled equations for rotational excitation and gave results for rotational excitation of H_2 in collision with H, He and H_2. It initiated an industry of such calculations by chemical physicists which continues today. Indeed, Alex is still involved, with calculations of rotational-vibrational molecular collisional transitions at ultralow temperatures,[12] exploring the Wigner regime.

1.4. *The Founding of Molecular Astrophysics*

Key to the founding of Molecular Astrophysics were Alex's studies of the radiative properties of molecular hydrogen beginning with the determination of the oscillator strengths and transition probabilities of the Lyman and Werner systems[13] which were essential elements in the interpretation of the observational data expected from the ultraviolet spectrometer on the Copernicus satellite.

2. The Modus Operandi

The four examples above perhaps serve to illustrate some of the hallmarks of Alex's successful *modus operandi*, which, if possible, may be summarized as follows:

(a) Identify the great variety of atomic and molecular quantum processes of significance to the physics and chemistry of the environment under consideration.
(b) Perform calculations to understand the reaction mechanisms and determine the rate coefficients either from existing theory or from theoretical methods he would invent for their study. Often this involved extending quantum mechanical theory for application to a diverse range of processes that operate in astrophysical environments where conditions are extreme.

(c) Construct models of the particular environment, and then,
(d) Conduct rigorous investigations.

The wide array of applications has included:

(1) Early Universe
(2) Planetary and Stellar Atmospheres
(3) Interstellar Clouds
(4) Shocked Gases
(5) Photon-dominated regions (PDR's), or photo-dissociation regions, and X-ray dominated regions (XDR's). Responses of molecular material to UV and X-rays and to magneto-hydrodynamic (MHD) shocks
(6) Planetary Nebulae, Planetary Atmospheres and Comets, Supernova ejecta (SN1987A).

3. The Result of the Transition

Alex has made landmark advances and provided major contributions not only to the development and application of Atomic, Molecular and Optical Physics but also to the interpretation of astrophysical phenomena from the early Universe to planetary atmospheres, covered within the array of applications above. His method of construction of models of interstellar clouds, for example, with the close interplay of molecular physics and astrophysics has become part of the fabric of molecular astrophysics. His combination of quantum-mechanical theory and analyses of astrophysical data with the identification of the relevant molecular processes and recognition of their consequences is molecular astrophysics at work. And so the evolution of *Molecular Astrophysics* into a unified discipline of inquiry continues to be dominated by Alex and his students. Sir David Bates (my mentor at Queen's and afterward) and George Victor summed it up quite nicely when they wrote[14] 20 years ago that "There is no greater figure than Alex in the history of Atomic Physics and its applications" Also, 10 years ago, Alex's pioneering contributions to Molecular Astrophysics were acknowledged[15] by describing him as "*The Father of Molecular Astrophysics* ".

Alex, now more than ever, has earned these accolades. This three-day Symposium is indeed testament to the fact that Alex's career has essentially embodied three spectacular careers, each in (a) Theoretical Atomic and Molecular Physics, (b) Aeronomy (the study of the upper atmosphere), Planetary Atmospheres and Comets and (c) Astrophysics. Any one of us would consider ourselves most fortunate to be successful in any one of these endeavors and to have, at most, a one-day celebration, as acknowledgment.

4. The Beginning and End of the Transition

And Alex did all of this without the benefit of any formal undergraduate courses/training in Physics, Chemistry or Astronomy. Alex graduated in 1947 with a degree in Mathematics at University College, London (UCL) and had a fellowship to do whatever he pleased for the year afterward. According to Alex's recent account[1,16] of this time at UCL, he considered further work in geometry, a subject he thoroughly enjoyed, almost, as entertainment. But he thought of geometry more as an intellectual exercise, as he readily solved hypothetical problems and puzzles. Because he wanted to solve real physical problems, he identified Physics as a possible area for him to apply his considerable mathematical skills. He choose an appropriate graduate Physics course for study at UCL. One day, the Department head, then Sir Harrie Massey, happened to meet Alex by chance in the corridor and asked him what he wanted to do after finishing his Physics course. Alex had no idea, until Massey suggested Atomic Physics, a subject Alex, at that time, knew little about. But to Alex, Atomic Physics "sounded different" and "it might be interesting".[16] Massey offered Alex a Fellowship to pursue his Ph.D (with advisor R. A. Buckingham, an expert in interatomic forces). And so, it was this serendipitous encounter with Massey in the corridor of UCL that turned out to be the key event that led Alex from the study of Mathematics to Physics and subsequently, in later years, to Aeronomy, Chemistry, Astronomy and Astrophysics. And David Bates, who was about to leave UCL in 1951 to reinstate the Dept. of Applied Mathematics at Queen's, offered Alex, on completion of his Ph.D, a faculty position at Queen's.

Bates later introduced him to Aeronomy and Planetary Atmospheres. Then, as they say, the rest is history. He occupied his chair at Queen's until 1967 when he accepted the invitation of Leo Goldberg and Fred Whipple to a chaired position at, what is now, the Harvard-Smithsonian Center for Astrophysics. I linked up again with Alex in 1968 at Harvard, ten years after we first met in Belfast and I spent three very productive years there under his wise counsel and tutorship. And so, we are now back to the beginning of my talk. The circle is completed. On reviewing all his monumental advances and contributions to these vast fields, researchers in each of these areas must now be all extremely grateful that Alex easily and successfully made the transition from gifted mathematician to astrophysicist with such astonishing rapidity and spectacular success. I consider Alex's role in molecular astrophysics to be analogous to the roles played by Bethe, Chandrasekhar and Fowler in stellar and nuclear astrophysics. Alex has established Molecular Astrophysics as a unified intellectual field of great scientific endeavor, impact and achievement.

References

1. A. Dalgarno, *Annu. Rev. Astro. Astrophys.* **46**, 1 (2008).
2. See Chapters 2, 3, 4, and 5 of the present volume, Proceedings of the Dalgarno Celebratory Symposium, J. F. Babb, K. Kirby, and H. Sadeghpour (eds.), (Imperial College Press, London, 2010).
3. See Chapters 22, 23, and 24 of the present volume, Ref. 2.
4. H. Goldstein, *Classical Mechanics* (Addison-Wesley, Reading, Massachusetts, 1950).
5. D. R. Bates (ed.), *Quantum Theory 1. Elements* (Academic Press, New York, 1961).
6. A. Dalgarno and J. T. Lewis, *Proc. R. Soc. London Ser. A* **233**, 70 (1955).
7. L. Schiff, *Quantum Mechanics*, 3rd edn. (McGraw-Hill, New York, 1968).
8. B. Pagel, *Mon. Not. Roy. Astron. Soc* **119**, 609 (1959).
9. A. Dalgarno, *Annales de Geophysique* **17**, 16 (1961).
10. A. Arthurs and A. Dalgarno, *Proc. R. Soc. London Ser. A* **256**, 540 (1960).
11. A. C. Allison and A. Dalgarno, *Proc. Phys. Soc* **90**, 609 (1967).
12. See Chapter 24 of the present volume, Ref. 2.
13. A. C. Allison and A. Dalgarno, *Atomic Data* **1**, 289 (1970).
14. D. R. Bates and G. A. Victor, in *Adv. Atom. Mol. Physics.* **24**, 1 (1988).
15. T. W. Hartquist and D. A. Williams (eds.), *Molecular Physics of Stars and Galaxies* (Oxford University Press, Oxford, 1998).
16. Listen to and view Harry Kreisler's Conversation with Alexander Dalgarno at: `http://globetrotter.berkeley.edu/people3/Dalgarno/dalgarno-con0.html`.

CALCULATIONS OF ATOMIC AND MOLECULAR PROPERTIES

G. W. F. DRAKE

Department of Physics, University of Windsor,
Windsor, Ontario N9B 3P4, Canada
E-mail: gdrake@uwindsor.ca
www.uwindsor.ca/physics

The main ideas and results of dipole response theory are summarized, and the application of pseudospectral methods are discussed. Recent progress in high precision variational methods for helium and lithium are summarized and applied to the calculation of isotope shifts as a means to determine the nuclear charge radius of halo nuclei.

Keywords: Dipole response theory, variational methods, pseudospectral methods, helium energies, lithium energies, isotope shifts, nuclear charge radii, halo nuclei.

1. Introduction

The topic of this chapter is of course immense, and the contributions of Alex Dalgarno and his students over the past 60+ years is equally immense. It is impossible to do justice to it all, or even a small part of it, but I will try to hit upon a few key ideas that may be helpful to the reader as an introduction to the literature. I will particularly focus on those topics that were very much under discussion at the time that I arrived at the Harvard Smithsonian Center for Astrophysics in the fall of 1967 to begin my postdoctoral fellowship with Alex. These topics have had a profound influence on my own scientific career since then, and especially the following lesson:

> *The key to successful theoretical work is always to have more than one way of doing the same thing.* –Alex Dalgarno

2. Dipole Response Theory: Basic Ideas

We begin with an introduction to dipole response theory describing how the interaction of an atom or molecule with electromagnetic radiation can be characterized in terms of basic atomic and molecular properties.

Consider an atom or molecule interacting with an oscillating electric field

$$\vec{\mathcal{E}} = \mathcal{E}\hat{\epsilon}\,\cos(\omega t) = \mathcal{E}\hat{\epsilon}\left(\frac{e^{i\omega t} + e^{-i\omega t}}{2}\right) \tag{1}$$

where $\mathcal{E}$ is the field strength, $\hat{\epsilon}$ is the polarization vector defining the field direction, and ω is the frequency of oscillation. The interaction energy with an atomic electron is

$$V(\mathbf{r}, t) = -e\mathcal{E}\hat{\epsilon}\cdot\mathbf{r}\left(\frac{e^{i\omega t} + e^{-i\omega t}}{2}\right) \tag{2}$$

$$\equiv \mathcal{E}\left[G_+(\mathbf{r}, t) + G_-(\mathbf{r}, t)\right] \tag{3}$$

The time-dependent Schrödinger equation is then

$$i\hbar\frac{\partial\Psi(\mathbf{r}, t)}{\partial t} = (H_0 + \mathcal{E}G_+ + \mathcal{E}G_-)\,\Psi(\mathbf{r}, t) \tag{4}$$

where H_0 specifies the field-free Hamiltonian. In order to develop a systematic method of solving this equation, substitute the perturbation expansion

$$\Psi(\mathbf{r}, t) = \Psi^{(0)}(\mathbf{r}, t) + \mathcal{E}\left[\Psi_+^{(1)}(\mathbf{r}, t) + \Psi_-^{(1)}(\mathbf{r}, t)\right] + O(\mathcal{E}^2) \tag{5}$$

Then, collecting coefficients of equal powers of $\mathcal{E}$, the perturbation equations are:

$$\begin{aligned} \text{Zeroth order :}\quad & \left(H_0 - i\hbar\frac{\partial}{\partial t}\right)\Psi^{(0)}(\mathbf{r}, t) = 0 \\ \text{First order :}\quad & \left(H_0 - i\hbar\frac{\partial}{\partial t}\right)\Psi_\pm^{(1)}(\mathbf{r}, t) = -G_\pm\Psi^{(0)}(\mathbf{r}, t) \end{aligned} \tag{6}$$

The stationary-state solutions $\Psi_n^{(0)}(\mathbf{r}, t) = \psi_n^{(0)}(\mathbf{r})e^{iE_nt/\hbar}$ to the zeroth order equation determine the energy level spectrum E_n, $n = 0, 1, \ldots$ for the atom or molecule in the absence of the oscillating electric field. Solutions to the first-order equation depend in general on the details of how the external field is switched on. There is of course great current interest in the interaction of atoms with short high-intensity laser pulses where the field may be switched on in a time short compared with the response time of the atom. However, let us begin with the simplest case of adiabatic switching where the field is assumed to be switched on very slowly according to the switching function $\lim_{\epsilon \to 0} e^{\epsilon t}$ as $t \to -\infty$. After transient effects have died away, we then look for for steady-state solutions of the form:

$$\begin{aligned} \text{Zeroth order}: \quad & \Psi^{(0)}(\mathbf{r},t) = \psi^{(0)}(\mathbf{r})e^{-iE_0 t/\hbar} \\ \text{First order}: \quad & \Psi^{(1)}_{\pm}(\mathbf{r},t) = \psi^{(1)}_{\pm}(\mathbf{r})e^{-iE_0 t/\hbar \pm i\omega t} \end{aligned} \tag{7}$$

The time-independent equations are then:

$$\begin{aligned} \text{Zero order}: \quad & (H_0 - E_0)\,\psi^{(0)}(\mathbf{r}) = 0 \\ \text{First order}: \quad & (H_0 - E_0 \pm \hbar\omega)\,\psi^{(1)}_{\pm}(\mathbf{r}) = -\tfrac{1}{2}\hat{\boldsymbol{\epsilon}}\cdot\mathbf{r}\,\psi^{(0)}(\mathbf{r}) \end{aligned} \tag{8}$$

From here, traditional approaches have followed two distinct but complementary paths:

- Solve the first-order equation directly. This can be done analytically for hydrogenic atoms or ions.[1] For many-electron atoms, a great deal of work has been done to develop an arsenel of approximation methods such as the time-dependent Hartree-Fock approximation (Dalgarno and Victor,[2] Jamieson,[3]), configuratiuon interaction (CI) (Krause *et al.*[4]), and time-dependent density functional theory (Chu and Dalgarno[5]), and variational methods (Stewart, Watson, and Dalgarno,[6]).
- Construct a representation in terms of an explicit sum over virtual intermediate states. To obtain this form, multiply through by the resolvent operator $(H_0 - E_0 \pm \hbar\omega)^{-1}$ and insert a complete set of states to obtain:

$$\psi^{(1)}_{\pm}(\mathbf{r},\omega) = -\frac{e}{2}\sum_{n\neq 0}^{\infty} \frac{\psi_n(\mathbf{r})\langle\psi_n|\hat{\boldsymbol{\epsilon}}\cdot\mathbf{r}|\psi_0\rangle}{E_0 - E_n \mp \hbar\omega} \tag{9}$$

In the limit of zero frequency, the second-order energy shift is related to the polarizability $\alpha_{\rm d}$ by

$$\mathcal{E}^2 \Delta E^{(2)} \equiv \frac{1}{2}\alpha_{\rm d}\mathcal{E}^2 \tag{10}$$

$$= e\mathcal{E}^2\langle\psi^{(0)}|\hat{\boldsymbol{\epsilon}}\cdot\mathbf{r}|\psi^{(1)}_{+} + \psi^{(1)}_{-}\rangle \tag{11}$$

This leads to the definition of the frequency-dependent polarizability

$$\begin{aligned}
\alpha_{\rm d}(\omega) &= 2e\langle\psi^{(0)}|\hat{\boldsymbol{\epsilon}}\cdot\mathbf{r}|\psi_+^{(1)}(\omega)+\psi_-^{(1)}(\omega)\rangle \qquad (12)\\
&= -e^2\sum_{n\neq 0}^{\infty}\!\!\!\!\!\!\int |\langle\psi_n|\hat{\boldsymbol{\epsilon}}\cdot\mathbf{r}|\psi_0\rangle|^2\left(\frac{1}{E_0-E_n+\hbar\omega}+\frac{1}{E_0-E_n-\hbar\omega}\right)\\
&= 2e^2\sum_{n\neq 0}^{\infty}\!\!\!\!\!\!\int \frac{(E_n-E_0)\,\langle\psi_0|\hat{\boldsymbol{\epsilon}}^{\,*}\cdot\mathbf{r}|\psi_n\rangle\langle\psi_n|\hat{\boldsymbol{\epsilon}}\cdot\mathbf{r}|\psi_0\rangle}{(E_n-E_0)^2-(\hbar\omega)^2}\\
&= \frac{\hbar^2e^2}{m_{\rm e}}\sum_{n\neq 0}^{\infty}\!\!\!\!\!\!\int \frac{f_{0,n}}{(E_n-E_0)^2-(\hbar\omega)^2} \qquad (13)
\end{aligned}$$

where the symbol $\sum\!\!\!\!\!\!\int$ denotes integration over the continuum as well as summation over discrete states, and

$$f_{0,n}=\frac{2m_{\rm e}}{\hbar^2}(E_n-E_0)|\langle\psi_n|\hat{\boldsymbol{\epsilon}}\cdot\mathbf{r}|\psi_0\rangle|^2 \qquad (14)$$

is the dimensionless oscillator strength for the $0\to n$ transition.

From here, there are again two distinct but complementary options:

- Use experimental data for the oscillator strengths.
- Introduce a discrete variational basis set to construct a pseudospectrum to represent the intermediate states. This second, very powerful technique is discussed in the following section.

3. Pseudospectral Theory

The actual spectrum of an atom such as hydrogen consists of an infinity of bound states together with a scattering continuum. The basic idea of pseudospectral methods is to span the actual spectrum of states with a basis set of variationally determined pseudostates that is entirely discrete, as illustrated in Fig. 1. As the basis set is progressively enlarged, the pseudostates move inexorably downward to form a sequence of progressively lower upper bounds to the exact eigenvalues (including the excited states), provided that the spectrum is bounded from below. This is known as the Hylleraas-Undheim-McDonald Theorem.[7] In the one-electron case, the discrete variational representation is entirely equivalent to a Sturmian basis set.[8] From a computational point of view, the pseudospectrum is obtained by diagonalizing H_0 in a discrete variational basis set of nonorthogonal functions χ_p, $p=0,\ldots,N-1$ such that the linear combinations $\phi_p=\sum_{q=0}^{N-1}a_{p,q}\chi_q$

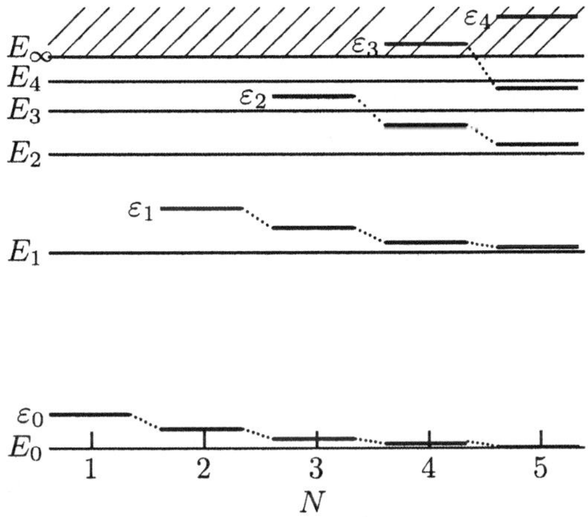

Fig. 1. Diagram illustrating the HUM theorem. The ε_p, $p = 0, \ldots, N-1$ are the variational eigenvalues for an N-dimensional basis set, and the E_i are the exact eigenvalues of H_0. The key point is that the old eigenvalues ε_p interleave the new as N is progressively increased. The highest ε_p lie in the continuous spectrum of H_0.

satisfy the generalized eigenvalue problem

$$\langle \phi_p | \phi_q \rangle = \delta_{p,q} \tag{15}$$

$$\langle \phi_p | H_0 | \phi_q \rangle = \varepsilon_p \, \delta_{p,q} \tag{16}$$

One then replaces the $\{\psi_n, \ E_n\}$, $n = 0, \ldots, \infty$ by $\{\phi_p, \ \varepsilon_p\}$, $p = 0, \ldots, N-1$ to obtain

$$\tilde{\alpha}_{\rm d}(\omega) = 2e^2 \sum_{p \neq 0}^{N-1} \frac{(\varepsilon_p - \varepsilon_0) \, \langle \phi_0 | \hat{\boldsymbol{\epsilon}}^* \cdot \mathbf{r} | \phi_p \rangle \langle \phi_p | \hat{\boldsymbol{\epsilon}} \cdot \mathbf{r} | \phi_0 \rangle}{(\varepsilon_p - \varepsilon_0)^2 - (\hbar\omega)^2}$$

Then $\tilde{\alpha}_{\rm d}(\omega) \to \alpha_{\rm d}(\omega)$ as $N \to \infty$, provided that the basis set is complete.

Variational Justification: Following Chan and Dalgarno,[9] write $\psi^{(1)} = \sum_{p=1}^{N-1} a_p \phi_p$, and construct $J_\pm(\omega) = \langle \psi^{(1)} | H - E_0 \pm \hbar\omega | \psi^{(1)} \rangle + 2\langle \psi^{(1)} | \hat{\boldsymbol{\epsilon}} \cdot \mathbf{r} | \psi_0 \rangle$. Then

$$\delta J_\pm(\omega) = \frac{\partial J_\pm(\omega)}{\partial a_p} \delta a_p = 0 \quad \Rightarrow \quad a_p = \frac{\langle \phi_p | \hat{\boldsymbol{\epsilon}} \cdot \mathbf{r} | \phi_0 \rangle}{E_0 - \varepsilon_p \mp \hbar\omega}$$

4. Sample Pseudostate Calculation for Hydrogen

As a simple but very instructive example,[10] consider a hydrogen atom in its $1s$ ground state and a static electric field $\mathcal{E}$ pointing in the z-direction.

The total Hamiltonian in polar coordinates is then

$$H = H^{(0)} + e\mathcal{E}r\cos\theta\,. \tag{17}$$

The first order perturbation equation can be solved analytically with the result

$$\psi^{(1)} = -\frac{1}{\sqrt{3}}(2r + r^2)e^{-r}Y_1^0(\hat{\mathbf{r}})\,. \tag{18}$$

$E^{(2)} = -\frac{9}{4}a_0^3$, and the dipole polarizability is $\alpha_{\rm d} \equiv -2E^{(2)} = \frac{9}{2}a_0^3$.

Suppose now that a variational solution is constructed of the form

$$\psi_{\rm tr}^{(1)} = -\frac{1}{\sqrt{3}}(b_1 r + b_2 r^2)e^{-\lambda r}Y_1^0(\hat{\mathbf{r}})\,,$$

where b_1 and b_2 are linear variational parameters multiplying the two basis functions $\chi_1 = -\frac{1}{\sqrt{3}}re^{-\lambda r}Y_1^0(\hat{\mathbf{r}})$ and $\chi_2 = -\frac{1}{\sqrt{3}}r^2e^{-\lambda r}Y_1^0(\hat{\mathbf{r}})$, and λ is an additional nonlinear variational parameter. This provides a two-dimensional basis set, with the exact solution being recovered for the case $\lambda = 1$. For $\lambda \neq 1$, the basis set provides the best variational representation of $\psi^{(1)}$. After orthogonalizing the basis set and diagonalizing H_0 to find b_1 and b_2, the expression for $\alpha_{\rm d}$ as a function of λ becomes

$$\alpha_{\rm d}(\lambda) = 6\lambda^5\left(\frac{2}{\lambda+1}\right)^{12}\frac{9\lambda^4 - 12\lambda^3 + 14\lambda^2 - 10\lambda + 5}{5\lambda^4 - 10\lambda^3 + 18\lambda^2 - 10\lambda + 5}\,.$$

Since $E^{(2)}$ is an upper bound, $\alpha_{\rm d}$ is a lower bound for any λ. It is clear from Fig. 2 that the exact value of $4.5\,a_0^3$ is recovered at $\lambda = 1$, where there is an absolute maximum. The remarkable point is that the entire spectrum of hydrogen is being well represented by just two pseudostates, neither of which corresponds to physical states of hydrogen.

5. Relationship of $\alpha(\omega)$ to Other Atomic Properties

A knowledge of the frequency-dependent polarizability leads directly to many other properties of atoms and molecules[1,11,12] as follows:

Index of Refraction: For a gas with number density N, the index of refraction $n(\omega)$ is

$$n(\omega) = 1 + 2\pi N\alpha(\omega)\,. \tag{19}$$

For molecules, $\alpha(\omega)$ is the averaged quantity

$$\alpha(\omega) = \frac{1}{3}\left[2\alpha_\perp(\omega) + \alpha_\parallel(\omega)\right] \tag{20}$$

where $\alpha_\perp(\omega)$ and $\alpha_\parallel(\omega)$ are the components of the polarizability parallel and perpendicular to the internuclear axis.

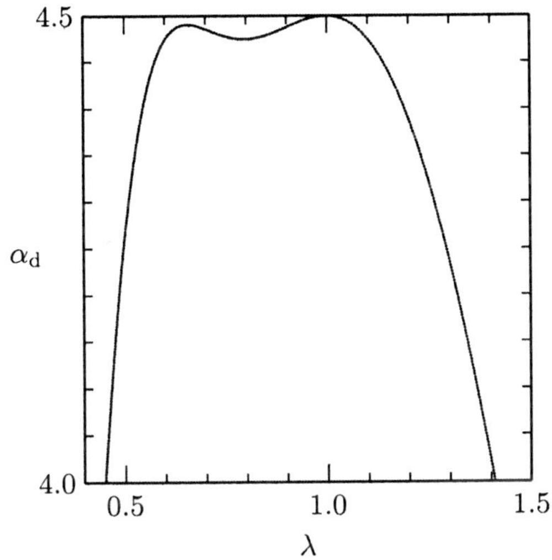

Fig. 2. Variational polarizability α_d of hydrogen, in units of a_0^3. The exact value at $\lambda = 1$ is $\alpha_d = 4.5\ a_0^3$.

Raman Scattering Cross Section: Far from resonance, the cross sections for scattering from an initial rotational state J are

$$\sigma(J \to J+2) = \frac{128\pi^5}{9\lambda^4} \frac{(J+1)(J+2)}{(2J+1)(2J+3)} \gamma(\omega)^2 \tag{21}$$

and

$$\sigma(J \to J-2) = \frac{128\pi^5}{9\lambda^4} \frac{(J-1)J}{(2J-1)(2J+1)} \gamma(\omega)^2 \tag{22}$$

where λ is the final wavelength, and $\gamma(\omega) = \alpha_{\|}(\omega) - \alpha_{\perp}(\omega)$ is the polarization anisotropy.

Rayleigh Scattering Cross Section:

$$\sigma(J \to J) = \frac{128\pi^5}{9\lambda^4} \left[3\alpha(\omega)^2 + \frac{2J(J+1)}{3(2J-1)(2J+3)} \gamma(\omega)^2 \right] \tag{23}$$

Rayleigh Depolarization Factor:

$$\rho_{\text{Ray}}(\omega) = \frac{6\gamma(\omega)^2}{45\alpha(\omega)^2 + 7\gamma(\omega)^2} \tag{24}$$

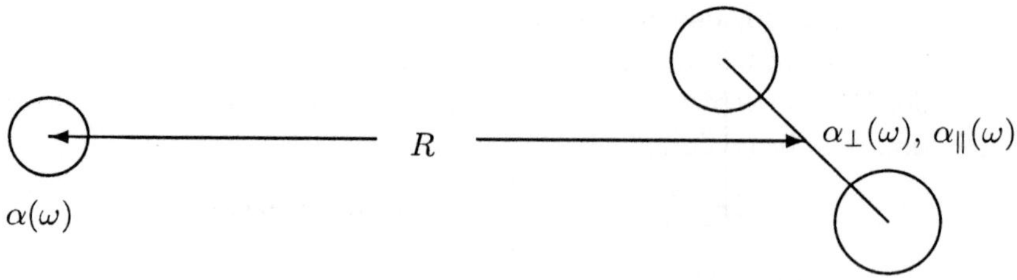

Fig. 3. Long-range interaction between an atom and a diatomic molecule.

Verdet Constant: The Verdet constant describes the strength of the Faraday rotation of a material according to the equation

$$V \simeq 2.98 \times 10^{-4} \omega \frac{dn(\omega)}{d\omega} \tag{25}$$

van der Waals Interactions: As shown in Fig. 3, the long range interaction potential between an atom and a diatomic molecule is

$$V(R) = -\frac{C_6}{R^6}\left[1 + \gamma P_2(\cos\theta)\right] \tag{26}$$

where

$$C_6 = \frac{1}{\pi}\int_0^\infty \left[\alpha_\parallel(i\omega) + 2\alpha_\perp(i\omega)\right]\alpha(i\omega)\,d\omega \tag{27}$$

and

$$\gamma = \frac{1}{\pi C_6}\int_0^\infty \left[\alpha_\parallel(i\omega) - \alpha_\perp(i\omega)\right]\alpha(i\omega)\,d\omega \tag{28}$$

6. Applications and Results

This section provides a few of the early applications as illustrative examples, and culminating with the most recent high-precision results to demonstrate the tremendous progress that has been made with the increasing speed and storage capacity of computers.

One of the earliest papers to draw attention to the power and versatility of variational methods and pseudospectral techniques was that of Chan and Dalgarno.[9] They obtained excellent convergence for the frequency-dependent polarizability and the index of refraction of helium with a simple 4-term variational basis set derived from the ground-state wave function of Hart and Herzberg. Their static polarizability already reached an accuracy of $\alpha_D = 1.378\ a_0^3$, $(2.04 \times 10^{-24}\ cm^3)$, in close agreement with the accurate nonrelativistic value $1.383\,192\ a_0^3$. This work was closely followed by the paper of Victor, Dalgarno and Taylor[13] on the dipole properties of the metastable states of helium, including frequency-dependent polarizabilities and van der Waals coefficients. This paper introduced fully correlated variational basis sets in Hylleraas coordinates of the form

$$\chi_{i,j,k} = r_1^i r_2^j r_{12}^k \exp(-\alpha r_1 - \beta r_2) \cos\theta_2 \pm \text{exchange} \qquad (29)$$

such as are still used today. Linear combinations of these functions then form the pseudostates ϕ_p as discussed in Sect. 3. As an important test, they also showed that the Thomas-Reiche-Kuhn oscillator strength sum rule

$$\sum_n f_{i\to n} = N \qquad (30)$$

where N is the number of electrons, is accurately satisfied when summed over the variational pseudostates in place of the actual spectrum. The most accurate calculation to date of the helium polarizability is that of Pachucki and Sapirstein.[14] Their result, including mass polarization, relativistic and quantum electrodynamic corrections, is

$$\alpha_{\rm D} = 1.383\,191(2)\ {\rm a}_0^3. \qquad (31)$$

This is but one example out of a huge literature on the polarizabilities of atoms and molecules.

There is an equally huge literature on the van der Waals coefficients for the long-range interactions between atoms and molecules. This is a topic of great current interest because of its importance in the formation of Bose-Einstein condensates. Calculations of the van der Waals coefficients have now reached a high degree of refinement, as illustrated by the paper by Zhu *et al.*[15] Starting from the dynamic polarizabilities, they obtained the C_6 and γ coefficients in Eq. (26) for long-range interactions between molecular hydrogen and alkali metal atoms to an accuracy of better than $\pm 2\%$.

7. Applications to the Study of Light Halo Nuclei

This brief review concludes with a survey of high-precision theory for helium as an illustration of what can be achieved for atomic systems more complicated than hydrogen. A topic of particular recent interest and significance is the use of high precision atomic theory and experiment to probe nuclear structure via the isotope shift. This technique provides a unique measurement tool to measure the nuclear charge radii of so-called ‘halo nuclei’ first

Table 1. Contributions to the energy and their orders of magnitude in terms of Z, $\mu/M = 1.370\,745\,624 \times 10^{-4}$, and $\alpha^2 = 0.532\,513\,5450 \times 10^{-4}$.

Contribution	Magnitude
Nonrelativistic energy	Z^2
Mass polarization	$Z^2\mu/M$
Second-order mass polarization	$Z^2(\mu/M)^2$
Relativistic corrections	$Z^4\alpha^2$
Relativistic recoil	$Z^4\alpha^2\mu/M$
Anomalous magnetic moment	$Z^4\alpha^3$
Hyperfine structure	$Z^3 g_I \mu_0^2$
Lamb shift	$Z^4\alpha^3 \ln\alpha + \cdots$
Radiative recoil	$Z^4\alpha^3(\ln\alpha)\mu/M$
Finite nuclear size	$Z^4\langle\bar{r}_c/a_0\rangle^2$

discovered by Tanihata about 30 years ago.[16,17] They observed that some isotopes of the lightest elements close to the neutron drip-line have extra planetary neutrons that make the matter (mass) radii substantially larger than the charge radii. The neutron wave functions extend far into the classically forbidden region far from the more tightly bound nuclear core. As reviewed by Tanihata,[18] the halo structure reveals itself most clearly as a difference between the nuclear mass radius and the nuclear charge radius. Since the neutrons are massive but uncharged, one expects the mass radius to be larger than the charge radius. Moreover, the change in the charge radius relative to the corresponding isotope without the halo neutrons provides a sensitive probe of the core structure, and the effective low energy nucleon-nucleon interaction potential.

The possibility of using the isotope shift to determine nuclear radii for helium was first suggested by Drake,[19] and applied to the stable isotope combinations ^{3}He/^{4}He and ^{6}Li$^+$/^{7}Li$^+$.[20] Short-lived halo nuclei are of course much more difficult to handle. As reviewed by Drake *et al.*,[21] two groups have succeeded in obtaining isotope shifts with a precision that is sufficient to extract nuclear charge radii. At Argonne National Lab the charge radius of ^{6}He was obtained from laser spectroscopy on helium atoms in a magneto-optical trap, while a collaboration from GSI Darmstadt and the University of Tübingen developed a method for high-resolution resonance ionization spectroscopy and measured the charge radii of ^{8,9}Li at GSI[22] and of ^{11}Li at the ISAC mass separator at TRIUMF.[23]

To determine the charge radius from the isotope shift, it is necessary to calculate the mass-dependent part of the isotope shift with high accuracy and reliability. The various contributions to the energy (and hence the isotope shift) are shown in Table 6 in the form of a double perturbation expansion in powers of $\alpha \simeq 1/137.035\,999\,679$ and the electron reduced mass ratio $\mu/M \simeq 10^{-4}$. An important feature is that the terms that are independent of μ/M cancel from the isotope shift and so do not contribute. The key problem to be overcome is the calculation of the mass-dependent part of the QED correction. Fortunately, this problem is now solved with the calculation of the Bethe logarithm part of the electron self-energy for helium[24] and lithium.[25] Since all the lower-order terms up to order α^3 Ry can now be calculated to very high precision, the dominant source of uncertainty comes from the QED corrections of order $\alpha^4\mu/M$ Ry or higher, where once again it is only the mass dependent part that contributes. Since $\alpha^4\mu/M \simeq 10^{-12}$Ry ($\sim$ 10 kHz), the uncertainty from these higher-order terms is much less than the finite nuclear size correction of about 1 MHz. The comparison between theory and experiment therefore provides a clear determination of the nuclear size.

7.1. *Solution to the Nonrelativistic Schrödinger Equation*

The foundation for the calculation of isotope shifts, including relativistic and QED corrections, is to find high-precision solutions to the nonrelativistic Schrödinger equation for infinite nuclear mass. The past 20 years have seen important advances in developing specialized techniques for doing this in the case of the three-body problem (heliumlike systems),[19,26,27] and more recently the four-body problem.[25,28–33] The usual methods of theoretical atomic physics, such as the Hartree-Fock approximation or configuration interaction methods, are not capable of yielding results of spectroscopic accuracy, and so specialized techniques are needed. The most successful calculations are those based on variational trial functions constructed from linear combinations of correlated functions in Hylleraas coordinates, such as the $\chi_{i,j,k}$ functions defined by Eq. (29). The result is a sequence of progressively lower upper bounds for the energy, exactly as shown in Fig. 1. As shown by Klahn and Bingle,[34] the Hylleraas basis set is complete in the limit that the number of powers tends to infinity. This important property ensures that the results converge to the correct answer, including all correlation effects.

The most studied example is the $1s^2\ ^1\mathrm{S}$ ground state of helium. As shown by Drake *et al.*,[35] convergence to 20 or more figures can be readily obtained, using conventional quadruple precision (32 decimal digit) arithmetic in FORTRAN. The very large calculation by Schwartz,[36] using 104-digit arithmetic, provides a benchmark for comparison. Recently, even higher accuracy has been obtained by Nakashima and Nakatsuji.[37] High precision results are available for all states of helium up to $n = 10$ and angular momentum $L = 7$ are available.[19,26,38] Combined with asymptotic expansion methods for high L and quantum defect methods for high n, this provides a complete coverage of the entire spectrum of singly-excited states for helium.[39]

The lithium problem is much more difficult than helium both because the integrals over fully correlated wave functions are more difficult, and because the basis set grows much more rapidly with increasing Ω. Nevertheless, there has been important progress in recent years,[25,28–33] and results of spectroscopic accuracy can be obtained for the low-lying states. For isotope shifts, the breakthrough came in 2000 when Yan and Drake succeeded in calculating the mass effect in transitions of neutral lithium with a relative accuracy of better than 5×10^{-6}.

7.2. *Finite Nuclear Mass Effects*

For high precision calculations, and especially for the isotope shift, it is necessary to include also the motion of the nucleus in the center-of-mass (CM) frame. A transformation to CM plus relative coordinates results in the additional $-(\hbar^2/M)\nabla_1 \cdot \nabla_2$ mass polarization term to be added to the Hamiltonian for infinite nuclear mass,[40] where M is the nuclear mass. Since our aim is to calculate isotope shifts for different values of μ/M, it is highly desirable to express all the results in the form of a perturbation expansion

$$E_M = E_\infty^{(0)} + \frac{\mu}{M} E_\infty^{(1)} + \left(\frac{\mu}{M}\right)^2 E_\infty^{(2)} + \cdots \tag{32}$$

where μ is the electron reduced mass, and to evaluate the expansion coefficients to sufficient accuracy.[26,27]

7.3. *Relativistic and Quantum Electrodynamic Corrections*

Relativistic and quantum electrodynamic corrections (especially the mass-dependent parts) play an important role in calculating the isotope shift to spectroscopic accuracy and applying it to the determination of nuclear charge radii. However, a detailed discussion of them would take us beyond the scope of this brief overview. Suffice it to say that these effects can be treated as perturbations once the nonrelativistic part of the problem has been solved to sufficient precision.

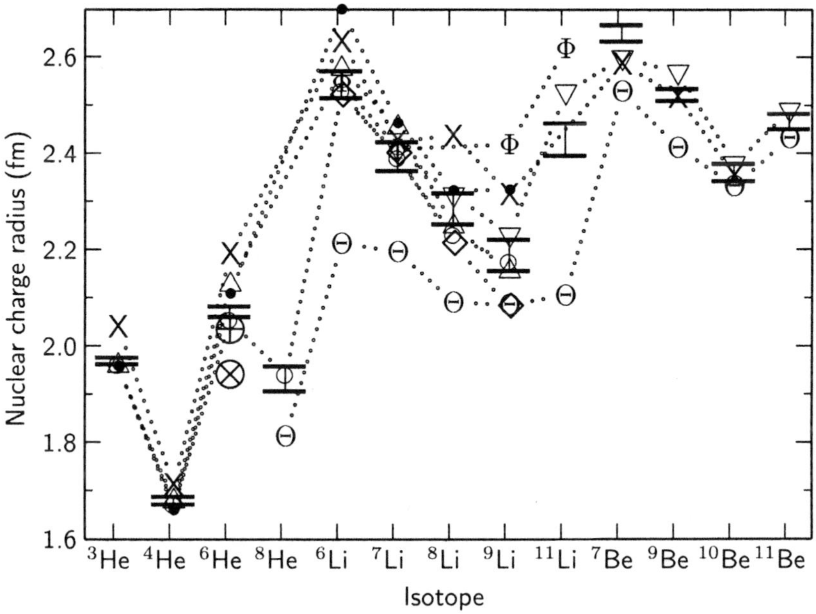

Fig. 4. Comparison of various nuclear structure theories with experiment for the rms nuclear charge radius r_c. The points with errror bars correspond to experimental values for the nuclear radii and the others are grouped as (⊗) variational microcluster calculations[41–43] and a no-core shell model;[44,45] (⊕) effective three-body cluster models;[46,47] (Θ) large-basis shell model;[48] (▽) stochastic variational multicluster;[49] (Φ) dynamic correlation model.[50] The remaining points are quantum Monte Carlo calculations[51,52] with various effective potentials as follows: (X) AV8'; (•) AV18/UIX; (O) AV18/IL2; (△) AV18/IL3; (◇) AV18/IL4 (for Li only).

7.4. *Results for Nuclear Charge Radii*

Figure 4 provides an overview of all the nuclear charge radii r_c that have been measured to date by the isotope shift method. There are three separate sequences of measurements for the isotopes of helium,[53] lithium[23] and beryllium,[54] with the stable isotopes ^{3}He, ^{7}Li and ^{9}Be taken as reference nuclei. The measurements are represented by the large vertical error bars, while the various sequences of theoretical predictions are connected by dotted lines to show their trends. Of particular importance are the two- and four-neutron halo nuclei ^{6}He, ^{8}He, the two-neutron halo nucleus ^{11}Li, and the single-neutron halo nucleus ^{11}Be. The last is the simplest example of a single-neutron halo. It is clear that, although the theoretical predictions show the correct overall trends, it is the halo nuclei that are best able to discriminate amongst the various theories of nuclear structure. It is for these cases that the various theoretical predictions show the greatest divergence.

Beginning first with the helium sequence, r_c is quite small for the tightly bound α-particle corresponding to the nucleus of ^{4}He, but then it rises dramatically for ^{6}He, indicating that the charge distribution of the α-particle becomes smeared out due to its recoil motion in the center-of-mass frame. It then drops again for ^{8}He as the halo of four neutrons becomes more spherically symmetric. A similar pattern of steadily decreasing values of r_c is apparent for both the Li and Be sequences, followed by a sharp increase for the last member of each sequence corresponding to the halo nucleus case. It is interesting that nuclear polarization by the atomic electrons has a significant effect on the calculated isotope shifts.[55]

8. Concluding Remarks

Atomic and molecular theory continues to provide key input data for astrophysical applications, such as transition frequencies, oscillator strengths, etc., as well as cooling and trapping through long-range interactions. However, what is perhaps the most exciting is the possibility of combining high precision theory and experiment to create new measurement tools, as illustrated by the previous section on halo nuclei. Other possibilities are improved measurements of fundamental importance such as the electron to proton mass ratio m_e/m_p, the fine structure constant α, and the cosmological variation of the fundamental constants. There also remain important challenges ahead in pushing high precision variational calculations beyond lithium to four and more electron atoms.

Acknowledgments

The work of Zong-Chao Yan has played an especially important role in advancing computational techniques for lithium and other three-electron problems. Research support by the Natural Sciences and Engineering Research Council is gratefully acknowledged.

References

1. M. Karplus and H.J. Kolker, J. Chem. Phys. **39**, 1493 (1963).
2. A. Dalgarno and G. A. Victor, Proc. Roy. Soc. (London) **A291**, 291 (1966).
3. M.J. Jamieson, J. Phys. B **8**, 537 (1975).

4. P. Krause, T. Klamroth, and P. Saalfrank, J. Chem. Phys. **127**, 034107 (2007).
5. X. Chu, and A. Dalgarno, J. Chem. Phys. **121**, 4083 (2004).
6. R.F. Stewart, D.K. Watson, and A. Dalgarno, J. Chem. Phys. **65**, 2104 (1976), and earlier references therein.
7. E.A. Hylleraas and B. Undheim, Z. Phys. **65**, 759 (1930); J.K.L. MacDonald, Phys. Rev. **43**, 830 (1933).
8. E.L. Ince, *Ordinary Differential Equations* (Dover, New York, 1956).
9. Y. M. Chan and A. Dalgarno, Proc. Phys. Soc. (London) **85**, 227 (1965).
10. G.W.F. Drake, in *Encyclopedia of Applied Physics*, Edited by George L. Trigg (Wiley-VCH, Weinheim and New York, 1998), Vol. 23, pp. 121–149; reprinted in *Mathematical Tools for Physicists*, Edited by George L. Trigg (Wiley-VCH, New York, 2005).
11. A. Dalgarno, Adv. Phys. **11**, 281 (1962).
12. Y.M. Chan and A. Dalgarno, Proc. Phys. Soc. **85**, 227 (1965); *ibid.* **86**, 777 (1965).
13. G. A. Victor, A. Dalgarno and A. J. Taylor, J. Phys. B **1**, 14 (1968).
14. K.Pachucki and J. Sapirstein, Phys. Rev. A **63**, 012504 (2000).
15. C. Zhu, A. Dalgarno, and A. Derevianko, Phys. Rev. A **65**, 034708 (2002).
16. I. Tanihata *et al.*, Phys. Rev. Lett. **55**, 2676 (1985).
17. I. Tanihata *et al.*, Phys. Lett. **160B**, 380 (1985).
18. I. Tanihata, J. Phys. G **22**, 157 (1996).
19. G.W.F. Drake, in *Long-range Casimir forces: Theory and recent experiments on atomic systems*, edited by F.S. Levin and D.A. Micha (Plenum, New York, 1993), pp. 107–217.
20. E. Riis, A.G. Sinclair, O. Poulsen, G.W.F. Drake, W.R. C.Rowley, A.P. Levick, Phys. Rev. A **49**, 207 (1994).
21. G.W.F. Drake, Z.-T. Lu, and W. Nörtershäuser, in *Precision Physics of Simple Atoms and Molecules*, Lecture Notes in Physics, edited by Sabvely G. Karshenboim **745**, 131–153 (Springer-Verlag, Berlin, Heidelberg, 2008).
22. G. Ewald, W. Nörtershäuser, A. Dax, S. Göte, R. Kirchner, H.-J. Kluge, Th. Kühl, R. Sanchez, A. Wojtaszek, B.A. Bushaw, G.W.F. Drake, Z.-C. Yan, and C. Zimmermann, Phys. Rev. Lett. **93**, 113002 (2004).
23. R. Sanchez *et al.*, Hyperfine Int. **171**, 181 (2006); see also R. Sanchez *et al.*, Phys. Rev. Lett. **96**, 033002 (2006)
24. G.W.F. Drake and S.P. Goldman, Can. J. Phys. **77**, 835 (2000).
25. Z.-C. Yan, W. Nörtershäuser, and G.W.F. Drake, Plys. Rev. Lett. **100** 243002 (2008).
26. G.W.F. Drake and Z.-C. Yan, Phys. Rev. A, **46**, 2378 (1992).
27. G.W.F. Drake, Adv. At. Mol. Opt. Phys. **31**, 1 (1993).
28. Z.-C. Yan and G.W.F. Drake, Phys. Rev. Lett. **81**, 774 (1998).
29. Z.-C. Yan and G.W.F. Drake, Phys. Rev. A, **66**, 042504 (2002).
30. Z.-C. Yan and G.W.F. Drake, Phys. Rev. Lett., **91**, 113004 (2003).
31. M. Puchalski, A.M. Moro, and K. Pachucki, Phys. Rev. Lett. **97**, 133001 (2006).
32. K. Pachucki, Phys. Rev. A **74**, 062510 (2006).

33. M. Puchalski and K. Pachucki, Phys. Rev. A **78**, 052511 (2008).
34. B. Klahn and W.A. Bingel, Theor. Chem. Acta **44**, 27 (1977); Int. J. Quantum Chem. **11**, 943 (1978).
35. G.W.F. Drake, M.M. Cassar, and R.A. Nistor, Phys. Rev. A **65**, 054501 (2002).
36. C. Schwartz, Int. J. Mod. Phys. E–Nucl. Phys. **15**, 877 (2006).
37. H. Nakashima and H. Nakatsuji, J. Chem. Phys. **128**, 154107 (2008); *ibid.*, Phys. Rev. Lett. **101** 240406 (2008).
38. G.W.F. Drake and Z.-C. Yan, Chem. Phys. Lett. **229**, 486–490 (1994).
39. G.W.F. Drake, in *Handbook of Atomic, Molecular and Optical Physics*, Edited by G.W.F. Drake (Springer, New York, 2005), p. 1355.
40. H.A. Bethe and E.E. Salpeter, *Quantum mechanics of one- and two-electron atoms*, (Springer-Verlag, New York, 1957).
41. K. Varga, Y. Suzuki, and Y. Ohbayasi. Phys. Rev. C **50**, 189 (1994).
42. A. Csótó. Phys. Rev. C **48**, 165 (1993).
43. J. Wurzer and H.M. Hofmann. Phys. Rev. C **55**, 688 (1997).
44. P. Navrátil, J.P. Vary, W.E. Ormand, and B.R. Barrett. Phys. Rev. Lett. **87**, 172502 (2001).
45. C. Forssén, P. Navrátil, and W.E. Ormond, Phys. Rev. C **71**, 044312 (2005).
46. S. Fuñada, H. Kameyama, and Y. Sakuragi, Nucl. Phys. A **575**, 93 (1994).
47. H. Esbensen, G.F. Bertsch, and K. Hencken, Phys. Rev. C **56**, 3054 (1997).
48. P. Navrátil and B.R. Barrett, Phys. Rev. C **57**, 3119 (1998).
49. Y. Suzuki, R.G. Lovas, and K. Varga, Prog. Theo. Phys. Suppl. **146**, 413 (2002).
50. M. Tomaselli, T. Kühl, W. Nörtershäuser, G. Ewald, R. Sanchez, S. Fritzsche, and S.G. Karshenboim. Can. J. Phys. **80**, 1347 (2002).
51. S.C. Pieper and R.B. Wiringa, Ann. Rev. Nucl. Part. Sci. **51**, 53 (2001).
52. S.C. Pieper, K. Varga, and R.B. Wiringa, Phys. Rev. C **66**, 044310 (2002).
53. P. Mueller *et al.*, Phys. Rev. Lett. **99**, 252501 (2008).
54. W. Nörtershäuser *et al.*, Phys. Rev. Lett. **102**, 062503 (2009).
55. K. Pachucki and A.M. Moro, Phys. Rev. A **75**, 032501 (2007).

AN ELASTIC APPROXIMATION FOR SPIN FLIPPING TRANSITIONS IN COLLISIONS OF HYDROGEN ATOMS

B. ZYGELMAN

Department of Physics and Astronomy, University of Nevada Las Vegas, Las Vegas, NV 89154, USA
E-mail: bernard@physics.unlv.edu

The elastic approximation, developed by Dalgarno for the calculation of spin exchange cross sections, is extended to predict the cross sections for spin flipping transitions induced by the anisotropic magnetic coupling between two hydrogen atoms.

Keywords: Spin exchange; Spin re-alignment collisions; Rotational coupling.

1. Introduction

In binary collisions of hydrogen atoms, the spin orientations of the electron pair can interchange provided that

$$\Delta m \equiv m_a + m_b - m'_a - m'_b = 0, \tag{1}$$

where m_i, m'_i are the initial and final azimuthal components of electronic spin in atom i. This process is called spin exchange and, since the electron spin is coupled via the hyperfine interaction to the spin of the proton, it can excite or quench the hyperfine levels of a hydrogen atom.[1–3] In 1961 Dalgarno[3] introduced a quantum mechanical formulation of the process and derived the expression

$$\sigma(F = 1 \to F = 0) = \frac{\pi}{4k^2} \sum_l (2l + 1) \sin^2(\delta_{lt} - \delta_{ls}) \tag{2}$$

for the cross section of a hydrogen atom to undergo a transition from hyperfine level $F = 1$ to its ground $F = 0$ level. Here, k is the wave number for the collision system and δ_{lt}, δ_{ls} are the triplet and singlet scattering partial wave phaseshifts. This expression was later generalized to include the effects of quantum exchange symmetry.[4,5] Because Eq. (2) incorporates elastic scattering phase shifts, this description is sometimes called the elastic approximation. We recently employed a multichannel theory[6] to calculate[7] hyperfine level changing cross sections in collisions of hydrogen atoms. For collision energies corresponding to gas temperatures $T > 1K$, we found that the predictions of the elastic approximation are in excellent agreement with those obtained with the multichannel theory. Experimental measurements[8,9] of hydrogen atom spin exchange cross sections are in good agreement with values calculated within the framework of the elastic approximation. The method has also been applied to calculate spin exchange cross sections in collisions of alkali atoms,[10] and to predict frequency shift parameters that are important in the operation of the hydrogen maser.[11–13] In addition, the elastic approximation has been utilized to calculate cross sections for the proton impact excitation of fine structure levels in ions.[14,15]

More recently, there is renewed interest in the mechanisms that determine the hyperfine level populations of hydrogen atoms in the early universe.[16] Astronomers hope to utilize a new generation of proposed radio arrays in order to map the $21cm$ line of hydrogen, which corresponds to the ground state $F = 1$ to $F = 0$ hyperfine transition, at high redshifts. Cosmologists have predicted[17] that the line could be observed in absorption of the background microwave radiation field for redshifts in the range $30 < z < 200$. They have argued that observation of this signal could provide tomographic maps that yield unprecedented information on primordial matter density fluctuations. It has been shown[7,17] that spin exchange is a primary mechanism that determines the spin temperature[2] of hydrogen in the dark age universe, and allows for the $21cm$ absorption features.

Electron spin orientation re-alignment can also occur via the long range magnetic dipolar interaction[18] between atoms. Because the dipolar interaction is weak spin exchange[7] dominates, however the former can induce spin flips in which selection rule Eq. (1) does not hold. So if the hydrogen gas is polarized, spin exchange collisions are ineffective in re-distributing the hyperfine level populations to their statistical equilibrium values. In addition, $\Delta m \neq 0$ transitions could be important in over-dense halo regions of the early universe. In this discussion we refer to $\Delta m \neq 0$ transitions as spin flip processes, in contrast to spin exchange where $\Delta m = 0$.

In a recent study[19] we calculated, using multichannel methods, $\Delta m \neq 0$ hydrogen collision cross sections, for an energy range corresponding to gas temperatures $1\,mK < T < 300K$. We compared those results with values predicted by the Born approximation and found that the latter provide a poor estimate for temperatures above $1K$. At higher energies many partial waves participate and the use of the multichannel method becomes increasingly prohibitive. Keeping in mind the efficiency of the elastic approximation, well suited for calculations at higher energies, we wondered if a similar description is possible for the calculation of $\Delta m \neq 0$ spin flip transitions. The main purpose of this paper is to report on progress in that direction.

In the first section we give a brief review of the multichannel theory. We then use the Born approximation to estimate the spin flip cross sections and compare those results with that obtained by the multichannel method. In order to understand the reason for the discrepancy between the results of the two approaches, we develop a partial wave expansion and express the cross sections in terms of elastic scattering phaseshifts. We propose to employ the latter method for the calculation of $\Delta m \neq 0$ spin flip cross sections at higher collision energies. We offer a few personal observations in the conclusion. Atomic units are used throughout unless otherwise stated.

2. Multichannel Theory

The multichannel Schroedinger equation for two atoms with angular momenta j_a and j_b is given by[20]

$$\boldsymbol{\nabla}^2 F_\Omega(\boldsymbol{R}) - 2\mu \sum_{\Omega'} V_\Omega^{\Omega'}(R,\theta,\phi) F_{\Omega'}(\boldsymbol{R}) + k^2 F_\Omega(\boldsymbol{R}) = 0. \tag{3}$$

$F_\Omega(\boldsymbol{R})$ is the wave amplitude for the pair of atoms in state Ω, μ is the nuclear reduced mass of the collision system, $k = \sqrt{2\mu E}$ is the wave number and $V_\Omega^{\Omega'}(R,\theta,\phi)$ is an anisotropic multichannel potential matrix.[20] In a previous study[20] we considered the case where j_a, j_b are the total electronic angular momenta of two complex atoms. Here we consider two ground state hydrogen atoms and $j_a \equiv F_a$, $j_b \equiv F_b$ are total spin (electronic + nuclear) angular momenta, i.e. hyperfine, quantum numbers. For this case expression (58) in reference[20] reduces to

$$\begin{aligned}
\underline{V}(R\theta\phi) &= \sum_{IM_I} \sum_{SM_S} \sum_{\Omega} [F,F',F_a,F'_a,F_b,F'_b]^{1/2}[S,I] \\
&\times \mathcal{D}^{F'}_{M'_F\Omega}(\phi,\theta,-\phi)\, \mathcal{D}^{F}_{\Omega M_F}(\phi,-\theta,-\phi) \begin{pmatrix} I & S & F \\ M_I & M_S & -\Omega \end{pmatrix} \begin{pmatrix} I & S & F' \\ M_I & M_S & -\Omega \end{pmatrix} \\
&\times \begin{Bmatrix} 1/2 & 1/2 & F_a \\ 1/2 & 1/2 & F_b \\ S & I & F \end{Bmatrix} \begin{Bmatrix} 1/2 & 1/2 & F'_a \\ 1/2 & 1/2 & F'_b \\ S & I & F' \end{Bmatrix} \epsilon_{S\,M_S}(R)
\end{aligned} \tag{4}$$

where

$$\mathcal{D}^{j'}_{\Omega_1,\Omega'}(\phi,\theta,-\phi) \equiv \langle j\Omega|\exp(-i\phi \boldsymbol{j}_z)\exp(-i\theta \boldsymbol{j}_y)\exp(i\phi \boldsymbol{j}_z)|j\Omega'\rangle \tag{5}$$

is a Wigner rotation matrix, F, M_F are the total, atom-pair, angular momentum quantum numbers, and $\epsilon_{S\,M_S}(R)$ are the Born-Oppenheimer (BO) energies for two atoms with total electronic spin angular momentum $S\,M_S$. Typically, the BO energies do not depend on the azimuthal component of the angular momentum but the magnetic anisotropic interaction leads to an M_S dependence.

For two ground state hydrogen atoms, we can approximate the interaction Hamiltonian with the expression

$$\begin{aligned} H_{AD} = &\left({}^3\Sigma(R) -{}^1 \Sigma(R)\right)\boldsymbol{S}_a \cdot \boldsymbol{S}_b + \frac{3\,{}^3\Sigma(R) +{}^1 \Sigma(R)}{4} \\ &+H_{DIP} + H_{FS} \end{aligned} \tag{6}$$

where ${}^3\Sigma(R) \equiv b^3\Sigma_u^+(R), {}^1\Sigma(R) \equiv X^1\Sigma_u^+(R)$ are, respectively, the ground state triplet and singlet Born-Oppenheimer (BO) potential curves for H_2, and $\boldsymbol{S}_a, \boldsymbol{S}_b$ are the electronic spin operators for atoms a, b respectively. H_{DIP} is the long-range magnetic electron spin-spin interaction[18] (Breit interaction) between the pair of atoms,

$$H_{DIP} = \frac{\alpha^2}{R^3}\left[\boldsymbol{S}_a \cdot \boldsymbol{S}_b - 3\frac{(\boldsymbol{S}_a \cdot \boldsymbol{R})(\boldsymbol{S}_b \cdot \boldsymbol{R})}{R^2}\right] \tag{7}$$

where α is the fine structure constant. H_{FS} is the hyperfine Hamiltonian. Because the 21 cm splitting is small compared to the collision energies of interest, we can ignore it in this discussion. We also do not consider interactions involving nuclear spin since the magnetic moments of the nuclei are about three orders of magnitude smaller than that of the electrons.

With those approximations, the resulting dynamics is simpler to analyze if we use the $|S\,M_S I M_I\rangle$ representation, where I, M_I and S, M_S are the angular momenta of the nuclei and electron pairs respectively. A unitary transformation into the $|F\,M_F\,F_a F_b\rangle$ representation can be applied after all collision amplitudes in the $|S\,M_S I M_I\rangle$ picture are calculated.[6] It is desirable to express the cross sections in the $|m_a m_b\rangle \otimes |I M_I\rangle$ basis, where m_a, m_b are the spin angular momenta for electrons in atom a and b respectively. For the sake of economy in notation, we drop explicit reference to the nuclear spin angular momenta since they do not participate in the process under consideration.

In this basis we obtain a simplified expression[19] for $\underline{\mathrm{V}}(R\theta\phi)$ that involves only four channels. We find that the BO eigenvalues for the Hamiltonian in Eq. (6) are given by[19]

$$\begin{aligned}
\epsilon_{S=1,M_S} &= {}^3\Sigma(R) - \frac{\alpha^2}{2R^3} \qquad |M_S| = 1 \\
\epsilon_{S=1,M_S} &= {}^3\Sigma(R) + \frac{\alpha^2}{R^3} \qquad M_S = 0 \\
\epsilon_{S=0,M_S} &= {}^1\Sigma(R).
\end{aligned} \tag{8}$$

In Eq. (8) the splitting of the triplet manifold arises from the anisotropic term H_{DIP}. Using standard re-coupling methods and introducing a partial wave expansion Eq. (3) leads to the coupled radial equations[19]

$$\frac{d^2\underline{G}^{JM}(R)}{dR^2} - \frac{\underline{L}(\underline{L}+1)}{R^2}\underline{G}^{JM}(R) - 2\mu\underline{W}\,\underline{G}^{JM}(R) + k^2\underline{G}^{JM}(R) = 0, \tag{9}$$

where $\underline{G}^{JM}(R)$ is the multichannel radial scattering function, $\underline{L}$ is a diagonal matrix whose entries contain the orbital angular momenta, l, for the collision partners, J is a total angular momentum quantum number and the relationship between J, l and the channel indices are itemized in Table 1.

Table 1. Channel quantum numbers for block J.

Channel	l	S
1	$J-1$	1
2	$J+1$	1
3	J	1
4	J	0

The coupled equations are block diagonal in J and with the definition of channel indices given above, the multichannel radial potential matrix, $\underline{W}(R)$, is given by

$$\begin{pmatrix}
{}^3\Sigma(R) + \frac{(J+2)\alpha^2}{2(2J+1)R^3} & -\frac{3}{2}\frac{\sqrt{J(J+1)}\alpha^2}{(2J+1)R^3} & 0 & 0 \\
-\frac{3}{2}\frac{\sqrt{J(J+1)}\alpha^2}{2(J+1)R^3} & {}^3\Sigma(R) + \frac{(J-1)\alpha^2}{2(2J+1)R^3} & 0 & 0 \\
0 & 0 & {}^3\Sigma(R) - \frac{\alpha^2}{2R^3} & 0 \\
0 & 0 & 0 & {}^1\Sigma(R)
\end{pmatrix}. \tag{10}$$

Numerical solution of these equations allows us to construct the radial multichannel T-matrix, $T^{lS}_{l'S'}(JM)$. The scattering amplitude for a pair of atoms, in internal state $|m_a m_b\rangle$ approaching the solid angle centered at

$\theta_i\phi_i$ to scatter into state $|m'_a m'_b\rangle$ and the solid angle centered at $\theta\phi$ is given by[19]

$$f^{m'_a m'_b}_{m_a m_b}(\theta\phi;\theta_i\phi_i) = \sum_{lm}\sum_{l'm'} Y_{lm}(\theta\phi)Y^*_{l'm'}(\theta_i\phi_i)\frac{2\pi i}{k}W^{m'_a m'_b}_{m_a m_b}(lm, l'm')$$

$$W^{m'_a m'_b}_{m_a m_b}(lm, l'm') \equiv \sum_{JM}\sum_{S,S'}\sum_{M_S,M'_S}\begin{pmatrix} \frac{1}{2} & \frac{1}{2} & S \\ m_a & m_b & -M_S \end{pmatrix}\begin{pmatrix} \frac{1}{2} & \frac{1}{2} & S' \\ m'_a & m'_b & -M'_S \end{pmatrix}$$
$$\times \begin{pmatrix} S & l & J \\ M_S & m & -M \end{pmatrix}\begin{pmatrix} S' & l' & J \\ M'_S & m' & -M \end{pmatrix}$$
$$\times [S,S']^{1/2}[J]\mathcal{T}^{S'l'}_{Sl}(JM). \tag{11}$$

The corresponding cross section is

$$\sigma(m_a m_b \to m'_a m'_b) \equiv \frac{1}{4\pi}\int d\hat{\Omega}_i \int d\hat{\Omega}\, |f^{m'_a m'_b}_{m_a m_b}(\theta\phi;\theta_i\phi_i)|^2$$
$$= \frac{\pi}{k^2}\sum_{lm}\sum_{l'm'}|W^{m'_a m'_b}_{m_a m_b}(lm, l'm')|^2. \tag{12}$$

According to Eq. (10) $\mathcal{T}^{S'l'}_{Sl}(JM)$ is diagonal if the anisotropic dipolar interaction is ignored, i.e. $\alpha = 0$. In that case,

$$\mathcal{T}^{S'l'}_{Sl}(JM) = \delta_{S,S'}\delta_{l,l'}\mathcal{T}(S,l)$$
$$\mathcal{T}(S,l) \equiv \mathcal{T}^{Sl}_{Sl}(JM). \tag{13}$$

We make use of the fact that $\mathcal{T}$ is not an explicit function of JM (it depends on J through the implicit relationship between l, l' i.e. $l = l(J), l' = l'(J)$. For a fixed value of l we can then contract

$$\sum_{JM}[J]\begin{pmatrix} S & l & J \\ M_S & m & -M \end{pmatrix}\begin{pmatrix} S & l & J \\ M'_S & m' & -M \end{pmatrix} = \delta_{m,m'}\delta_{M_S,M'_S} \tag{14}$$

to obtain,

$$\sigma(m_a m_b \to m'_a m'_b) = \frac{\pi}{k^2}\sum_l (2l+1)|W^{m'_a m'_b}_{m_a m_b}(l)|^2$$

$$W^{m'_a m'_b}_{m_a m_b}(l) = \sum_{SM_S}[S]\begin{pmatrix} \frac{1}{2} & \frac{1}{2} & S \\ m_a & m_b & -M_S \end{pmatrix}\begin{pmatrix} \frac{1}{2} & \frac{1}{2} & S \\ m'_a & m'_b & -M_S \end{pmatrix}\mathcal{T}(S,l). \tag{15}$$

Since

$$\mathcal{T}(S,l) = 1 - \exp(2i\delta_{Sl}) \tag{16}$$

where δ_{Sl} is the l'th partial phase shift for elastic scattering in the triplet $S = 1$ and singlet $S = 0$ states of the hydrogen molecule. Evaluation of Eq. (15) leads to an expression for the electron spin exchange cross section

$$\sigma\left(m_a = \frac{1}{2};\, m_b = -\frac{1}{2} \to m_a' = -\frac{1}{2};\, m_b' = \frac{1}{2}\right) = \frac{\pi}{k^2}\sum_l (2l+1)\,\sin^2\Big(\delta_{sl} - \delta_{tl}\Big). \tag{17}$$

Transforming into the $|F_a M_a F_b M_b\rangle$ representation, we recover the classic result of Dalgarno[3]

$$\sigma = \frac{\pi}{4k^2}\sum_l (2l+1)\,\sin^2\Big(\delta_{sl} - \delta_{tl}\Big). \tag{18}$$

For spin exchange, selection rules in Eq. (15) require that $m_a + m_b = m_a' + m_b'$. The selection rule also holds in the hyperfine basis $|F_a M_a F_b M_b\rangle$. The expression given above illustrates the essence of the elastic approximation, the amplitude for each atom to undergo a spin flip (as long as $\Delta m = m_a + m_b - m_a' - m_b' = 0$) can simply be expressed in terms of the elastic scattering phase shifts. Quantum exchange symmetry can be incorporated by a straightforward extension of the above discussion.[19]

In a recent study[19] we calculated the $\Delta\, m \neq 0$ spin flipping transition cross section in collisions of hydrogen atoms. The results for the $\Delta\, m \neq 0$ transitions are shown in Fig. 1. At lower temperatures, where the elastic approximation is no longer valid, we used a fully multichannel theory to calculate the cross sections. The solid blue and red lines correspond to $\Delta F = 1$ transitions and $\Delta F = 2$ cross sections are given by the green line. At energies around $1\,K$ the $\Delta F = 1, 2$ cross sections merge but exchange symmetry effects persist and lead to two distinct cross sections at thermal energies. Because the dipolar interaction H_{DIP} is weak the Born approximation is sometimes employed[7,21] to approximate spin flipping transition cross sections.

2.1. *The Born Approximation*

In the Born approximation the dipolar induced hyperfine changing cross sections, are given by the expression[19]

$$\sigma(F_a M_a F_b M_b \to F_a' M_a' F_b' M_b') = \frac{16\pi\xi}{15}\mu^2\alpha^4\Big[|\mathcal{D}|^2 + |\mathcal{E}|^2 + 2h(\xi)\,\mathcal{D}\,\mathcal{E}\Big], \tag{19}$$

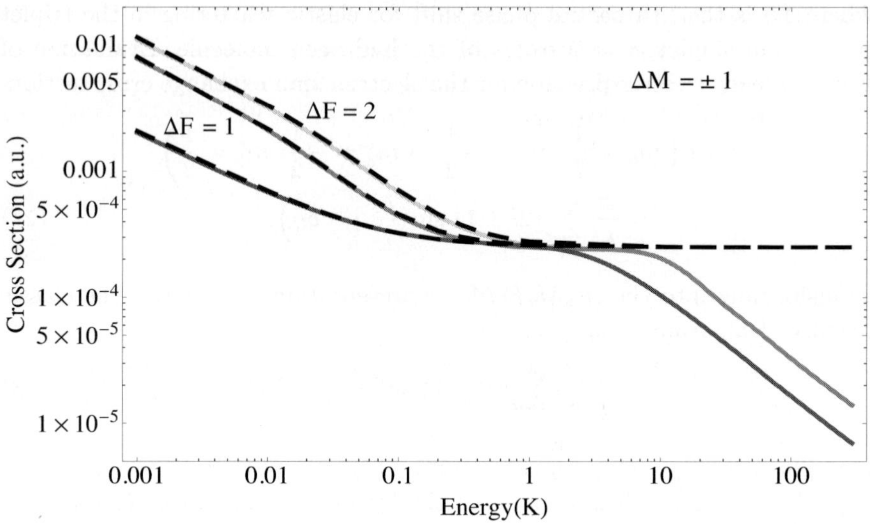

Fig. 1. Solid lines represent results for the multichannel calculation of $\Delta F = 1, 2$ hyperfine level quenching cross sections. The dashed lines represent results obtained using the Born approximation.

in which

$$\xi \equiv \frac{k_f}{k_i}$$

and where

$$\begin{aligned} \mathcal{D} &\equiv \langle F_a M_a F_b M_b | S_Q^{(2)} | F_a' M_a' F_b' M_b' \rangle \\ \mathcal{E} &\equiv \langle F_a M_a F_b M_b | S_Q^{(2)} | F_b' M_b' F_a' M_a' \rangle \end{aligned} \tag{20}$$

are direct and exchange atomic matrix elements of the rank two spin operator[22] $S_Q^{(2)}$, and $Q = \Delta M \equiv M_a' + M_b' - M_a - M_b$. k_i and k_f are the wavenumbers for the system in the initial and final hyperfine levels, and

$$h(\xi) = \frac{3\left(\xi^2 - 1\right)^2 \log\left(\frac{(\xi+1)^2}{(\xi-1)^2}\right)}{8\xi\left(\xi^2+1\right)^2} - \frac{1}{2}.$$

Evaluating Eq. (19) for the cases $\Delta M = \pm 1$, $\Delta F = 1, 2$, we obtain the two distinct expressions

$$\begin{aligned} \sigma_a &\equiv \frac{\pi}{15}\frac{k_f}{k_i}(\mu\alpha^2)^2\frac{1}{2} \\ \sigma_b &\equiv \frac{\pi}{15}\frac{k_f}{k_i}(\mu\alpha^2)^2\frac{1}{2}\Big[2 + 2h(\xi)\Big]. \end{aligned} \tag{21}$$

The subscripts a, b are state dependent and denote the exchange symmetry properties of the target states. Those symmetries lead to even or odd partial wave contributions in the corresponding cross sections.

The cross sections calculated using Eq. (21) are illustrated in Fig. 1 by the dashed lines. At higher collision energies $k_f \approx k_i$, and $h(\xi) \to -1/2$. In this limit, the cross sections tend to the common value given by

$$\sigma \longrightarrow \frac{\pi(\mu\alpha^2)^2}{30}. \tag{22}$$

Inspection of Fig. 1 shows that the predictions of the Born approximation are in harmony with the results of the fully coupled channel approach for collision energies $E < 1\,K$. However, at higher energies the Born approximation fails, predicting the constant cross sections given by Eq. (22).

Though the multichannel approach provides for an accurate description, its application becomes increasingly prohibitive at higher energies where many partial waves contribute. Keeping in mind the efficiency and utility of the elastic approximation for the calculation of $\Delta m = 0$ transition cross sections, it is worthwhile to investigate the feasibility of extending such an approach to the case of $\Delta m \neq 0$ processes. In the section below we make use of Eq. (11) to develop such a theory. We are interested in collision energies for $T > 1K$ and so we can neglect hyperfine splittings, and employ the basis discussed above to simplify the algebra. In addition, we keep our discussion simple by focusing on the $H - D$ system which does not exhibit the additional complications arising from exchange symmetry effects.

2.2. *Partial Wave Analysis*

Evaluating Eq. (11) we find

$$\sigma(m_a m_b \to m'_a m'_b) = \frac{\pi}{k^2} \sum_{lm} \Big(|W^{m'_a m'_b}_{m_a m_b}(lm, lm')|^2 + |W^{m'_a m'_b}_{m_a m_b}(lm, l+2\,m')|^2 + |W^{m'_a m'_b}_{m_a m_b}(lm, l-2\,m')|^2 \Big) \tag{23}$$

where we have used the selection rules, $l' = l, l' = l \pm 2$ and $m' = \Delta m = \Delta M$. We first investigate the contribution due to the diagonal $l = l'$ term. Though the radial T-matrix contains off-diagonal contributions they are small and to first order we can approximate

$$T^{Sl}_{Sl}(J) \approx 1 - \exp(2i\delta_l(J)) \tag{24}$$

where $\delta_l(J)$ is the elastic scattering phase shift for a given l and J. They are obtained solving Eq. (9) but including only the diagonal entries in the radial potential Eq. (10). Evaluating the $l = l'$ contribution, we get

$$\sum_m |W^{m'_a m'_b}_{m_a m_b}(lm, lm')|^2 = \Big[\frac{(2l-1)(l-1)}{15l}\sin^2(\delta_2 - \delta_3) + \frac{2(3+2l)(2l-1)}{15(1+2l)}\sin^2(\delta_1 - \delta_2) + \frac{(2+l)(3+2l)}{15(1+l)}\sin^2(\delta_1 - \delta_3)\Big] \quad (25)$$

where we used the shorthand notation δ_i to represent the elastic scattering phaseshift for channel i, as itemized in Table 1, for a given block J. The $\Delta l = 0$ component to the total cross section is then given by,

$$\sigma(\Delta l = 0) = \frac{\pi}{k^2}\sum_l \Big[\frac{(2l-1)(l-1)}{15l}\sin^2(\delta_2 - \delta_3) + \frac{2(3+2l)(2l-1)}{15(1+2l)}\sin^2(\delta_1 - \delta_2) + \frac{(2+l)(3+2l)}{15(1+l)}\sin^2(\delta_1 - \delta_3)\Big]. \quad (26)$$

We now turn our attention to the $l' = l \pm 2$ contribution. According to Eq. (11)

$$\sum_m |W^{m'_a m'_b}_{m_a m_b}(lm, l+2\, m')|^2 = \frac{(3+2l)}{20}|T^{Sl+1}_{Sl-1}(JM)|^2 = \frac{(3+2l)}{20}|T^2_1|^2 \quad (27)$$

where $J = l$, and in the second equality we used the channel indices given in Table 1 to label the radial T - matrix. Because the T matrix element is off-diagonal we cannot, at first glance, express it in terms of an elastic scattering phase shift. We estimate T^2_1 by solving the coupled pair of equations,

$$\frac{d^2 f_1}{dR^2} - \frac{l_1(l_1+1)}{R^2}f_1 - 2\mu V_{BO} f_1 - k^2 f_1 = \frac{3\mu\alpha^2\sqrt{J(J+1)}}{(2J+1)R^3}f_2$$
$$\frac{d^2 f_2}{dR^2} - \frac{l_2(l_2+1)}{R^2}f_2 - 2\mu V_{BO} f_2 - k^2 f_2 = \frac{3\mu\alpha^2\sqrt{J(J+1)}}{(2J+1)R^3}f_1 \quad (28)$$

where we used Eq. (9) and Eq. (10), V_{BO} is the Born-Oppenheimer energy for the $^3\Sigma$ state, $l_1 = J-1$, $l_2 = J+1 = l_1+2$, and we ignored the diagonal components proportional to α^2.

To estimate f_1, f_2 we appeal to the distorted wave approximation, in which f_i on the r.h.s. of Eq. (28) is replaced by $f_i^{(0)}$ a solution to the homogeneous equations

$$\frac{d^2 f_i^{(0)}}{dR^2} - \frac{l_i(l_i+1)}{R^2} f_i^{(0)} - 2\mu V_{BO} f_i^{(0)} - k^2 f_i^{(0)} = 0. \tag{29}$$

As $R \to \infty$ they have the limit

$$f_i^{(0)}(R) \to \sin(kR - l_i \frac{\pi}{2} + \delta_{BO}) \tag{30}$$

where δ_{BO} is the phase shift due to $V_{BO}(R)$. In this approximation, the off-diagonal element of the T matrix is given by

$$\mathcal{T}_1^2 \approx -2iK \exp(2i\delta_{BO}) \tag{31}$$

where

$$K \equiv -\frac{3\mu\alpha^2\sqrt{J(J+1)}}{(2J+1)} \frac{1}{k} \int_0^\infty \frac{dR}{R^3} f_1^{(0)}(R) f_2^{(0)}(R), \tag{32}$$

and we make use of the fact $|K| << 1$. Consider the following inhomogeneous equation,

$$\frac{d^2 f_1}{dR^2} - \frac{l_1(l_1+1)}{R^2} f_1 - 2\mu V_{BO} f_1 - k^2 f_1 = \frac{3\mu\alpha^2\sqrt{J(J+1)}}{(2J+1)R^3} f_2^{(0)} \tag{33}$$

Our solution to Eq. (33) has the asymptotic limit

$$f_1(R) \to \sin(kR - l_1 \frac{\pi}{2} + \delta_{BO}) + K \cos(kR - l_1 \frac{\pi}{2} + \delta_{BO}) \tag{34}$$

where K is given by Eq. (32). Or,

$$\begin{aligned} & f_1(R) \sim C \sin(kR - l_1 \frac{\pi}{2} + \delta_{BO} + \delta_4) \\ & \tan(\delta_4) = K. \end{aligned} \tag{35}$$

Therefore, we can express $\mathcal{T}_1^2$ in terms of the phase shift δ_4 obtained solving the inhomogeneous Eq. (33). Thus

$$\sum_m |W_{m_a m_b}^{m_a' m_b'}(lm, l+2\, m')|^2 = \frac{(3+2l)}{20} 4\sin^2(\delta_4) \tag{36}$$

and

$$\sigma(\Delta l = 2) = \frac{4\pi}{k^2} \sum_l \frac{(3+2l)}{20} \sin^2(\delta_4). \tag{37}$$

Similarly, we find that $\sigma(\Delta l = -2) = \sigma(\Delta l = 2)$.

We now need to calculate the partial wave phase shifts δ_i defined above. For $i = 1, 2, 3$ we solve the elastic scattering equations with the potentials V_i given by the diagonal elements of potential matrix Eq. (10), and δ_4 is obtained from solution to the inhomogeneous equation Eq. (33). In the Born approximation we ignore V_{BO} and obtain,

$$\delta_i^B = -\frac{1}{k}\int_0^\infty dR\, f_i^{(0)}(R)V_i(R)f_i^{(0)}(R) \tag{38}$$

for $i = 1, 2, 3$, and $f_i^{(0)}(R) = (kR)j_{l_i}(kR)$ are the regular Bessel-Ricatti functions. Inserting $V_i(R)$ from Eq. (10) (and excluding V_{BO}) we obtain, for $(l > 0)$

$$\begin{aligned}\delta_1^B &= -\frac{k\,\alpha^2\mu}{2(1+l)(3+2l)}\\ \delta_2^B &= -\frac{k\,\alpha^2\mu}{2l(2l-1)}\\ \delta_3^B &= -\frac{k\,\alpha^2\mu}{2l(l+1)}.\end{aligned} \tag{39}$$

Inserting the δ_i^B into Eq. (26) we find

$$\begin{aligned}\sigma(\Delta l = 0) &= \frac{\pi}{k^2}\sum_{l=1}^{\infty}\frac{3k^2(2l+1)\alpha^4\mu^2}{10l(l+1)(2l-1)(2l+3)}\\ &= \pi(\alpha^2\mu)^2\left[\frac{2}{5}\log(2) - \frac{1}{6}\right]\end{aligned} \tag{40}$$

Also,

$$\begin{aligned}\delta_4^B &= 3\mu\alpha^2\frac{\sqrt{(l+1)(l+2)}}{(2l+3)}k\int_0^\infty \frac{dR}{R}\,j_l(kR)j_{l+2}(kR)\\ &= \mu\alpha^2\frac{k}{2\sqrt{(1+l)(2+l)}(3+2l)}\end{aligned} \tag{41}$$

and

$$\begin{aligned}\sigma(\Delta l = 2) &= \frac{\pi}{k^2}\sum_l \frac{k^2\alpha^4\mu^2}{20(l+1)(l+2)(2l+3)}\\ &= \frac{\pi}{20}(\alpha^2\mu)^2\,(3 - 4\log(2)).\end{aligned} \tag{42}$$

The total $\Delta m = \pm 1$ cross section is given by

$$\begin{aligned}\sigma &= \sigma(\Delta l = 0) + \sigma(\Delta l = 2) + \sigma(\Delta l = -2)\\ &= \frac{2\pi}{15}(\alpha^2\mu)^2.\end{aligned} \tag{43}$$

Recognizing that the transformation from the $|m_a m_b\rangle \otimes |IM_I\rangle$ to the $|F_a M_a F_b M_b\rangle$ basis involves a factor of 1/4 we find that Eq. (43) is in harmony with the previous expression given in Eq. (22).

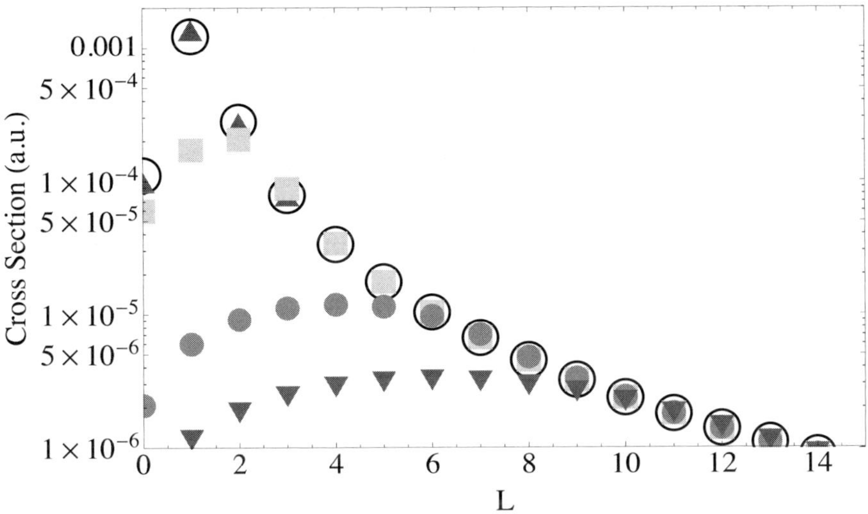

Fig. 2. Solid open circles denote the Born approximation results. The solid markers represent the cross sections obtained using the elastic approximation. For collision energies of; 1 K - triangles, 10K - squares, 100K - circles, and the inverted triangles represent the 300K data.

In Fig. 2 we plot the partial wave Born approximation cross sections as a function of angular momentum l by the open circles. They are independent of the collision energy. In that figure we superimpose the results obtained when V_{BO} was included in the determination of the partial wave phase shifts $\delta_i(l)$. The blue triangles represent the results obtained for the collision energy that corresponds to the temperature $T = 1K$. The cross sections are in good agreement with those obtained by the partial wave Born approximation. The green squares are cross sections for the collision energy $T = 10K$. A discrepancy between the Born approximation values is apparent at lower values of l. Those discrepancies become more dominant for increasing l as the collision energy is increased. The Born approximation partial phase shifts are proportional to

$$\delta^B \approx k \int_0^\infty \frac{dR}{R} j_l(kR) j_{l'}(kR) \tag{44}$$

and, due to the scaling property of the $1/R^3$ potential, the integral is independent of k. The effect of including V_{BO} in calculation of the distorted wave $f_i^{(0)}(R)$ is an effective lower limit cutoff, due to the repulsive inner wall of the $^3\Sigma$ BO potential, in the radial integrals that determine $\delta_i(l)$. The reason for the failure of the Born approximation is now clear: it overestimates the contributions of close atom encounters.

3. Summary, Conclusion and Personal Reflections

Equation (26) and Eq. (37) provide an accurate expression, useful at higher energies, for the calculation of $\Delta m = \pm 1$ spin flip cross sections of the $H-D$ collision system. A similar analysis is available for $\Delta m = \pm 2$ transitions and for the inclusion of exchange symmetry effects in $H - H$ collisions. In such a description, spin flips occur due to interference of different phase histories, induced by rotational couplings, in the $^3\Sigma$ sector.

Consider this process in the adiabatic (or molecular gauge[20]) picture. In it, the scattering equations have the form[20]

$$-\frac{1}{2\mu}\left(\boldsymbol{\nabla} - i\underline{\boldsymbol{A}}\right)^2 \underline{\Psi}(\boldsymbol{R}) + \underline{V}(R)\underline{\Psi}(\boldsymbol{R}) = E\underline{\Psi}(\boldsymbol{R}) \tag{45}$$

where the scalar matrix potential $\underline{V}(R)$ is diagonal with the BO energies Eq. (8) as its entries. $\underline{\boldsymbol{A}}$ is a matrix vector potential.[20] Equation (3) is obtained from it by a unitary gauge transformation.[20] Suppose we ignore the vector potential in Eq. (45) and proceed to solve for the amplitudes $\underline{\Psi}(\boldsymbol{R})$. In the $^3\Sigma$ sector the amplitudes can be parameterized by two distinct phase histories, δ_A, δ_B, that result from the two components $^3\Sigma - \alpha^2/2R^3$, $^3\Sigma + \alpha^2/R^3$ of the triplet manifold. The triplet BO splitting is independent of the internuclear orientation and it is a consequence of the magnetic interaction for the two configurations shown in Fig. 3. Can the $\Delta m \neq 0$ cross sections be expressed solely in terms of δ_A and δ_B? Our analysis in the previous section suggests that the answer is no. In addition to the dynamic phases generated by the BO potentials, the vector potential $\underline{\boldsymbol{A}}$ also contributes, and in our formulation it is incorporated implicitly through the rotational couplings terms given in Eq. (10).

Alex's landmark paper on spin exchange is representative of the outstanding contributions that he has made to atomic, molecular and optical science. The paper is emblematic of his scientific style; austere, succinct and elegant. It has been noted at this symposium that Alex is a maestro of minimalism. Perhaps, but he uses that esthetic to great effect.

Among his students and colleagues, Alex is know for his accessibility. Alex has that unique ability to guide and mentor by the application of the gentlest of perturbations. As a former postdoc, I remember the daily queue of students, postdocs and visitors outside his office waiting for their turn

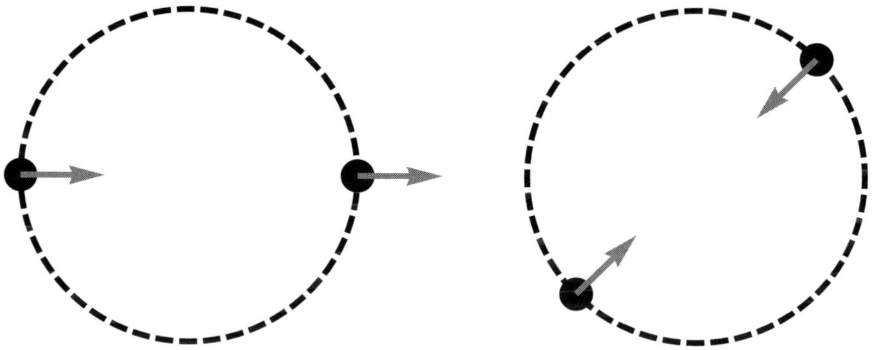

Fig. 3. Spin orientation configurations corresponding to the two components of the $^3\Sigma$ BO energy given in Eq. (8).

to see him. It was clear to me how seriously Alex took this responsibility. Everyone obtained the individualized attention that was needed. I re-collect one of those meetings in particular.

Alex is always full of research ideas, and in that session he proposed that I should tackle, what he suggested is, an unresolved problem in AMO physics. Excited, I returned to my office and immediately set out to research the existing literature on the proposed topic. I soon became frustrated since the ostensibly authoritative treatments in those texts gave no hint of deficiencies in the theory that Alex alluded to. However, I kept thinking about the problem. After twenty years, I now fully comprehend and appreciate Alex's prescient observations, and have a renewed admiration of his deep and profound mastery of his field. Those, long-ago, late afternoon discussions still serve as a well-spring for me today. That's Alex, what you see is only a fraction of what you get.

Acknowledgments

This work was supported by NSF grant PHY-0758140 and, in part, by a NASA EPSCoR grant awarded to the state of Nevada.

References

1. J. P. Wittke and R. H. Dicke, *Phys. Rev.* **103**, 620 (1956).
2. E. M. Purcell and G. B. Field, *ApJ* **124**, 542 (1956).
3. A. Dalgarno, *Royal Society of London Proceedings Series A* **262**, 132 (1961).
4. F. J. Smith, *Planet. Space Sci.* **14**, 929 (1966).
5. A. C. Allison and A. Dalgarno, *ApJ* **158**, 423 (1969).

6. B. Zygelman, A. Dalgarno, M. J. Jamieson and P. C. Stancil, *Phys. Rev. A* **67**, 042715 (2003).
7. B. Zygelman, *ApJ* **622**, 1356 (2005).
8. M. Desaintfuscien and C. Audoin, *Phys. Rev. A* **13**, 2070 (1976).
9. C. Baumgarten, B. Braun, M. Capiluppi, G. Ciullo, P. F. Dalpiaz, H. Kolster, P. Lenisa, H. Marukyan, A. Nass, D. Reggiani, M. Stancari and E. Steffens, *European Physical Journal D* **48**, 343 (2008).
10. A. Dalgarno and M. R. H. Rudge, *Royal Society of London Proceedings Series A* **286**, 519 (1965).
11. L. C. Balling, R. J. Hanson and F. M. Pipkin, *Phys. Rev.* **133**, 607 (1964).
12. S. B. Crampton and H. T. M. Wang, *Phys. Rev. A* **12**, 1305 (1975).
13. A. C. Allison, *Phys. Rev. A* **5**, 2695 (1972).
14. T. G. Heil, S. Green and A. Dalgarno, *Phys. Rev. A* **26**, 3293 (1982).
15. T. G. Heil, K. Kirby and A. Dalgarno, *Phys. Rev. A* **27**, 2826 (1983).
16. S. R. Furlanetto, S. P. Oh and F. H. Briggs, *Physics Reports* **433**, 181 (2006).
17. A. Loeb and M. Zaldarriaga, *Phys. Rev. Lett.* **92**, 211301 (2004).
18. H. A. Bethe and E. E. Salpeter, *Quantum Mechanics of One- and Two-Electron Atoms* (Quantum Mechanics of One- and Two-Electron Atoms, New York: Academic Press, 1957, 1957).
19. B. Zygelman, *Phys. Rev. A (to be submitted)* (2009).
20. B. Zygelman, A. Dalgarno and R. Sharma, *Phys. Rev. A* **49**, 2587 (1994).
21. S. Hensler, J. Werner, A. Griesmaier, P. O. Schmidt, A. Görlitz, T. Pfau, S. Giovanazzi and K. Rzażewski, *Applied Physics B: Lasers and Optics* **77**, 765 (2003).
22. B. Zygelman, *arXiv:physics/0209052* (2002).

MATRIX ELEMENT SUMS EVALUATED *VIA* DIFFERENTIAL EQUATIONS IN CALCULATIONS OF ATOMIC AND MOLECULAR PROPERTIES

MICHAEL JAMIESON

Department of Computing Science, University of Glasgow, Glasgow, G12 8QQ, UK.
E-mail: mjj@dcs.gla.ac.uk

Following the establishment of quantum mechanics around 80 years ago, many formulae that describe how atoms and molecules interact with each other and with external electromagnetic fields were derived. While such formulae provide prescriptions for calculating atomic and molecular properties many of them are inconvenient to evaluate because of the need to calculate the sums of a large, often infinite, number of expressions constructed from various matrix elements and excitation energies. Alex Dalgarno through his major work in perturbation theory pioneered the method, now standard, in which these sums are calculated from the solutions of finite numbers of differential equations. This article is a brief summary of the method and its use.

Keywords: Differential equations; perturbation theory; sum rules.

1. Introduction

I have had the pleasure of a friendship with Alex Dalgarno, and of working with him, since I started as one of his graduate students at the Queen's University of Belfast 42 years ago. I was very fortunate in being invited by him to complete my thesis at the Harvard College Observatory/ Smithsonian Astrophysical Observatory, now the CfA. I am very pleased to have the privilege of contributing to this Celebratory Symposium in his honour.

As is clear from this meeting, the scientific contribution made by Alex Dalgarno is enormous, covering Atomic and Molecular Physics, Planetary Atmospheres and Astrochemistry. His scholarship is manifest in many papers; those referred to here comprise only around 5% of the total. I have chosen to review an important, effective and valuable technique that Alex pioneered for calculations of atomic and molecular properties.

The formulae of quantum mechanics that quantify how atoms and molecules interact with each other and with external influences can be viewed as prescriptions for calculating atomic and molecular properties. They can be computationally impractical; often they involve the sums of many matrix elements that must be calculated first. Alex Dalgarno pioneered the method, now standard, in which such sums, often of an infinite number of terms, are calculated from the solutions of finite numbers of differential equations. Usually the differential equations must be solved numerically. The solutions can be obtained by variational methods with numerical optimisation and by finite-step methods. Alex, after using the marvellously named "Whirlwind" electronic computer at MIT [1] albeit for a different problem, advocated the idea that the increasing availability of computers would make the numerical solution of such differential equations routine. It would be impossible to do justice to Alex's contribution in this area alone in a single article and therefore I attempt to convey the spirit of the method rather than give a full exposition; I cite only Alex's original contributions since an enumeration of the publications of other investigators who use Alex's work would cover most of this entire book.

2. The Differential Equation Method

2.1. *Perturbation Theory*

Many of the applications of perturbation theory to investigations in atomic and molecular physics were developed by Alex Dalgarno. The formalism of the perturbation method was published at the start of a very fruitful period of its use by Dalgarno and Stewart [2]. When a system with Hamiltonian H in state $|0\rangle$ with energy E_0 is subjected to a perturbing potential v it undergoes a change in energy. The first order (in v) change is the expectation value $\langle 0|v|0\rangle$. The second order change is the sum over states $|p\rangle$ with energies E_p

$$\mathcal{E}_2 = -\mathsf{S}' \frac{|\langle 0|v|p\rangle|^2}{E_p - E_0} = -\mathsf{S}' \langle 0|v|p\rangle \frac{1}{E_p - E_0} \langle p|v|0\rangle \,. \tag{1}$$

Throughout S' will denote the sum over all states $|p\rangle$ including the continuum but excluding the state $|0\rangle$. Expression (1) is inconvenient to evaluate; neither the infinite number of discrete states nor the continuum can be summed and the sum must be truncated. Furthermore each retained term requires considerable effort for its evaluation. The more efficient method introduced by Alex is to cast (1) into a finite sum of terms dependent on the solutions of a finite number of differential equations. This was done by

Dalgarno and Lewis [3,4] in a study of long-range forces. The sum in Eq. (1) is the part of the resolvent $(H - E_0)^{-1}$ that is orthogonal to the state $|0\rangle$. Therefore

$$\mathcal{E}_2 = \langle 0|v|\chi\rangle \tag{2}$$

where $|\chi\rangle$ satisfies the differential equation

$$(H - E_0)\,|\chi\rangle + v\,|0\rangle - \langle 0|v|0\rangle\,|0\rangle = 0; \tag{3}$$

the term $\langle 0|v|0\rangle\,|0\rangle$ preserves the orthogonality to the state $|0\rangle$. Dalgarno and Lewis [3] showed that use of Eq. (3) is equivalent to use of a variational method where the expectation value of the energy is minimised with respect to a trial wave function, written as the product of the unperturbed wave function and a function dependent on parameters to be determined. This result started the development of variation-perturbation methods.

2.2. *Static and Dynamic Polarizabilities*

The differential equation method for calculating polarizabilities follows from the analysis outlined above. The static dipole polarizability of an N-electron atom with Hamiltonian H and dipole moment $\mathbf{m} = -e\sum_{i=1}^{N}\mathbf{r}_i$, in which e denotes the *magnitude* of the charge of an electron and $\mathbf{r}_i$ denotes the position of the i'th electron, is

$$\alpha = 2\left\langle 0\left|\hat{\mathbf{z}}.\mathbf{m}\left[\mathsf{S}'\,|p\rangle\frac{1}{E_p - E_0}\langle p|\right]\mathbf{m}.\hat{\mathbf{z}}\right|0\right\rangle \tag{4}$$

where $\hat{\mathbf{z}}$ is a unit vector in the direction of an arbitrarily chosen z-axis, The sum in Eq. (4) is the projection of $(H - E_0)^{-1}$ orthogonal to $|0\rangle$ and the polarizability is

$$\alpha = 2\,\langle 0\,|\hat{\mathbf{z}}.\mathbf{m}|\,\chi\rangle \tag{5}$$

where $|\chi\rangle$ satisfies the differential equation

$$(H - E_0)\,|\chi\rangle = \mathbf{m}.\hat{\mathbf{z}}\,|0\rangle - \langle 0|\mathbf{m}.\hat{\mathbf{z}}|0\rangle\,|0\rangle\,. \tag{6}$$

Of course Eq. (4) is derived *via* Eq. (6) by perturbation analysis of the effect of an electric field $F\hat{\mathbf{z}}$ ($v = -\mathbf{m}.\hat{\mathbf{z}}F$) but it was Alex Dalgarno who pointed out that solving Eq. (6) *directly* would prove fruitful. Eq. (6) can be solved variationally by minimising the functional

$$J = \langle\chi|(H - E_0)|\chi\rangle - 2\,\langle\chi|\mathbf{m}.\hat{\mathbf{z}}|0\rangle + 2\,\langle 0|\mathbf{m}.\hat{\mathbf{z}}|0\rangle\,\langle\chi|0\rangle \tag{7}$$

with respect to the parameters of a trial solution. In formulations that are separable in the electronic coordinates Eq. (6) can be replaced by linear second order ordinary differential equations in the radial coordinates; many finite difference techniques exist for a numerical solution of such equations. Dalgarno and Parkinson [5], in calculating the polarizabilities of atoms from boron to neon, used finite difference methods; they were probably the first investigators to do so[a]. They also presented variational calculations.

The dynamic dipole polarizability at frequency ν is

$$\alpha(\nu) = \mathsf{S}' \left(\frac{1}{E_p - E_0 + h\nu} + \frac{1}{E_p - E_0 - h\nu} \right) | \langle 0|\hat{\mathbf{z}}.\mathbf{m}|p\rangle |^2. \tag{8}$$

where h is Planck's constant. Chan and Dalgarno [6] replaced Eq. (8) by

$$\alpha(\nu) = \langle 0|\hat{\mathbf{z}}.\mathbf{m}|\chi(\nu) + \chi(-\nu)\rangle \tag{9}$$

where $|\chi(\pm\nu)\rangle$ satisfy the differential equations

$$(H - E_0 \pm h\nu)\,|\chi(\pm\nu)\rangle = \mathbf{m}.\hat{\mathbf{z}}\,|0\rangle - \langle 0\,|\mathbf{m}.\hat{\mathbf{z}}|\,0\rangle\,|0\rangle \pm h\nu\,\langle 0|\chi(\pm\nu)\rangle\,. \tag{10}$$

Eq. (8) implies that the contribution from state $|p\rangle$ is magnified when $h\nu \approx E_p - E_0$. The higher excited states of an atom are Rydberg states whose energies are close to each other; thus if $h\nu$ is near $E_p - E_0$ it is also near several other excitation energies and terms to at least the p'th must be included in a direct summation. The spatial extent of a state increases with its energy. If $h\nu \approx E_p - E_0$ the function $|\chi(-\nu)\rangle$ must take account of the state $|p\rangle$. A finite difference method to find $|\chi(-\nu)\rangle$ when applied over a range of radial coordinate that is not large enough to 'contain' $|p\rangle$ and its neighbours effectively truncates the sum in Eq. (8). The range must be carefully chosen but this usually requires less effort than attempts to find convergence in Eq. (8).

It is difficult to estimate the contribution from excitation into the continuum; in variational methods it is often replaced by the contribution of a 'fictitious' discrete state. The matrix elements and the excitation energies in Eq. (8) constitute the dipole spectrum. The spectrum is often approximated by a model consisting of n, say, effective excitation energies ΔE_p and *oscillator strengths* f_p; Eq. (8) is replaced by

$$\alpha(\nu) = e^4 a_0 \sum_1^n \frac{f_p}{\Delta E_p^2 - h^2\nu^2} \tag{11}$$

where a_0 is the Bohr radius. Chan and Dalgarno [6,7], in calculating the refractive index of helium, solved Eqs. (10) variationally. They reflected the

[a]The 'DEUCE' computer of Short Brothers and Harland, Belfast was used; this machine, and the similar model at the Queen's University that was acquired largely at the instigation of Alex, were the first digital computers in Northern Ireland.

resonant structure of Eq. (11) by including terms with coefficients $(\nu \pm \nu_t)^{-1}$ for ν_t to be determined. The same procedure was used by Victor, Dalgarno and Taylor [8]. They calculated a model dipole spectrum and used it and Eq. (11) to obtain the polarizabilities at imaginary frequencies that they needed to find the leading van der Waals coefficient C_6 of the long-range interatomic potential.

The advantages of the differential equation method are clear; the excited states and energies are not required explicitly, all the discrete terms in Eqs. (1), (4) and (8) are included automatically and the continuum part, which is otherwise awkward to evaluate, is included. Has 'something for nothing' been obtained? The answer is affirmative when calculating static polarizabilities but care is needed when calculating dynamic polarizabilities.

2.3. *Sum Rules*

The replacement of sums of functions of matrix elements and energies by Eqs. (2), (5) and (9) are examples of sum rules. Dalgarno [9] used Eqs. (8), (9) and (10) to derive and generalise the Thomas-Kuhn sum rule for oscillator strengths; these sum rules are useful in testing the accuracy of various calculations.

Dalgarno and Lewis [3] showed that the sum in Eq. (1) can be replaced by $\mathsf{S}' \langle 0|v|p\rangle \langle p|f|0\rangle$ where f satisfies a differential equation; substitution of the unit projector for $\mathsf{S}\,|p\rangle\langle p|$ reduces $\mathsf{S}' \langle 0|v|p\rangle \langle p|f|0\rangle$ to two terms. In a separable model this sum rule can be expressed in terms of functions of a single electronic coordinate. Dalgarno and Lewis derived the sum rule

$$\mathsf{S}' \frac{\langle 0|g|p\rangle \langle p|g|0\rangle}{E_0 - E_p} = \langle 0|fg|0\rangle - \langle 0|f|0\rangle \langle 0|g|0\rangle \tag{12}$$

where f satisfies the differential equation, in which $\psi_0 \equiv |0\rangle$,

$$2\nabla f.\nabla\psi_0 + \psi_0\nabla^2 f = g\psi_0. \tag{13}$$

Thus the problem of summing an infinite number of elements is reduced to that of solving the single differential equation (13). Care is needed when finding a finite-step solution of Eq. (13) in the neighbourhood of a zero of ψ_0 where it is possible for the solution f to have a pole.

Dalgarno and Lewis [3] used the sum rule (12) to find the second order long-range interaction between hydrogen atoms; they were able to solve Eq. (13) analytically. They derived another sum rule which they used in a study of the third order interaction. It allows a complicated infinite sum to be evaluated from the solution of only one differential equation.

3. Applications

3.1. *Polarizabilities, Shielding and Relativistic Energies*

Dalgarno and Stewart [10,11] solved Eq. (13) analytically for a hydrogen-like atom and used the solution in an analysis of properties of the $1s\sigma$, $2p\sigma$ and $2p\pi$ states of HeH^{2+}. They considered the ion to consist of a hydrogen-like atom perturbed by a nucleus at distance R from the atom's nucleus and they expanded the perturbing potential in inverse powers of R. They solved various versions of Eq. (13) in the region of large R where ψ_0 has no zeros and they obtained inverse power series in R for the potentials, the kinetic energies and the dipole and quadrupole polarizabilities. Dalgarno and Lynn [12] studied the long-range interaction in a diatomic molecular ion; they expressed Eq. (13) as a partial differential equation in confocal elliptic coordinates which they solved analytically.

Dalgarno and Lewis [13] derived another sum rule,

$$\mathsf{S}'(E_0 - E_p)\,\langle 0|f|p\rangle\,\langle p|f|0\rangle = \langle 0|fg|0\rangle\,. \tag{14}$$

They used it and Eq. (12) to simplify an expression for the second order energy of a perturbed atom or molecule and found semi-empirical formulae for the dipole and quadrupole polarizabilities of atoms and diatomic molecules. They computed dipole polarizabilities of the alkali-metals, the helium isoelectronic sequence and molecular hydrogen.

Dalgarno and Stewart [14] investigated the effect of a perturbing potential on diagonal matrix elements. They found the first order change of the ground state expectation value of a dynamical variable L to be

$$\mathcal{L}_1 = \langle 0|L|0\rangle + 2\left(\langle\chi|L|0\rangle - \langle 0|L|0\rangle\,\langle\chi|L|0\rangle\right) \tag{15}$$

where the response $|\chi\rangle$ satisfies Eq. (3); the first order correction, which would otherwise be expressed as an infinite expansion, is a single term dependent on the solution of one differential equation. They used Eqs. (3) and (15) in an extensive investigation in which the potential is split into a potential for which the eigenstate $|0\rangle$ and energy E_0 are known, and a balancing potential; subsequent analysis allows properties that depend on diagonal matrix elements to be evaluated to first order in the balancing potential. Dalgarno and Stewart used the method to calculate polarizabilities, diamagnetic susceptibilities, and relativistic and radiative contributions to the energies of the members of the helium isoelectronic sequence. Dalgarno and Pengelly [15] used a similar method to evaluate the dipole polarizabilities of the alkali-metals; they solved Eq. (3) analytically at large nuclear separations where they were able to use Whittaker functions in the representation of the unperturbed state $|0\rangle$. Chen and Dalgarno [16] derived a procedure to correct matrix elements evaluated with approximate *wave functions* that satisfy an eigenvalue equation with a potential that is not necessarily known; no infinite expansion is required. They solved Eq. (3) variationally.

Dalgarno and Stewart [17–20] used Eq. (15) and an extension of the theory to second order to evaluate oscillator strengths of HeH^{2+}, and energies with relativistic and radiative corrections, dipole and quadrupole polarizabilities, magnetic shielding factors and diamagnetic susceptibilities for the helium isoelectronic sequence.

Dalgarno, Davison and Stewart [21] evaluated the nuclear shielding factors and quadrupole polarizabilities for the helium isoelectronic sequence. In calculating the shielding factors they considered an atom perturbed by two external influences, an electric field and an electric field gradient. They introduced two first order functions like the response $|\chi\rangle$ of Eq. (3); the resulting expression for the shielding factor contains few terms, again dependent on the solutions of differential equations which they solved variationally.

3.2. *Static and Time-dependent Hartree-Fock Theory*

Hartree-Fock theory provides examples of studies in which an approximate potential is used initially and the calculated properties are subsequently corrected, as described in Sec. 3.1. Alex Dalgarno explored the application of perturbation theory within the Hartree-Fock model; his development used the differential equation formalism.

Evaluation of the Hartree-Fock energy requires solutions of the Hartree-Fock eigenvalue equations. Dalgarno [22] and Cohen and Dalgarno [23] calculated the Hartree-Fock energy by taking the Coulomb electron-nucleus interaction as the unperturbed potential and the Hartree-Fock average electron-electron interaction as the perturbation; they found the correction by solving Eq. (3) and its higher order counterparts as expansions in powers of the nuclear charge Z. Cohen and Dalgarno [24] used Eqs. (3) and (15) to derive the theorem that Hartree-Fock theory yields the expectation value of a sum of one-electron operators correct to first order in the difference between the Hartree-Fock and exact Hamiltonians, *i.e.* correct to first order in correlation; they demonstrated that the expectation values of certain one-electron operators for helium are accurate to within 3% and most to within 1%.

The studies described above are twofold; they concern the difference between an *approximation* to the Hartree-Fock Hamiltonian and the Hartree-Fock Hamiltonian itself, and they concern the difference between the Hartree-Fock Hamiltonian and the exact Hamiltonian; the differences are regarded as perturbations. The next step in the development of Hartree-Fock perturbation theory was the introduction of an external influence such as an electric field to explore, for example, the accuracy of the Hartree-Fock account of polarizabilities. Dalgarno [25], in an exposition of static Hartree-Fock perturbation theory, derived the Hartree-Fock equivalent of Eq. (3); the Hartree-Fock self-consistent potential introduces extra terms into the counterpart of Eq. (3) and its generalisations to higher orders, and the resulting equations are integro-differential equations. Although the equations contain integrals of their solutions they are straightforward to solve variationally or by finite-step methods. Dalgarno and McNamee [26] used static Hartree-Fock perturbation theory to calculate the dipole and quadrupole polarizabilities of the helium atom and the lithium ion. Their calculated dipole polarizability of helium was within 4.3% of the experimental value.

Dalgarno and Victor [27] used Hartree-Fock perturbation theory to calculate the lifetime of metastable helium against two photon decay. The two-photon transition matrix element involves an infinite sum that can be replaced by two terms involving the solution of the Hartree-Fock counterparts of Eqs. (10). The lifetime calculated by Dalgarno and Victor is of the same order of magnitude as the accurate value. Their theory is a hybrid of static Hartree-Fock theory and time-dependent perturbation theory. To overcome the shortcomings of this approach Dalgarno and Victor [28] formulated time-dependent Hartree-Fock perturbation theory which is logically more consistent and is more accurate. It is equivalent to the Random Phase Approximation. Expressions like (9) hold for the dynamic polarizabilities but perturbation equations for the responses $|\chi(\pm\nu)\rangle$ and the higher order responses fall into groups of integro-differential equations for positive and negative frequencies that are coupled *via* the electron-electron interaction. In Hartree-Fock theory the diagonal elements of single-electron operators are correct to first order in correlation [24]. Time-dependent Hartree-Fock theory can be regarded as adding the external perturbation as single electron operators to the electron-nucleus interactions. In time-dependent Hartree-Fock theory single-electron properties, which essentially involve products of the diagonal elements of single-electron operators and energies, are correct to first order in correlation; the theory accounts partially for the two-electron interaction whereas unperturbed Hartree-Fock theory only averages it. Dalgarno and Victor [28] solved the perturbation equations variationally. I, as a graduate student of Alex, studied time-dependent Hartree-Fock theory. I explored the structure of the dipole spectrum by solving the coupled perturbation equations numerically at frequencies that were varied. Dalgarno and Victor [29] used time-dependent Hartree-Fock theory to calculate the van der Waals dispersion coefficients C_6, C_8 and C_{10} for interacting helium atoms. They determined model multipole spectra from variational solutions, with trial resonant parameters as described in Sec. 2.2, of the coupled perturbation equations and they used the model spectra to calculate the required multipole polarizabilities at imaginary frequencies.

Within time-dependent Hartree-Fock theory, Kaveeshwar, Dalgarno and Hurst [30] calculated the frequency-dependent dipole shielding factors for members of the beryllium isoelectronic sequence up to N^{2+} at frequencies below the first transition frequency. They derived an explicit expression in terms of response functions for the shielding factor. The response functions satisfy a few differential equations which they elected to solve variationally; the shielding factor can be expressed as an infinite sum but use of the solutions of the differential equations is more convenient. Kaveeshwar, Dalgarno and Hurst derived a relation between the shielding factor and the polarizability which they used to reduce numerical errors in their evaluations of the dynamic polarizabilities. Time-dependent Hartree-Fock theory for these four-electron atoms and ions is considerably more complicated than it is for helium.

Stewart, Watson and Dalgarno [31] extended the application of time-dependent Hartree-Fock theory to four-electron molecules; they studied LiH and the molecular ion BeH^+. They solved the response equations variationally and for each system they were able to determine, in a similar way to that used by Dalgarno and Victor [29], model dipole spectra and the parallel and perpendicular dynamic dipole polarizabilities.

Relativistic effects are important in heavy systems and in highly ionised systems. Time-dependent Hartree-Fock theory was extended to include relativistic effects by Johnson, Lin and Dalgarno [32] who applied the theory in a study of highly ionised members of the helium isoelectronic sequence and by Shorer and Dalgarno [33,34] who applied it in calculating excitation energies and oscillator strengths of the zinc and magnesium isoelectronic sequences. The latter studies are complicated by the need to consider the angular momentum coupling scheme. The response equations were solved numerically by a Green's function method.

3.3. *Predissociation*

Predissociation shifts and broadens an absorption line. Du, Dalgarno and Jamieson [35] used the sum rules of Dalgarno and Lewis [3] to find a simple expression for the shift, providing another example of an infinite sum being evaluated from the solution of a single differential equation. Dalgarno, with Cheung, Child, Chiu, Finch, Jamieson, Parkinson and Yoshino, [36–38] calculated shifts in a search for parameter-dependent analytic expressions to describe the interactions that predissociate the $B^3\Sigma_u^-$ state of molecular oxygen; values of the shifts and widths calculated with trial parameters were compared with measurements of absorption spectra in the Schumann-Runge bands that were obtained by experimentalists (including Cheung, Chiu, Parkinson and Yoshino) at the Harvard-Smithsonian Center for Astrophysics; the trial parameters were adjusted to yield a good overall match.

3.4. *Rayleigh and Raman Scattering*

Dalgarno and Williams [39] calculated the Rayleigh cross sections for scattering of light by molecular hydrogen at large wavelengths; they used oscillator strength sum rules in calculating the dipole polarizabilities of molecular hydrogen at frequencies below the first vibrational transition frequency.

Equation (8) has a resonant structure; care is thus needed when using Eqs. (9) and (10) at shorter wavelengths corresponding to frequencies greater than the first vibrational transition frequency. Sadeghpour and Dalgarno [40], in a study of Rayleigh and Raman scattering, modified the differential method as follows. At non-resonant frequencies Eqs. (10) are solved as they stand. Widths Γ_p are introduced for the resonances at the energies E_p in Eq. (8). Near the resonance at energy E_p the state $|p\rangle$ is projected out of Eq. (10) for $|\chi(-\nu)\rangle$, the frequency ν is replaced by $\nu + i\Gamma_p/2h$ and the resulting equations for the real and imaginary parts of $|\chi(-\nu)\rangle$ are solved.

In Rayleigh scattering the atom or molecule is left in its initial state $|0\rangle$; in Raman scattering it is left in a different state $|f\rangle$, say. The Rayleigh cross section is obtained directly from the polarizability; to evaluate the Raman cross section an infinite sum, similar to that in Eq. (8) but with the energy E_0 of the second denominator replaced by the final energy E_f, is needed. The sum is essentially that used in the two-photon decay study of Dalgarno and Victor [27] and it can be reduced to two single terms dependent on the solutions of differential equations and evaluated in the manner described above. In the different context of investigations of photoabsorption processes (in helium and the hydrogen negative ion) the procedure near resonance, of extracting the resonant term from the response (or from the infinite sum) and treating it separately by inserting an imaginary shift to give the resonance a width, was used by Dalgarno, Doyle and Oppenheimer [41,42]. Simbotin, Marinescu, Sadeghpour and Dalgarno [43] used this method to study resonant Raman scattering by molecular oxygen in the Schumann-Runge absorption bands. The analysis is the same in principle but is complicated by the need to account for rotation. However the wavefunctions and the response functions $|\chi(\pm\nu)\rangle$ are separable in vibrational and angular coordinates and angular momentum algebra provides much simplification. Simbotin, Jamieson and Dalgarno [44] used the same method to extend these calculations to Rayleigh scattering by molecular oxygen in the Schumann-Runge absorption bands and at longer wavelengths; the cross sections are useful in interpreting observations of Rayleigh scattering in the atmosphere. Van der Loo, Groenenboom, Jamieson and Dalgarno [45] used this method to study a Raman association process in which a pair of hydrogen atoms, approaching in the ground electronic state of the molecule, collide with a photon which removes enough kinetic energy from the atoms to leave them in a bound vibrational state of the molecule; the photon undergoes Raman scattering. This process is important in investigations of conditions prevailing in the early universe.

3.5. *Recent Calculations of Long-range Forces*

The influence of long-range forces is greater in a slower atomic collision because the outcome of any collision depends on the forces the atoms experience throughout and a slow collision takes a relatively long time. The comparatively recent development of experimental techniques to study the behaviour of atoms and molecules that move slowly in ultra-cold traps has increased interest in calculations of long-range forces.

In calculating dispersion coefficients polarizabilities at imaginary frequencies are needed. When calculating C_6 Victor, Dalgarno and Taylor [8] derived a model spectrum from their dipole polarizabilities calculated at real frequencies, substituted imaginary frequencies into Eq. (11) and made an analytic evaluation of the integral in C_6. Recently calculations were made of dynamic polarizabilities of various atoms at real and imaginary frequencies by solving Eqs. (10) and of the dispersion coefficients for their interactions with helium atoms. Zhu, Dalgarno and Porsev [46] made these calculations for alkali-metal atoms; the calculations are of interest in interpreting the spectra of stars, and planets outside the solar system. Xi, Dalgarno and Groenenboom [47,48] made the calculations for Sc, Ti and rare-earth-metal atoms; the calculations are of interest in studies of these atoms cooled and trapped in a buffer gas of helium atoms.

In calculations of molecular structure it is assumed that the nuclear and electronic motions can be separated. In limited cases it is possible to make an exact correction to counterbalance this assumption. For example Taylor, Dalgarno and Babb [49] used a generalisation to molecules of the response equation (6) to calculate the static dipole polarizability of the hydrogen and deuterium molecular ions including the coupling of the nuclear and electronic motions by solving their generalisation of Eq. (6) variationally using a functional like (7).

It is often possible to make corrections that account for the non-separability of nuclear and electronic motions by perturbation theory. This was first discussed by Dalgarno and McCarroll [50]. An important correction to the long-range interaction between alkali-metal atoms was made by Marinescu and Dalgarno [51]. Marinescu and Dalgarno found so-called non-adiabatic corrections to the van der Waals dispersion terms. To see how their evaluation was performed by replacing infinite sums by quadratures consider the additional term in the long-range potential at atomic separation R that depends on R^{-8}; they expressed it as $D_8 R^{-8}$ and, following considerable analysis, they found an expression for D_8 in terms of an infinite sum of matrix elements but with squares of the differences of energies in the denominators. By ingenious use of the relation

$$\frac{1}{(a+b)^2} = \frac{2}{\pi}\int_0^\infty \left[\mathrm{Real}\left\{\frac{1}{b-i\omega}\right\}\mathrm{Real}\left\{\frac{1}{(a-i\omega)^2}\right\}\right] \mathrm{d}\omega \tag{16}$$

they separated the sum into independent contributions from each atom and cast the infinite sums from each atom as quadratures over ω of integrals involving the response functions calculated at imaginary frequencies $i\omega/2\pi$. They found the real and imaginary parts of the response functions by separating the differential equations that they satisfy into pairs of coupled differential equations which they solved by finite-step methods; this procedure was developed by Marinescu, Sadeghpour and Dalgarno [52] in a calculation of the Van der Waals dispersion coefficients C_6, C_8 and C_{10}.

Using the same method, Marinescu and Dalgarno [53] calculated dispersion coefficients (and also long-range electronic transition dipole moments) of alkali-metal atoms interacting *via* excited states of the corresponding dimers, and Marinescu, Babb and Dalgarno [54] calculated long range potentials that include retardation for pairs of two ground state alkali-metal atoms; the atoms considered in these calculations are lithium, sodium, potassium, rubidium and caesium.

Jamieson, Dalgarno and Wolniewicz [55] explored the effect on a diatomic system, of a two-centre perturbation expressible as the sum of products of single-centre perturbations, using an integral like (16) to separate the contributions made by each atom to the first and second order perturbed wave functions. The techniques described in Sec. 2.3 were used to simplify the influence of a two-centre perturbation on the expectation value of an operator that is effective on each centre separately. The Dirac contribution to the relativistic energy of two interacting ground-state hydrogen atoms influences cold collisions. It can be expressed as the expectation value of the sum of delta-functions; Jamieson, Dalgarno and Wolniewicz used their two-centre analysis to find the long range Dirac relativistic energy. One of the inhomogeneous differential equations that arise from application of the methods of Dalgarno and Lewis [3] was solved analytically. A leading correction to the long-range potential in R^{-6} was found; it is much smaller than the R^{-6} van der Waals term because it contains the square of the fine-structure constant. The importance of this relativistic correction arises from the availability of very precise adiabatically corrected *ab initio* Born-Oppenheimer potentials for ground state molecular hydrogen and the need to interpret increasingly precise measurements that can be made in ultra-cold environments.

4. Conclusion

The differential equation technique for performing calculations and developing analysis in studies of atomic and molecular physics, which is one of the very many scientific contributions of Alex Dalgarno, is extremely valuable. It has been adopted by many others and has become a standard technique. It has a timeless quality. As this review has demonstrated, Alex introduced it near the middle of the last century and now, more than 50 years later, it is still being used in increasingly novel and sophisticated applications.

Acknowledgments

I prepared this article, in part, during a visit to the Harvard-Smithsonian Center for Astrophysics (CfA). I thank the Institute for Theoretical Atomic and Molecular Physics at CfA for support; the Institute is supported by a grant from the National Science Foundation. I also acknowledge, with thanks, the support of the Chemical Sciences, Geosciences and Biosciences Division of the Office of Basic Energy Sciences, Office of Science, US Department of Energy.

References

1. A. Dalgarno and M. R. C. McDowell, *Proc. Phys. Soc.* **69**, 615 (1956).
2. A. Dalgarno and A. L. Stewart A L, *Proc. Roy. Soc.* **238**, 269 (1956).
3. A. Dalgarno and J. T. Lewis, *Proc. Roy. Soc.* **233**, 70 (1955).
4. A. Dalgarno and J. T. Lewis, *Proc. Phys. Soc.* **628**, 56 (1956).
5. A. Dalgarno and D. Parkinson, *Proc. Roy. Soc.* **250**, 422 (1958).
6. Y. M. Chan and A. Dalgarno, *Proc. Phys. Soc.* **85**, 227 (1965).
7. Y. M. Chan and A. Dalgarno, *Proc. Phys. Soc.* **86,** 777 (1965).
8. G. A. Victor, A. Dalgarno and A. J. Taylor, *J. Phys. B* **1**, 13 (1968).
9. A. Dalgarno, *Rev. Mod. Phys.* **35**, 522 (1963).
10. A. Dalgarno and A. L. Stewart, *Proc. Roy. Soc.* **238**, 276 (1956).
11. A. Dalgarno and A. L. Stewart, *Proc. Roy. Soc.* **240**, 274 (1957).
12. A. Dalgarno and N. Lynn, *Proc. Phys. Soc.* **70**, 223 (1957).
13. A. Dalgarno and J. T. Lewis, *Proc. Roy. Soc.* **240**, 284 (1957).
14. A. Dalgarno and A. L. Stewart, *Proc. Roy. Soc.* **247**, 245 (1958).
15. A. Dalgarno and R. M. Pengelly, *Proc. Phys. Soc.* **89**, 503 (1966).
16. J. C. Y. Chen and A. Dalgarno, *Proc. Phys. Soc.* **85**, 399 (1965).
17. A. Dalgarno and A. L. Stewart, *Proc. Phys. Soc.* **76**, 467 (1960).
18. A. Dalgarno and A. L. Stewart, *Proc. Phys. Soc.* **75**, 441 (1960).
19. A. Dalgarno and A. L. Stewart, *Proc. Roy. Soc.* **254**, 570 (1960).
20. A. Dalgarno and A. L. Stewart, *Proc. Roy. Soc.* **257**, 534 (1960).
21. A. Dalgarno, W. D. Davison and A. L. Stewart, *Proc. Roy. Soc.* **257**, 115 (1960).
22. A. Dalgarno, *Proc. Phys. Soc.* **75**, 439 (1960).
23. M. Cohen and A. Dalgarno, *Proc. Phys. Soc.* **77**, 165 (1960).
24. M. Cohen and A. Dalgarno, *Proc. Phys. Soc.* **77**, 748 (1961).
25. A. Dalgarno, *Proc. Roy. Soc.* **251**, 282 (1959).
26. A. Dalgarno and J. M. McNamee, *Proc. Phys. Soc.* **77**, 673 (1960).
27. A. Dalgarno and G. A. Victor, *Proc. Phys. Soc.* **87**, 371 (1966).
28. A. Dalgarno and G. A. Victor, *Proc. Roy. Soc.* **291**, 291 (1966).
29. A. Dalgarno and G. A. Victor, *Proc. Phys. Soc.* **90**, 605 (1967).
30. V. G. Kaveeshwar, A. Dalgarno and R. P. Hurst, *J. Phys. B* **2** 984, (1969).
31. R. F. Stewart, D. K. Watson and A. Dalgarno, *J. Chem. Phys.* **63**, 3222 (1975).
32. W. R. Johnson, C. D. Lin and A. Dalgarno, *J. Phys. B* **9**, L303 (1976).
33. P. Shorer and A. Dalgarno, *Phys. Rev. A* **15**, 154 (1977).
34. P. Shorer and A. Dalgarno, *Phys. Scripta* **21**, 432 (1980).
35. M. L. Du, A. Dalgarno and M. J. Jamieson, *J. Chem. Phys.* **91**, 2980 (1989).
36. S. S-L. Chiu, A. S-C. Cheung, M. Finch, M. J. Jamieson, K. Yoshino, A. Dalgarno and W. H. Parkinson, *J. Chem. Phys.* **97**, 1787 (1992).
37. A. S-C. Cheung, D. K-W. Mok, M. J. Jamieson, M. Finch, K. Yoshino, A. Dalgarno and W. H. Parkinson, *J. Chem. Phys.* **99**, 1086 (1993).
38. A. S-C. Cheung, D. K-W. Mok, K. Yoshino, W. H. Parkinson, M. J. Jamieson, A. Dalgarno and M. S. Child, *J. Chem. Phys.* **103**, 2369 (1995).
39. A. Dalgarno and D. A. Williams, *Proc. Phys. Soc.* **85**, 605 (1965).
40. H. R. Sadeghpour and A. Dalgarno, *J. Phys. B* **25**, 4801 (1992).

41. A. Dalgarno, H. Doyle and M. Oppenheimer, *Phys. Rev. Lett.* **20**, 1051 (1972).
42. H. Doyle, M. Oppenheimer and A. Dalgarno, *Phys. Rev. A* **11**, 909 (1975).
43. I. Simbotin, M. Marinescu, H. R. Sadeghpour and A. Dalgarno, *J. Chem. Phys.* **107**, 7057 (1997).
44. I. Simbotin, M. J. Jamieson and A. Dalgarno, *J. Geophys. Res. - Atmospheres* **109**, D13302 (2004).
45. M. P. J. van der Loo, G. C. Groenenboom, M. J. Jamieson and A. Dalgarno, *Trans. Faraday Soc.* **133**, 43 (2006).
46. C. Zhu, A. Dalgarno, S. G. Porsev and A. Derevianko, *Phys. Rev. A* **70,** 032722 (2004).
47. C. Xi, A. Dalgarno and G. C. Groenenboom, *Phys. Rev. A* **72**, 032703 (2005).
48. C. Xi, A. Dalgarno and G. C. Groenenboom, *Phys. Rev. A* **75**, 032723 (2007).
49. J. M. Taylor, A. Dalgarno and J. F. Babb, *Phys. Rev. A* **60**, R2630 (1999),
50. A. Dalgarno and R. McCarroll, *Proc. Roy. Soc.* **237**, 383 (1956).
51. M. Marinescu and A. Dalgarno, *Phys. Rev. A* **57**, 1821 (1998).
52. M. Marinescu M, H. R. Sadeghpour and A. Dalgarno, *Phys. Rev. A* **49**, 982 (1994).
53. M. Marinescu and A. Dalgarno, *Phys. Rev. A* **52**, 311 (1995).
54. M. Marinescu, J. F. Babb and A. Dalgarno, *Phys. Rev. A* **50**, 3096 (1994).
55. M. J. Jamieson, A. Dalgarno and L. Wolniewicz, *Phys. Rev. A* **61**, 042705 (2000).

LINEAR RESPONSE TIME DEPENDENT DENSITY FUNCTIONAL THEORY FOR DISPERSION COEFFICIENTS BETWEEN ATOMIC PAIRS

XI CHU

Department of Chemistry, The University of Montana,
Missoula, MT 59801, USA
E-mail: Xi.Chu@mso.umt.edu
www.umt.edu

GERRIT C. GROENENBOOM

Theoretical Chemistry, Institute for Molecules and Materials, Radboud University Nijmegen, Heyendaalseweg 135, 6525 AJ Nijmegen, The Netherlands
E-mail: gerritg@theochem.ru.nl

Dispersion coefficients C_6 can be calculated from the dynamic polarizabilities of the interacting monomers at imaginary frequencies. This article describes a linear response time dependent density functional approach that has been recently developed for evaluating these polarizabilities. Using a spherical tensor expansion, both the isotropic and anisotropic components of the C_6 coefficient can be obtained for a set of molecular states. Isotropic C_6 coefficients for the interactions between a pair of transition metals are presented.

Keywords: Dispersion coefficient; dynamic polarizability; time dependent density functional theory.

1. Introduction

Long-range interactions between atomic pairs are crucial for the elastic and inelastic scattering cross sections in gases at low temperatures, which control the trap loss in buffer gas loading experiments.[1–4] For two neutral atoms, one of which is in an S state, such as helium, the leading term of their long range interaction is the dispersion or van der Waals interaction. The leading term in the long-range interaction of neutral non-S-state atoms ($\propto R^{-5}$) is the quadrupole-quadrupole interaction.

It is difficult to obtain the long range interactions accurately by treating the interacting atoms as one molecule. Theoretically, it usually requires high level, wave function based quantum chemical calculations. Numerically, such calculations are subject to the basis set superposition error (BSSE) which is difficult to correct.

It is an area of active current research to formulate density functional theory (DFT), such that they can be properly represented.[5] The accuracy of density functional based approaches depends on the exchange-correlation energy functional E_{xc}. Most popular functionals, however, do not describe the dispersion energy well.

The alternative is to relate the long-range interactions to the monomer properties. In a recent paper,[6] we have adopted an expression for the electrostatic, dispersion, and induction interactions in terms of the multipole moments and dynamic polarizabilities of the interacting atoms . A spherical tensor expansion of these interactions is provided for this expression.[6] The long-range Born-Oppenheimer potentials of various molecular states for the the collisions of atoms A and B with angular momentum L_A and L_B are related to the tensorial expansion coefficients.

We have developed a linear response time-dependent density functional theory (TDDFT) approach for calculating dynamic polarizabilities of atoms.[7–10] A spherical tensor expansion can also be applied to the dynamic polarizabilities for imaginary frequencies to obtain their scalar and tensor components, which can be related to the scalar and tensor components of the dispersion coefficients.[6] We have calculated the isotropic C_6 dispersion coefficients for the like atomic pairs of many atomic species and estimated their accuracy.[7] The isotropic dispersion coefficients between atomic pairs have been employed for the calculation of the dispersion interaction between molecules.[11] Furthermore, we calculated both the isotropic and anisotropic C_6 coefficients for atoms with non-zero angular momentum interacting with helium atoms.[8–10] The effectiveness of the buffer gas loading technique[2–4,12] in creating ultra-cold gases of atoms depends on the relative efficiency of elastic and inelastic collisions of the atoms with ^{3}He atoms at temperatures below 0.4 K. Inelastic collisions lead to trap loss. The inelastic cross sections are determined by the energy separations of the interaction potentials of the different symmetries of the molecular system formed by the approaching atoms,[13] which is dominated by the anisotropic term of the dispersion interactions. The ratio of isotropic to anisotropic dispersion coefficients, therefore, were used to estimate if an atomic species is a candidate for helium buffer gas cooling.

2. Perturbation Theory for Dispersion Interactions

In a recent paper we discussed the multipole expansion of the long-range interaction between two neutral atoms in first and second order perturbation theory.[6] The first order interaction matrix elements of the electrostatic Hamiltonian $\hat{H}(R)$ are given by

$$E_{L_A M_A M'_A L_B M_B M'_B}(R) = \langle L_A M_A L_B M_B | \hat{H}(R) | L_A M'_A L_B M'_B \rangle, \quad (1)$$

in which L_A is the electronic orbital angular momentum of atom A, M_A its projection onto the interatomic axis, and the same notations applies to atom B. These matrix elements can be expanded as

$$E_{L_A M_A M'_A L_B M_B M'_B}(R) = -\sum_{n=5}^{\infty} \frac{C_n(L_A, M_A, M'_A, L_B, M_B, M'_B)}{R^n}. \quad (2)$$

Only odd powers n contribute in first order. When both atoms have nonzero quadrupole moments, the leading term is $-C_5/R^5$. Second order interactions, i.e., dispersion and induction, can be expanded in the same form, but only even powers n contribute. The leading term in second order is $-C_6/R^6$, regardless of the angular momenta of the atoms. If one of the atoms, say B, is in an S state, only diagonal elements in Eq. (1) are nonzero and the C_6 coefficients are given by

$$C_6(L_A, M_A, M_A, 0, 0, 0) = C_{6,0}(L_A) - \frac{3M_A^2 - L_A(L_A+1)}{(2L_A-1)(2L_A+3)} C_{6,2}(L_A), \quad (3)$$

where $C_{6,0}$ and $C_{6,2}$ are the scalar and tensor dispersion coefficient, respectively. If neither atom is in an S state the C_6 coefficients are given by

$$C_n(L_A, M_A, M'_A, L_B, M_B, M'_B) = \sum_{k_1 k_2 k} C_{6(k_1 k_2)k} (-1)^{L_A - M'_A + L_B - M'_B}$$
$$\times \sum_q \langle L_A M_A L_A {-} M'_A | k_1 q \rangle \langle L_B M_B L_B {-} M_B{}' | k_2 {-} q \rangle \langle k_1 q k_2 {-} q | k 0 \rangle. \quad (4)$$

It follows from the Clebsch-Gordan coefficients $\langle ....|..\rangle$ that nonzero contributions can only arise when $|M_A - M'_A| \leq k_1$, $|M_B - M'_B| \leq k_2$, and $M_A + M_B = M'_A + M'_B$. There are three isotropic ($k = 0$) terms, $C_{6(0,0)0}$, $C_{6(1,1)0}$, and $C_{6(2,2)0}$, four rank 2 tensor terms, $C_{6(1,1)2}$, $C_{6(0,2)2}$, $C_{6(2,0)2}$, and $C_{6(2,2)2}$, and one rank 4 term, $C_{6(2,2)4}$. The C_6 tensor dispersion coefficients can be evaluated using the following expression:

$$C_{6(k_1k_2)k} = \frac{15\langle 2\,0\,2\,0|k\,0\rangle}{\pi} \begin{Bmatrix} 1 & 1 & k_1 \\ 1 & 1 & k_2 \\ 2 & 2 & k \end{Bmatrix} \sqrt{(2k_1+1)(2k_2+1)}$$

$$\times \int_0^\infty \alpha^{(A)}_{(1,1)k_1}(i\omega)\alpha^{(B)}_{(1,1)k_2}(i\omega)d\omega, \tag{5}$$

where $\alpha^{(A)}_{(1,1)k_1}(i\omega)$ is the dynamic polarizability of atom A at imaginary frequency $i\omega$ and the symbol in curly brakets is a 9–j symbol. For $L_A = L$ and $L_B = 0$, one must have $k_2 = 0$ and the scalar and tensor dispersion coefficients in Eq. (3) are related to $C_{6(k,0)k}$ through

$$C_{6,0}(L) = (2L+1)^{-\frac{1}{2}} C_{6(0,0)0} \tag{6}$$

and

$$C_{6,2}(L) = -\sqrt{\frac{5(2L-1)(2L+3)}{L(L+1)(2L+1)}}\, C_{6(2,0)2}. \tag{7}$$

In this case the dispersion coefficients may also be written as

$$C_{6,0} = \frac{3}{\pi}\int_0^\infty \alpha_0^{(A)}(L;i\omega)\alpha_0^{(B)}(i\omega)d\omega \tag{8}$$

and

$$C_{6,2} = -\frac{3(2L+3)}{2\pi L}\int_0^\infty \alpha_2^{(A)}(L;i\omega)\alpha_0^{(B)}(i\omega)d\omega \tag{9}$$

with

$$\alpha_0(L) = -\frac{\alpha_{(1,1)0}(L)}{\sqrt{3(2L+1)}} \tag{10}$$

and

$$\alpha_2(L) = \sqrt{\frac{10L(2L-1)}{3(L+1)(2L+1)(2L+3)}}\,\alpha_{(1,1)2}(L), \tag{11}$$

where $\alpha_0(L)$ and $\alpha_2(L)$ are defined such that

$$\alpha_z(L,M) = \alpha_0(L) + \frac{3M^2 - L(L+1)}{L(2L-1)}\alpha_2(L), \tag{12}$$

$$\alpha_x(L,M) = \alpha_0(L) - \frac{3M^2 - L(L+1)}{2L(2L-1)}\alpha_2(L), \tag{13}$$

where α_z is the polarizability along the z axis, which is in the direction of the internuclear axis of AB and α_x is the polarizability perpendicular to the z axis.

From linear response TDDFT calculations, we obtain $\alpha_z(L, M; i\omega)$ and $\alpha_x(L, M; i\omega)$ for both A and B with a series of M values. Through Eqs. (12) and (13), we acquire the scalar and rank 2 polarizabilities $\alpha_k(L, i\omega)$. For two interacting non-S state atoms also odd-rank polarizability tensors are required, as explained in reference.[6] We then use Eq. (5) to calculate the dispersion coefficients in tensor form and with Eq. (4) we can calculate coefficients corresponding to different L and M quantum numbers of A and B. If one of the two atoms is in S state, we can use Eqs. (8), (9), and (3) instead.

3. TDDFT Calculation of Dynamic Polarizabilities

Combining linear response theory and TDDFT, the dynamic polarizabilities are calculated as

$$\alpha_f(L, M; i\omega) = \sum_\sigma \iint \chi_{\sigma,L,M}(\mathbf{r}, \mathbf{r}', i\omega)\delta v^f_{\sigma,L,M}(\mathbf{r}', i\omega) f d\mathbf{r}d\mathbf{r}', \quad (14)$$

where f is either x or z, σ is the spin index, $\chi_{\sigma,L,M}$ is the response function of σ spin state electrons of the atom in state $|LM\rangle$, $\mathbf{r}$ and $\mathbf{r}'$ are electron position vectors, $i\omega$ is the photon frequency, and δv is the perturbation of the potential due to an applied field. To calculate α_f, we apply a field in the direction of the f axis, and

$$\begin{aligned}\delta v^f_{\sigma,L,M}(\mathbf{r}, i\omega) = f &+ \int K_{\sigma\uparrow,L,M}(\mathbf{r}, \mathbf{r}', i\omega)\delta v^f_{\uparrow,L,M}(\mathbf{r}', i\omega)d\mathbf{r}' \\ &+ \int K_{\sigma\downarrow,L,M}(\mathbf{r}, \mathbf{r}', i\omega)\delta v^f_{\downarrow,L,M}(\mathbf{r}', i\omega)d\mathbf{r}', \quad (15)\end{aligned}$$

in which

$$\begin{aligned}K_{\sigma\sigma,L,M}(\mathbf{r}, \mathbf{r}', i\omega) = &\int \frac{\chi_{\sigma,L,M}(\mathbf{r}', \mathbf{r}'', i\omega)}{|\mathbf{r} - \mathbf{r}''|} d\mathbf{r}'' \\ &+ \frac{\partial V_{xc}}{\partial \rho_\sigma} \chi_{\alpha,L,M}(\mathbf{r}, \mathbf{r}', i\omega), \quad (16)\end{aligned}$$

$$K_{\uparrow\downarrow,L,M}(\mathbf{r}, \mathbf{r}', \omega) = \int \frac{\chi_{\downarrow,L,M}(\mathbf{r}', \mathbf{r}'', \omega)}{|\mathbf{r} - \mathbf{r}''|} d\mathbf{r}''. \quad (17)$$

In Eq. (16), V_{xc} is the exchange-correlation potential and ρ_σ is the spin density. The adiabatic approximation is employed here and the functional derivative $\partial V_{xc}/\partial \rho_\sigma$ is evaluated with solutions of the static Kohn-Sham (KS) equation

$$\left[-\frac{\hbar^2\nabla^2}{2m_e} + V_\sigma(\mathbf{r})\right]\phi_{i\sigma}(\mathbf{r}) = \epsilon_{i\sigma}\phi_{i\sigma}(\mathbf{r}), \tag{18}$$

where

$$V_\sigma(\mathbf{r}) = -\frac{Ze^2}{r} + \int d\mathbf{r}' \frac{\rho(\mathbf{r}')}{|\mathbf{r}-\mathbf{r}'|} + V_{xc,\sigma}(\mathbf{r}). \tag{19}$$

Here Z is the nuclear charge, e is the electron charge, and

$$\rho_\sigma(\mathbf{r}) = \sum_{i=1}^{N_\sigma} \rho_{i\sigma}(\mathbf{r}) = \sum_{i=1}^{N_\sigma} |\phi_{i\sigma}(\mathbf{r})|^2, \tag{20}$$

where N_σ is the number of orbitals whose spin quantum number is σ and $\rho(\mathbf{r}) = \rho_\uparrow(\mathbf{r}) + \rho_\downarrow(\mathbf{r})$.

The response functions are also obtained from solutions of the KS equation as

$$\chi_{\sigma,L,M}(\mathbf{r},\mathbf{r}',i\omega) = \sum_{k,j}\left[\frac{\phi^*_{k\sigma}(\mathbf{r})\phi_{j\sigma}(\mathbf{r})\phi_{k\sigma}(\mathbf{r}')\phi^*_{j\sigma}(\mathbf{r}')}{i\omega + (\epsilon_{j\sigma} - \epsilon_{k\sigma})} - \frac{\phi^*_{j\sigma}(\mathbf{r})\phi_{k\sigma}(\mathbf{r})\phi_{j\sigma}(\mathbf{r}')\phi^*_{k\sigma}(\mathbf{r}')}{i\omega - (\epsilon_{j\sigma} - \epsilon_{k\sigma})}\right], \tag{21}$$

where $\phi_{k\sigma}$ and $\phi_{j\sigma}$ are occupied and unoccupied orbitals respectively, and $\epsilon_{k\sigma}$ and $\epsilon_{j\sigma}$ are their energies.

To evaluate $\alpha_f(L,M;i\omega)$ we first solve the KS equations with proper XC functionals to obtain χ using Eq. (21), then we solve Eq. (15) followed by Eq. (14).

4. Numerical Method

To numerically solve KS equations and evaluate the dynamic polarizabilities, we used the generalized pseudospectral (GPS) method which we developed for spherical polar, cylindrical, and spheroidal coordinates. These methods are all basis-set free. They all generate and employ spacial grid points that are very dense where the Coulomb potential is steep and sparse where it is not. These methods also take advantage of the symmetry of a system to reduce the computation.

In the program that uses spherical coordinate, we only input the nuclear charge Z and the electron configuration $l_{i\sigma}, \sigma = \uparrow, \downarrow, i = 1, ..., N_\sigma$, and assume $m_{i\sigma}, \sigma = \uparrow, \downarrow, i = 1, ..., N_\sigma$ are such that the electron density of an atom is spherically symmetric. For instance, for atom C, we input its electron configuration as $1s^1_\uparrow 1s^1_\downarrow 2s^1_\uparrow 2s^1_\downarrow 2p^{\frac{2}{3}}_{-1,\uparrow} 2p^{\frac{2}{3}}_{0,\uparrow} 2p^{\frac{2}{3}}_{1,\uparrow}$. Because of this configuration, what we calculate is only α_0, with which we can calculate $C_{6(0,0)0}$ of a pair of atoms. Since the electronic structure calculation is only one dimensional with a minimum number of grid points, the calculation is extremely fast. In reference, we tabulated the $C_{6(0,0)0}$ value of 42 pairs of like atoms.

The anisotropy of the polarizabilities and, subsequently, the C_6 coefficients can be resolved with the GPS method with cylindrical or spheroidal coordinates. In these methods we input $m_{i\sigma}, \sigma = \uparrow, \downarrow, i = 1, ..., N_\sigma$ and $M = \sum_\sigma \sum_i m_{i\sigma}$. Since solutions of the KS equations provide us the ground state of a given symmetry, we can obtain $\alpha(L, M)$ with this method if it is the lowest energy state among states with the same M value. A state $|LMS\rangle$ is not always representable by one configuration. States $|LLS\rangle$ and $|LL-1S\rangle$, however, always are. Polarizabilities of these states are sufficient for us to calculate α_0 an α_2 using Eqs. (12) and (13), from which dispersion coefficients involving any $|LMS\rangle$ state can be obtained.

5. Exchange-Correlation Functionals

The choice of XC functionals are essential for density functional based methods. Since this method is not a supermolecular approach, the XC functional we choose does not have to properly describe the dispersion interaction. From Eq. (21), we see that orbital energies are important input for evaluating polarizabilities. The accuracy for evaluating Eq. (21) will be improved if the absolute values of the orbital energies are closer to the ionization potentials and the difference between occupied and unoccupied energies is closer to transition energies. Energies of the highest occupied molecular orbital (HOMO) and the lowest few virtual orbitals are particular important for polarizability calculations. It is demonstrated in reference[14] that the HOMO energies for atoms are greatly improved by applying a self-interaction correction (SIC) procedure. In that calculation, electron densities for atoms are assumed to have spherical symmetry. Here in Table 1 we compare HOMO energies of three diatomic molecules calculated with different methods and the measured vertical ionization energies. The LDA with and without SIC calculations are performed with our 2D GSP method with 21×18 grid points. The B3LYP and MP2FU values with very large basis sets are from the NIST computational chemistry comparison and benchmark database (CCCBDB). This comparison shows that LDA+SIC gives much improved HOMO energies compared to LDA or B3LYP. For HF and CO, its predictions are also better than MP2FU. For BF its value is only slightly worse.

Table 1. Comparison of HOMO energies calculated with different methods and the measured vertical ionization energies in Hartree.

Molecule	HF	BF	CO
LDA	0.3109	0.2096	0.2886
B3LYP+aug-cc-pVQZ*	0.4241	0.2884	0.3866
LDA+SIC	0.6108	0.3905	0.5028
MP2FU*	0.6503[a]	0.4045[b]	0.5509[c]
Experiment*	0.5924	0.4084	0.5149

Note: brackets * from CCCBDB
[a]The basis set is aug-cc-pVQZ.
[b]The basis set is cc-pCVTZ.
[c]The basis set is cc-pVDZ.

The accuracy of transition dipole moments are also important for the polarizabilities. Because the transition dipole moments are not well defined in DFT, we estimate their accuracies by comparing dipole moment calculations instead. In Table 2 we compare the dipole moments calculated with different methods and the measured values. Here B3LYP gives the best values.

Table 2. Comparison of dipole moments calculated with different methods and the measured values in ea_0.

	HF	BF	CO
HF+aug-cc-pVQZ*	−0.741	0.381	−0.054
LDA	−0.6969	0.4434	0.1160
B3LYP+aug-cc-pVQZ*	−0.711	0.385	0.040
LDA+sic	−0.7925	0.4426	0.06758
Experiment*	−0.72	0.20	0.04

Note: * from CCCBDB

In Table 3 we also compare values of quadrupole moments. Those of LDA+SIC are slightly better than the rest.

Table 3. Comparison of quadrupole moments calculated with different methods and the measured values in ea_0^2.

Molecule	HF	BF	CO
HF+aug-cc-pVQZ*	1.547	−3.177	−1.581
LDA	1.593	−2.455	−1.528
B3LYP+aug-cc-pVQZ*	1.581	−2.666	−1.517
LDA+sic	1.630	−3.007	−1.673
Experiment*	1.739		−2.111

Note: * from CCCBDB

6. Isotropic C_6 Coefficients

We have compared the static polarizabilities calculated with the LDA+SIC functional for many atomic species with what we estimated as the best values available.[7] This comparison allows us to estimate a correction factor for the C_6 values. In Table 4 the C_6 coefficients for transition metal atoms interacting with each other are listed. A larger correction factor for Cr is adopted here than in reference.[7] For the Sc-Cr interaction, $C_{6,0}$ is estimated to be 1366 a.u. using SAPT[15] and we compute 1009 a.u.

Table 4. C_6 coefficient of selective atomic pairs.

	Sc	Ti	V	Cr	Mn	Fe	Co	Ni	Cu	Zn
Sc	1383	1199	1067	1009	862.1	805.3	737.2	686.6	581.3	606.9
Ti	1199	1044	930.8	879.4	754.7	705.4	646.7	603.0	509.3	534.1
V	1067	930.8	831.7	785.5	676.0	632.2	580.1	541.4	456.6	480.3
Cr	1009	879.4	785.5	743.0	638.2	596.7	547.6	511.1	431.9	453.4
Mn	862.1	754.7	676.0	638.2	551.4	515.9	474.2	443.0	372.9	393.9
Fe	805.3	705.4	632.2	596.7	515.9	482.8	443.8	414.8	349.0	368.9
Co	737.2	646.7	580.1	547.6	474.2	443.8	408.3	381.8	321.0	339.8
Ni	686.6	603.0	541.4	511.1	443.0	414.8	381.8	357.2	300.2	318.2
Cu	581.3	509.3	456.6	431.9	372.9	349.0	321.0	300.2	253.0	267.1
Zn	606.9	534.1	480.3	453.4	393.9	368.9	339.8	318.2	267.1	283.7

Acknowledgment

This work was under the mentorship and guidance of Professor Alex Dalgarno.

References

1. A. L. Migdall, J. V. Prodan, W. D. Phillips, T. H. Bergeman and H. J. Metcalf, *Phys. Rev. Lett.* **54**, p. 2596 (1985).
2. J. M. Doyle, B. Friedrich, J. Kim and D. Patterson, *Phys. Rev. A* **52**, p. R2515 (1995).
3. J. D. Weinstein, R. deCarvalho, T. Guillet, B. Friedrich and J. M. Doyle, *Nature* **395**, p. 148 (1998).
4. J. D. Weinstein, R. deCarvalho, K. Amar, A. Boca, B. C. Odom, B. Friedrich and J. M. Doyle, *J. Chem. Phys.* **109**, p. 2656 (1998).
5. Y. Zhao and D. G. Truhlar, *Acc. Chem. Res.* **41** (2008).
6. G. C. Groenenboom, X. Chu and R. V. Krems, *J. Chem. Phys.* **126**, p. 204306 (2007).
7. X. Chu and A. Dalgarno, *J. Chem. Phys.* **121**, p. 4083 (2004).
8. X. Chu and A. Dalgarno, *Adv. At. Mol. Opt. Phys.* **51**, p. 83 (2005).
9. X. Chu, A. Dalgarno and G. C. Groenenboom, *Phys. Rev. A* **72**, p. 032703 (2005).
10. X. Chu, A. Dalgarno and G. C. Groenenboom, *Phys. Rev. A* **75**, p. 032723 (2007).
11. A. Tkatchenko and M. Scheffler, *Phys. Rev. Lett.* **102**, p. 073005 (2009).
12. C. I. Hancox, S. C. Doret, M. T. Hummon, L. Luo and J. M. Doyle, *Nature* **431**, p. 281 (2004).
13. R. V. Krems, G. C. Groenenboom and A. Dalgarno, *J. Phys. Chem. A* **108**, p. 8941 (2004).
14. X. M. Tong and S. I. Chu, *Phys. Rev. A* **55**, p. 3406 (1997).
15. L. Rajche, P. S. Zuchowski, J. Klos, M. M. Szczesniak and G. Chalasinski, *J. Chem. Phys.* **127**, p. 244302 (2007).

Session II

Astrochemistry

Charles Alcock, Alex Dalgarno

SOME THOUGHTS ON INTERSTELLAR CHEMISTRY

WILLIAM KLEMPERER

Department of Chemistry and Chemical Biology, Harvard University, Cambridge, MA 02138

It is indeed a pleasure to start our symposium celebrating Alec Dalgarno with the topic of astrochemistry. This subject, so intimately connected to radioastronomy, continues to provide new molecular species in a variety of astronomical sources, molecular clouds, circumstellar shells, and planetary nebula. A currently maintained catalog is online.[1] The majority of species observed are not commercially available stable gas-phase species. Thus their laboratory production and the production of their rotational spectra is a most significant challenge. There is a pressure to make observations at increasingly higher frequencies, for reasons of higher signal strength. This further challenges the laboratory astrophysicist. For *semi-rigid species* the rotational energy may be well expressed in powers of angular momentum by three rotational constants (quadratic) and five centrifugal distortion constants (quartic).[2] Thus a quite finite number of rotational transitions, at frequently convenient laboratory frequencies, allows the reasonably accurate estimation of the higher frequency spectrum. For larger molecules, especially those with large amplitude motions such as internal rotation or low frequency bending motions, the semi-rigid model is likely inadequate and a more complete laboratory spectrum is essential for reliable carrier identification.

The classes of species whose rotational transitions have been observed in radioastronomical sources have been importantly enlarged with the assignment of previously unidentified lines to negative ions. The first, C_6H^- (now identified) is linear and semi-rigid, with rotational constant 1% greater than the neutral radical C_6H.[3] This series of polyacetylenic radicals and anions is observed for C_nH, $n = 4, 6$, and 8 as well as for the isoelectronic species C_mN, $m = 3, 5$.[4] The laboratory observation of C_6H^- was the Rosetta Stone for the laboratory observation of these two sequences of negative ions. As pointed out,[5] these species have closed shell electronic structure

$^1\Sigma$. Thus, their rotational spectra consist of unsplit single lines, quite close in frequency to the more complex spectra of their neutral parent.

The variety of observed cations is large, and well known. The apparently largest class is that of protonated observed neutral species. The examples are HCO^+ and COH^+ (CO), N_2H^+ (N_2), HCS^+ (CS), H_3^+ (H_2), H_3O^+ (H_2O), $HCNH^+$ (HCN, HNC), $HOCO^+$ (CO_2), H_2COH^+ (H_2CO), HC_3NH^+, (HC_3N). Somewhat rarer are the class of positive ions of neutral molecules represented by CH^+, CO^+, SO^+ and CF^+. For CH^+, CO^+, and SO^+ the neutral is observed, while for CF^+ only the ion is observed. (We note that HCO is observed). The observation of rotational transitions of CO^+ and CH^+ may, at first, be regarded as surprising since both react with H_2 at a collision frequency given by the Langevin rate. These species are observed in emission in planetary nebula.[6,7] The suggested formation of CO^+ is by the reaction $C^+ + OH \rightarrow CO^+ + H$, a reaction 4 eV exothermic. Since the electric dipole moments of both CO^+ and CH^+ are appreciable the radiative lifetime of the excited rotational levels is short compared to collision times of these species with H_2. This indicates that the signal strength of the rotational transitions is not indicative of their abundance. The suggested formation[8] of CF^+ is by the analogous reaction $C^+ + HF \rightarrow CF^+ + H$. The reaction of CF^+ with H_2 producing HCF^+ is almost certainly endothermic since CF^+ is isoelectronic with CO and the H—C bond in HCO is extremely weak (1.1 eV). The ion CF^+ is likely to be stable in a molecular hydrogen atmosphere. In like manner the destruction of SO^+ by reaction with H_2 is unlikely since the closely related reaction $O_2^+ + H_2 \rightarrow HO_2^+ + H$ is 1.6 eV endoergic.

There are a number of ions whose absence is somewhat surprising. H_2CO^+ does not react[9] with H_2, and would appear to be producible by radiative association of $H + HCO^+$. NO has an extremely low ionization potential, 9.26 eV, and extremely strong bond ($D_0 = 10.85$ eV) and the destruction of NO^+ by the reaction $NO^+ + H_2 \rightarrow HNO^+ + H$ is quite endoergic. A search[10] for NO^+ in Sgr B2 yielded quite low limits, $NO^+/NO < 3 \times 10^{-4}$. While there are presently no high resolution spectra for H_2CO^+, the rotational constants have been estimated from photoelectron spectroscopy and *ab initio* calculations.[11,12] The accuracy of the constants is estimated at 1%.

I have sketched some thoughts on interstellar chemistry to fit into this wonderful occasion celebrating the birthday of our colleague, Alex Dalgarno. I have greatly enjoyed my collaborations with Alex.

References

1. http://www.astro.uni-koeln.de/site/vorhersagen/molecules.
2. See H. Meyer, Ann. Rev. Phys. Chem. **53**, 141 (2002), which provides a discussion and reference to the theory of J. K. G. Watson.
3. M. C. McCarthy, C. A. Gottlieb, H. Gupta and P. Thaddeus, *Ap. J. Lett.* **652**, p. L141 (2006).
4. See references in the catalog of Ref. 1.
5. Frequent private discussions with M. C. McCarthy, C. A. Gottlieb, H. Gupta, and P. Thaddeus.
6. W. B. Latter, C. K. Walker and P. R. Maloney, *Ap. J. Lett.* **419**, p. L97 (1993).
7. J. Cernicharo, X.-W. Liu, E. Gonzalez-Alfonso, P. Cox, M. J. Barlow, T. Lim and B. M. Swinyard, *Ap. J. Lett.* **483**, p. L65 (1997).
8. D. A. Neufeld, P. Schilke, K. M. Menten, M. G. Wolfire, J. H. Black, F. Schuller, H. S. P. Müller, S. Thorwirth, R. Güsten and S. Philipp, *Astron. Astrophys.* **454**, p. L37 (2006).
9. $H_2CO^+ + H_2$.
10. D. T. Halfen, A. J. Apponi and L. M. Ziurys, *Ap. J.* **561**, p. 244 (2001).
11. R. T. Wiedmann, M. G. White, K. Wang and V. McKoy, *J. Chem. Phys.* **100**, p. 4738 (1994).
12. P. J. Bruna, M. R. J. Hachey and F. Grein, *Mol. Phys.* **94**, p. 917 (1998).

ASTROCHEMISTRY: BUILDING ON DALGARNO'S LEGACY

EWINE F. VAN DISHOECK

Leiden Observatory, Leiden University,
2300 RA Leiden, The Netherlands
E-mail: ewine@strw.leidenuniv.nl
and Max-Planck-Institute für Extraterrestrische Physik,
Giessenbachstrasse 1, Garching, 85748, Germany

An overview of some developments in astrochemistry over the last 40 years is given, with special emphasis on topics that Alex Dalgarno has opened up (which is nearly all of astrochemistry!). The development of astrochemistry into an integral part of modern astrophysics is illustrated with recent examples. The bright future of astrochemistry is discussed in the light of new observational facilities and the need for continued studies of basic molecular processes under interstellar conditions is emphasized.

Keywords: Astrochemistry; Interstellar Molecules; Basic Molecular Processes; Molecular clouds; Star formation.

1. Introduction

Astrochemistry, also known as molecular astrophysics, is 'the study of the formation, destruction and excitation of molecules in astronomical environments and their influence on the structure, dynamics and evolution of astronomical objects' (Dalgarno 2008). Implicit in this definition is the fact that molecules are excellent diagnostics of the physical conditions and processes in the regions where they reside, both through their excitation as well as their abundance which is sensitive to temperature and density. Astrochemistry also provides an inventory of the molecules – both simple and complex – available in regions in which new stars and planets are formed. From a chemical physics perspective, interstellar space provides a unique laboratory in which basic molecular processes can be studied under conditions different from those on Earth, making it doubly interesting (and fun!) for researchers working at the boundary of both fields. Indeed, as stated by Alex, it is a 'blending of astronomy and chemistry in which each area enriches the other in a mutually stimulating interaction'.

Molecules are found in an increasing variety of regions in space. CO has been detected in several distant dusty galaxies up to a record redshift $z = 6.42$ when the universe was only 5% of its current age (Walter et al. 2003). Both the molecules themselves as well as the dust emission from the molecular clouds are prime tracers of the mass, temperature, dynamics and star formation activity of these earliest galaxies. At lower redshifts, pure rotational lines of CO, HCN and HCO^+ at millimeter wavelengths are commonly seen in ultra-luminous galaxies containing active galactic nuclei and undergoing bursts of star formation at a rate at least 100 times higher than that in our own Milky Way (Gao et al. 2007). Even closer to home, large scale maps in CO are used to make an inventory of molecular clouds and star-forming sites throughout the Milky Way and nearby galaxies (e.g., Dame et al. 2001). In all these studies, CO is used as a proxy for the dominant molecule in interstellar clouds, H_2, which does not have strong rotational emission lines because it lacks a permanent electric dipole moment.

Astrochemistry has traditionally focussed on clouds in the solar neighborhood of our galaxy, where a myriad of molecules have been found, more than 130 different ones not including isotopes (see www.astrochemystry.org for an up-to-date list). Several of these are complex organic, perhaps even prebiotic, molecules (Herbst & van Dishoeck 2009). Together, they are used to study the detailed physics, chemistry and dynamics of the diffuse gas, of shocks and clouds exposed to intense UV and X-rays, of dense star-forming regions (e.g., van Dishoeck 2006, Ceccarelli et al. 2007, Bergin & Tafalla 2007) and of disks in which planets may form (Bergin et al. 2007). Excitingly, simple molecules are now even detected in the atmospheres of exoplanets (Tinetti et al. 2007). Molecules also provide insight into the fate of evolved stars and their mass return to the interstellar medium (e.g., Pardo et al. 2007).

Astrochemistry is an inherently interdisciplinary field, and requires an integrated approach in which astronomical observations are combined with astrochemical models and studies of basic molecular processes to answer questions (Fig. 1). The basic processes form the foundation on which the rest of astrochemistry (and even astronomy as a whole) is built. Indeed, in Alex' view, 'astrophysics is almost entirely applied Atomic, Molecular and Optical Physics' (Dalgarno 2008).

This review focusses on a number of topics in astrochemistry in which Alex has been particularly active. The historical context and Alex's contributions are briefly sketched and then put in light of recent developments.

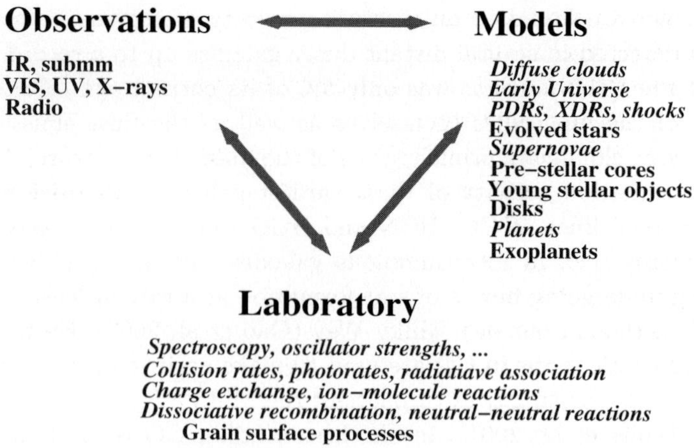

Fig. 1. The triangle of observations, models and laboratory studies needed to address problems in astrochemistry. Examples of each category are given, with areas in which Alex Dalgarno has been particularly active highlighted in italics. Note that 'laboratory' implies both experimental and theoretical work, with Alex focussing on the latter.

Early universe chemistry is not included since it is covered elsewhere in this volume. Alex himself has written several fine reviews recently, which are highly recommended as complementary literature (Dalgarno 2006a,b).

2. Diffuse Cloud Chemistry

Observations of molecules in diffuse clouds started around 1940 (in Anno Dalgarno (AD) ~10) with the identification of the first simple molecules: CH, CH^+ and CN. Diffuse clouds have visual extinctions $A_V \sim 1$ mag so that the atoms and molecules can be studied by optical absorption lines superposed on the spectra of bright background stars. They have typical densities $n_H = n(\mathrm{H}) + 2n(\mathrm{H_2})$ of a few hundred cm^{-3} and temperatures ranging from 100–200 K at the edge to about 30 K in the center of the cloud. Even earlier, the first diffuse interstellar bands (DIBs) had been discovered, a set of broad lines for which the carriers have still not been identified. The first models trying to explain the observed abundances in diffuse clouds were made by Kramers & ter Haar (1946) and Bates & Spitzer (1951). The heart of both studies consists of trying to identify the primary molecular processes that form and destroy molecules under the cold and tenuous conditions of space. In particular, inspired by studies of aeronomy, Bates & Spitzer noted that dissociative recombination of ions, $AB^+ + e \rightarrow$

A + B, can be a significant process thus making the high abundances of interstellar CH^+ more puzzling.

In 1968 (AD$\sim$40), the field of astrochemistry fully opened up with the discovery of the first polyatomic molecule, NH_3 (Cheung et al. 1968). A flood of new detections followed as soon as the newly-built millimeter telescopes became operational, including some quite complex and exotic species. This was the time when Alex had just moved to the Harvard-Smithsonian Center for Astrophysics and entered the field. Indeed, a flurry of activity in astrochemistry ensued at Harvard in the early 1970's, with Alex and his group focussing on diffuse cloud chemistry and Bill Klemperer and Eric Herbst concentrating on the foundations of dense cloud chemistry. Other major contributions at that time were made by Bill Watson (Illinois) and Phil Solomon (Stony Brook).

2.1. *Interstellar CH and Related Molecules*

The models by Bates & Spitzer (1951) and Solomon & Klemperer (1972) failed to reproduce the abundances of the simplest observed molecules in diffuse clouds. Following the detection of H_2 at ultraviolet wavelengths, Black & Dalgarno (1973a) realized that reactions with H_2 may be more important than previously thought. In diffuse clouds, most of the carbon is in the form of C^+ because the ionization potential of C (11 eV) is less than that of H (13.6 eV). However, the reaction of C^+ with H_2 to form CH^+ cannot proceed at low temperatures because it is endoergic by 4640 K. Instead, Black & Dalgarno realized that the radiative association reaction $C^+ + H_2 \rightarrow CH_2^+ + h\nu$ (with rate coefficient k_{ra}), followed by dissociative recombination of CH_2^+ to form CH, is rapid enough to explain the observed CH abundances. This reaction is faster than other radiative association reactions because it can proceed through an excited electronic state from which radiative stabilization is orders of magnitude more rapid than by emission within the vibrational manifold of the electronic ground state.

Black & Dalgarno (1977) subsequently made a comprehensive model of the diffuse cloud toward ζ Ophiuchi with which they could explain the observed CH column densities with their best estimates of the rate coefficients. Their postulated value of $k_{ra} \approx 5 \times 10^{-16}$ cm^3 s^{-1} can be viewed as the first astronomical determination of the rate of a basic molecular process which is very difficult to measure in the laboratory. Indeed, even 30 years later, the best estimate of k_{ra} is still uncertain by about 50% but close to the above value.

This same model, which was successful for CH, continued to fail to reproduce the observed CH^+ column densities. A number of other puzzling observations have appeared since, most notably the comprehensive surveys of polyatomic molecules by Lucas & Liszt using millimeter absorption lines against quasars behind diffuse clouds (Liszt et al. 2005). The high abundances of HCO^+ and HCN may well be linked to that of CH^+ since reactions of CH^+ and CH_2^+ with O an N can lead to more complex hydrocarbons. An overview of diffuse cloud studies and associated problems has been given by van Dishoeck (1998). For most simple molecules, including CH^+, the chemical processes are now so well understood that the problem must lie in our poor understanding of the physical structure. As noted by Dalgarno (2008), such 'failures are often more instructive than successes'.

2.2. H_3^+ *and the Cosmic-Ray Ionization Rate*

H_3^+ was recognized by Herbst & Klemperer (1973) to be the cornerstone of the ion-molecule chemistry in dense clouds, which allows molecules to be formed rapidly even at the very low temperatures of interstellar space. However, since it is a symmetric molecule with no dipole moment, it does not have strong millimeter emission lines. H_3^+ also does not have strong electronic transitions in the optical or UV below 13.6 eV, so that the only way to observe it in cold clouds is through vibrational absorption lines at infrared wavelengths. When sensitive infrared detectors became available, the molecule was finally detected by Geballe & Oka (1996) in dense star-forming regions. More recently, Indriolo et al. (2007) observed H_3^+ in several diffuse clouds. Since H_3^+ is formed by cosmic-ray ionization of H_2 and destroyed primarily by dissociative recombination in diffuse clouds, its chemistry is exceedingly simple and H_3^+ can be used to directly measure the cosmic ray ionization rate ζ. Alex and co-workers had previously used HD and OH as more indirect probes to constrain ζ (Black & Dalgarno 1973b, Hartquist et al. 1978) and earlier estimates were in the range of a few $\times 10^{-17}$ to 10^{-16} s^{-1} (see also van Dishoeck & Black 1986). With the new, much higher measured value of the H_3^+ dissociative recombination rate (McCall et al. 2003), however, larger average values of $\zeta \approx 2 \times 10^{-16}$ s^{-1} are inferred, and comparable rates are found for some dark clouds. The various determinations have been summarized by Dalgarno (2006b), who notes that 'the interesting question may be not why they are so different but why they are so similar'.

3. H_2 Excitation and Chemistry

3.1. *H_2 Photodissociation*

H_2 is the dominant molecule in the universe so understanding its formation, destruction and excitation is of paramount importance in astrochemistry. In contrast with many other interstellar molecules like OH and CH, H_2 does not have dissociative electronic states below 13.6 eV to which dipole-allowed transitions are possible. Thus, photodissociation can only occur indirectly by absorption of a UV photon into a bound excited electronic state (either the B $^1\Sigma_u^+$ or $C^1\Pi_u$ state) followed by emission into the vibrational continuum of the ground state (see Fig. 2). This process, proposed by Solomon & Wickramasinghe (1969) and Stecher & Williams (1967) was quantified by Stephens & Dalgarno (1972), whose values for the dissociation efficiencies (i.e., the fraction of absorptions into an excited state that lead to dissociation) are still used today. The calculations predicted a detailed line and continuum spectrum peaking around 1600 Å and continuing below 912 Å, which was beautifully confirmed by measurements by G. Herzberg in a joint paper by Dalgarno et al. (1970).

Excitation of H_2 by secondary electrons following cosmic ray ionization produces a similar emission spectrum, first proposed by Prasad & Tarafdar (1983) and computed quantitatively by Sternberg et al. (1987) and Gredel et al. (1989). This cosmic-ray induced radiation field is present even inside dense clouds, where it can drive specific chemical reactions. Compilations of the photodissociation cross sections of various molecules, necessary to compute the photodissociation rates, have been given by van Dishoeck (1988), Roberge et al. (1991) and van Dishoeck et al. (2006).

3.2. *H_2 Infrared Emission*

Most of the UV emission from the B and C electronic states ends up in vibrationally excited states $v > 0$ of the ground $X^1\Sigma_g^+$ state which relax down to $v = 0$ through emission of infrared photons at low densities. The first quantitative calculation of this spectrum was made by Black & Dalgarno (1976) for a diffuse cloud exposed to the normal interstellar radiation field. Black & van Dishoeck (1987) extended these calculations to denser clouds (up to few $\times 10^4$ cm^{-3}) exposed to orders of magnitude more intense UV fields, whereas Sternberg & Dalgarno (1989) considered even higher density clouds where collisional (de-)excitation becomes important in determining the populations of the ro-vibrational levels. With increasing density, the infrared spectrum of H_2 changes from that typical of UV fluorescence to

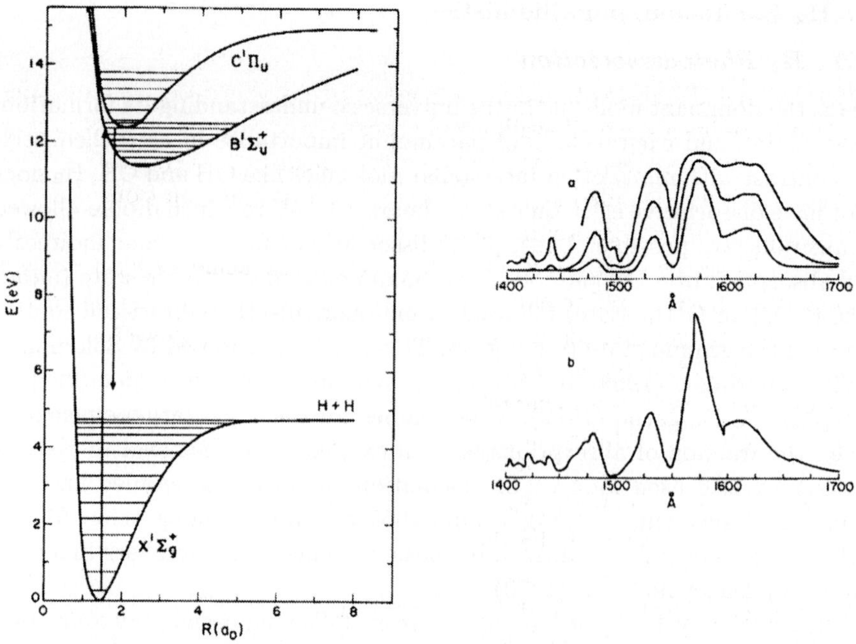

Fig. 2. Left: H_2 potential energy curves involved in the excitation and dissociation of H_2 in interstellar clouds. Right: Comparison of measured (top) and calculated (bottom) H_2 UV continuous emission spectrum following absorption into the B and C electronic states (Dalgarno et al. 1970).

that typical of thermal emission at high temperatures. Gredel & Dalgarno (1995) and Tiné et al. (1997) extended these calculations to exposure by X-rays in which higher excited electronic states start to contribute. Today, H_2 near-infrared lines are routinely observed in a wide variety of regions in our own and distant galaxies with the above models serving as the basis for their interpretation.

3.3. *H_2 Formation*

Because radiative association of H + H to form H_2 is exceedingly slow, H_2 must be formed largely on grains in interstellar clouds. In the early universe, when no grains yet exist, the only alternative is through reactions of H^+ and H^- with H, described by others in this volume. The $H^- + H \rightarrow H_2 + e$ dissociative attachment reaction was actually first studied by Dalgarno (1954) but its importance in astrophysics was realized only later by others.

3.4. *HD Chemistry and [D]/[H]*

With depth into an interstellar cloud, the H_2 column rapidly becomes so large that the UV absorption lines become optically thick: the H_2 molecules lying at the edge absorb all available photons at the line wavelengths so that H_2 molecules lying deeper into the cloud are shielded from the dissociating radiation, a process called self-shielding. Thus, the transition from $H \rightarrow H_2$ is very sharp with the H_2 abundance increasing rapidly at an $A_V \approx 0.1$ mag. Black & Dalgarno (1973b) realized that HD never achieves high enough columns to become self-shielding in diffuse clouds so that it is destroyed at orders of magnitude higher rates than H_2. This is compensated by the orders of magnitude higher formation rate of HD in the gas compared with H_2 through the charge-transfer reaction $H^+ + D \rightarrow H + D^+$, followed by $D^+ + H_2 \rightarrow HD + H^+$. This results in a HD/H_2 ratio which is generally within an order of magnitude of the overall [D]/[H] deuterium abundance in diffuse clouds. Since the HD abundance depends on both the cosmic ray ionization rate (producing H^+) and the deuterium abundance, HD can be used to constrain either of these two parameters if the other can be determined independently. Hartquist et al. (1978) used HD to infer [D]/[H]$\approx 2 \times 10^{-5}$.

Using the *Far Ultraviolet Spectroscopic Explorer* (FUSE), HD has recently been observed toward many more sightlines including some translucent clouds with higher reddening. Lower limits on the inferred [D]/[H] ratio range from $4 \times 10^{-7} - 4 \times 10^{-6}$ (Lacour et al. 2005). The HD pure rotational lines have been detected in several dense photon-dominated regions and shocks indicating [D]/[H]$\approx (5-7) \times 10^{-6}$, a factor of 2–3 lower than the canonical value of 1.5×10^{-5} (Wright et al. 1999, Neufeld et al. 2006). Variations in [D]/[H] indeed now appear well established in surveys of atomic D by Linsky et al. (2006), who proposes $(2.31 \pm 0.24) \times 10^{-5}$ as the representative value for the total gas + dust [D]/[H] ratio within 1 kpc of the Sun, with deuterium depleted onto grains in the denser colder parts.

3.5. *H_2 Cooling*

In a seminal review on the heating and ionization of the diffuse interstellar medium, Dalgarno & McCray (1972) computed the cooling rates in various pure rotational lines of H_2 and HD. At high temperature (> 100 K), H_2 is the more efficient coolant in units of erg s^{-1} H_2^{-1}, but at lower temperatures HD becomes more effective in spite of its 10^5 times lower abundance. This is because its levels lie lower in energy, with J=1 at 128 K compared with J=2 of H_2 at 510 K. Also, ΔJ=1 transitions can occur because HD has a small dipole moment, whereas H_2 can only cool through the slower $\Delta J = 2$ quadrupole transitions.

H_2 and HD emission is particularly important in the early universe where there are few heavy elements which can contribute to the cooling. This results in a lower Jeans mass for collapse and the formation of the first stars. Recent cosmological simulations indicate the formation of clumps with typical masses of 10^3 $M_\odot$ with conditions set by the H_2 cooling. The final masses of the first stars are thought to be typically 30–100 $M_\odot$. Indeed, in an early review and subsequent lively discussion, Dalgarno & Lepp (1987) noted that early universe clouds will 'fragment into stellar-sized objects with a tendency to heavier stars'.

Dalgarno & McCray (1972) also pointed out the importance of the [C II] 158 μm fine structure line for cooling of diffuse clouds in our Galaxy, a prediction which was confirmed by far-infrared observations in the late 1970's. Its dominance as a coolant was convincingly demonstrated by the COBE-FIRAS integrated spectrum of the Galaxy in which [C II] 158 μm is by far the strongest line (Wright et al. 1991). The [C II] line has now been detected in high-redshift galaxies out to z=6.4, and will likely become a prime diagnostic for the gas in the highest redshift galaxies (> 7) for which the CO lines have shifted out of the millimeter bands.

4. Cold Molecular Cores

4.1. *H_2D^+ and Extreme Deuteration*

Cold molecular clouds have long been known to harbor high abundances of deuterated molecules such as DCO^+ and DCN, with DCO^+/HCO^+ and DCN/HCN ratios at least three orders of magnitude higher than the overall [D]/[H] ratio. The origin of this huge fractionation lies in the fact that the zero-point vibrational energy of the deuterated molecules is slightly lower than that of their normal counterparts because of their higher mass.

In cold cores, most of the deuteration is initiated by the H_3^+ + HD $\rightarrow$ H_2D^+ + H_2 reaction which is exoergic by about 232 K. Proof of this mechanism comes from direct observations of H_2D^+ in cold clouds. Its pure rotational spectrum was first predicted by Dalgarno et al. (1973), who computed the lowest ortho transition $1_{10} - 1_{11}$ to lie at 344 GHz, compared with the measured value of 372.421 GHz. The line was finally detected 10 years ago toward a low-mass young stellar object (Stark et al. 1999) and subsequently even stronger toward cores with no embedded protostars (Caselli et al. 2003). Even D_2H^+ has been observed toward these sources

(Vastel et al. 2004), which also show high abundances of doubly- and triply-deuterated molecules such as D_2CO and ND_3. The observed ND_3/NH_3 ratio is about 0.001, indicating a deuterium enhancement of a factor of $\sim 10^{12}$ compared with that expected from a normal [D]/[H] ratio.

The high deuterium fractionation requires an additional explanation than just gas-phase reactions. In the centers of cold cores, there is now convincing evidence that most of the heavy elements, including CO, are frozen out onto the grains forming an icy mantle. This is inferred from the absence of gas-phase emission (e.g., Bergin et al. 2002) as well as from direct observations of the solid-state features (Pontoppidan 2006). Since CO is the main destroyer of both H_3^+ and H_2D^+, their abundances are even further enhanced when CO is removed from the gas. Indeed, the reaction of H_3^+ with HD then becomes its main destruction and the chemistry rapidly proceeds to a highly deuterated state in which D_3^+ can even become comparable to H_3^+ in abundance (Roberts et al. 2003).

The DCO^+/HCO^+ abundance ratio is also commonly used to infer the ionization fraction of cold cores, with typical values of a few $\times 10^{-9}$ – few $\times 10^{-8}$ in the densest regions. The ionization fraction controls the coupling of the gas to the magnetic field and thus the ability of the cloud to maintain itself against collapse.

4.2. *Negative Ions*

The formation and chemistry of negative ions in space was considered in models as early as those by Dalgarno & McCray (1973). Subsequent studies elaborated on the chemistry (e.g., Herbst 1981, Petrie & Herbst 1997) and considered the effects of large negative ions such as PAH^- on the overall chemistry and heating (Lepp et al. 1988, Lepp & Dalgarno 1988). Negative ions are also thought to be present in interstellar ice mantles, with a strong feature at 4.62 μm assigned to OCN^-. In ices, acid-base reactions of, for example, NH_3 with HNCO are known to readily lead to the ions NH_4^+ and OCN^- down to temperatures as low as 15 K (e.g., van Broekhuizen et al. 2005).

Dalgarno mentioned in a 1999 review that 'for carbon chains, a significant fraction may exist as anions'. Neutral carbon chains are prominently seen in cold clouds through their millimeter emission and are candidates for the DIBs. Tulej et al. (1999) pointed out that some DIBs coincide with lines of anions, most notably C_7^-, but this identification was not confirmed. The first gaseous negative ion, C_6H^- was detected only a few years ago by McCarthy et al. (2006) in the cold core TMC-1 and several other negative ions have followed since, both in TMC-1 and in the envelope of the evolved star IRC +10216. The larger ions are formed more readily because the increased electron binding energy with size of the neutral precursor makes simple radiative attachment quite efficient (Herbst & Osamura 2008).

5. Warm and Dense Molecular Clouds

5.1. *Dense PDRs and XDRs*

Photon-dominated regions (PDRs) are clouds exposed intense ultraviolet radiation, in which the UV photons control the chemistry and heating of the gas. The diffuse clouds discussed above can be viewed as the low density, low UV subset of PDRs. As described by Sternberg in this volume, the photon heating creates a hot (~1000 K) layer at the edge of the cloud where reactions with energy barriers can proceed, most notably $O + H_2 \rightarrow OH + H$ and $C^+ + H_2 \rightarrow CH^+ + H$ (Sternberg & Dalgarno 1995). Reactions of C^+ with OH leads to the characteristic PDR tracer CO^+.

X-rays also ionize and heat the gas, as explored in a number of papers by Alex and his group (e.g., Lepp & Dalgarno 1996, Yan 1997, Dalgarno et al. 1999). One difference of XDRs with PDRs is that doubly-ionized ions are created by X-ray ionization, some of which undergo slow charge-transfer to singly-ionized species (e.g., Ne^{2+} and S^{2+}, Butler et al. 1980). Also, X-rays penetrate much further into a cloud due to the lower absorption cross sections of hydrogen at high energies.

X-ray chemistry is currently being explored in the context of galactic nuclei (in particular to distinguish AGN- from starburst-dominated nuclei) and in the earliest stages of star formation and protoplanetary disks. Pre-main sequence stars are known to be powerful X-ray emitters, but the mechanism of X-ray production (magnetic reconnection along protostar-disk field lines vs. turbulent convective stellar dynamo) and the evolutionary stage at which they turn on are still under debate. Chemistry can potentially be a powerful diagnostic of X-rays in the deeply embedded phase when no X-rays can yet escape the source for direct observation and detailed models have been developed by Stäuber et al. (2005, 2006).

The [Ne II] fine-structure line at 12.8 μm can be a tracer of X-ray and EUV radiation as well and has recently been detected in a number of protoplanetary disks (e.g., Lahuis et al. 2007), with fluxes consistent with recent models (e.g., Glassgold et al. 2007).

5.2. *Shocks and Outflows*

High temperatures are also found in shocks created by outflows or supernovae remnants running into dense molecular clouds. Early work by Hartquist et al. (1980) indicated that H_2S, SO and SiO could be particularly enhanced, partly because of the large OH abundances. In a key paper, Draine, Roberge and Dalgarno (1983) studied the chemistry in non-dissociative, $C-$type shocks and showed that all oxygen is driven into H_2O for a wide variety of conditions. This prediction was beautifully confirmed by direct detection of H_2O lines using the *Infrared Space Observatory* (ISO) (Harwit et al. 1998) and the *Submillimeter Wave Astronomical Satellite* (SWAS) (Melnick et al. 2000), opening up the study of water in the universe. Most recently, highly excited mid-infrared lines of H_2O and OH have been unexpectedly detected by the *Spitzer Space Telescope* toward a low-mass deeply embedded protostar, perhaps originating in the accretion shock onto the growing disk (Watson et al. 2007).

Neufeld & Dalgarno (1989) performed a comprehensive study of the chemistry and cooling in dissociative shocks, in which all molecules are dissociated into atoms at the very high initial temperatures ($> 10^5$ K) and subsequently reform as the gas cools.

6. Chemistry in Star- and Planet-Forming Regions

Much of astrochemistry currently focusses on star-forming regions where the young star heats the surrounding gas and dust passively through its accretion luminosity and actively through UV radiation and X-rays, and through jets and outflows which create shocks in the surrounding envelope (see van Dishoeck 2008 for recent overview). In the latter cases, the PDR, XDR and shock models developed by Alex and his co-workers form the basis for interpretation. Over the last two decades, however, it has become clear that gas-grain interactions control the bulk of the chemical characteristics of embedded young stellar objects (YSOs). Molecular processes involving grains is one of the few areas of astrochemistry that Alex has hardly touched.

6.1. *Ices: Solid-state Astrochemistry*

Ices are primarily observed by their vibrational bands at mid-infrared wavelengths. The development of large format infrared detector arrays and the launch of *ISO* opened up the field of infrared spectroscopy unhindered by the Earth's atmosphere (van Dishoeck 2004). This allowed a full inventory of interstellar ices seen as broad absorption bands against the continuum from embedded YSOs or background stars. The dominant ice species are H_2O, CO, CO_2, CH_4, NH_3 and CH_3OH, i.e., those molecules produced by hydrogenation and oxidation of the dominant atoms (O, C and N) and molecules (CO) arriving from the gas on the grains at low temperatures (Tielens & Hagen 1982). More recent *Spitzer* surveys have extended such studies to large samples of low-mass YSOs, which have a remarkably similar overall composition as their high-mass counterparts.

The observations have stimulated the new field of solid-state astrochemistry, in which modern surface science techniques are used to quantitatively study processes such as thermal desorption (e.g., Collings et al. 2004, Bisschop et al. 2006), surface chemistry (e.g., Watanabe et al. 2004, Ioppolo et al. 2008) and photodesorption (e.g., Öberg et al. 2007). The resulting binding energies and rates are directly incorporated into astrochemical gas-grain models. Theoretical molecular dynamics techniques can also be usefully extended to solid-state processes, as demonstrated by the recent study of the photodissociation and photodesorption of water ice (Andersson et al. 2006).

6.2. *Hot Cores: Complex Molecules*

Saturated complex organic molecules such as CH_3OCH_3 and CH_3CN are prominently observed toward many (but not all) high-mass YSOs as well as some low-mass sources (see Herbst & van Dishoeck 2009 for review). Some of them are prebiotic, i.e., molecules involved in the origin of life. Most of these molecules are now thought to be produced on the surfaces of grains, in the cold ice formation phase (zeroth generation, e.g., CH_3OH) and during the subsequent warm-up phase when radicals produced by photodissociation become mobile on the surface (first generation, e.g., $HCOOCH_3$).

When the dust temperature reaches more than 100 K, the most strongly bound molecules including H_2O and CH_3OH evaporate from the grains and drive a hot gas-phase chemistry producing more complex organics (second generation). Recent determinations of reaction rate coefficients and branching ratios for dissociative recombination show that this route is not as important as thought before but it can still contribute for a period of $\sim 10^5$ yr after evaporation.

6.3. *Protoplanetary Disks and the Water Trail*

Because water is a dominant molecule in the oxygen chemistry, its abundance and partitioning between the gas and grains controls the chemistry of many other species. As a strong coolant, it also regulates the energy balance in warm and dense regions. In disks, its presence as water ice (the 'snow-line') may help the coagulation process that ultimately produces planets. Liquid water is a likely prerequisite for the emergence of life and much of the water in our oceans on Earth likely stems from the impact of asteroids and comets containing ice. The water molecules themselves are mostly formed on the surfaces of grains in the cold interstellar clouds. Following the water trail from dense cloud cores through collapsing envelopes to planet-forming disks is therefore a major goal of modern astrochemistry.

The ISO, SWAS and ODIN satellites have provided insight into the distribution of water in star-forming regions on larger scales, especially in high-mass star-forming regions. Recent *Spitzer* and ground-based observations have revealed surprisingly strong mid-infrared lines of hot ($\sim$ 800 K) water lines originating in the inner $\sim$1 AU of disks (Salyk et al. 2008, Carr & Najita 2008). Future observations with the *Herschel Space Observatory*, to be launched later in 2009, will allow observations of water from clouds to disks for Sun-like young stars.

6.4. *Exo-Planetary Atmospheres*

The detection and characterization of exo-planets is one of the fastest growing fields in astronomy. With the discovery of a number of transiting exo-planets, their atmospheres are now starting to be studied through accurate measurements of the flux of the parent star when the planet is in front and behind it (Seager 2008). Particularly exciting is the recent detection of H_2O in the atmosphere of the exo-planet around HD 189733b (Tinetti et al. 2007). One of the main science goals of future facilities is to study the composition of exo-planetary atmospheres, not only of Jupiter-like planets but down to the Earth-like regime. It brings together two subjects which are at the core of Alex' research – astronomy and aeronomy.

7. Concluding Remarks

In 1986, Dalgarno asked the provoking question "Is astrochemistry useful"? The answer 20 years later is a much more convincing "yes", as has been shown in the previous sections (see also Hartquist, this volume). At the same time, astrochemistry can only thrive by close interactions with neighboring fields, making it also highly rewarding intellectually. To stimulate such interactions worldwide, Alex established in 1984 the working group on Astrochemistry under Division VI of the International Astronomical Union

(IAU). Its main purpose is to ensure that IAU symposia on astrochemistry are regularly organized world wide, with a broad coverage of all aspects of astrochemistry including basic processes. The large and growing attendance at the IAU symposia in Goa (1985), Brazil (1991), Leiden (1996), Korea (1999) and Asilomar (2005) illustrates the need for such conferences.

The future of astrochemistry is bright, with several new and powerful infrared and submillimeter facilities on the horizon, including *Herschel*, the Atacama Large Millimeter/Submillimeter Array (ALMA), the *Stratospheric Observatory for Infrared Astronomy* (SOFIA), the *James Webb Space Telescope* (JWST) and in the more distant future ground-based optical Extremely Large Telescopes (ELTs). With increased sensitivity, spatial resolution and wavelength coverage these observatories will undoubtedly open up new areas in astrochemistry and astronomy, and stimulate the development of a new generation of astrochemical models. Complementary 'laboratory' measurements and calculations will remain essential to interpret the data from these billion $/Euro/Yen facilities. As shown repeatedly by Alex' work, even a very modest investment can result in a much more in-depth interpretation and understanding of the astronomical data than possible otherwise. This also requires a new generation of well-trained astrochemists, who understand the basic atomic and molecular physics underlying such laboratory measurents or computations. This is arguably Alex' largest legacy to astronomy: the Dalgarno scientific family tree.

Acknowledgments

On behalf of all the participants of this symposium, our warmest thanks to Alex for training us so well, for caring continuously about our careers, and for making astrochemistry such an enjoyable subject. Astrochemistry in Leiden is supported by a Spinoza grant from the Netherlands Organization for Scientific Research (NWO) and the Netherlands Research School of Astronomy (NOVA).

References

1. S. Andersson, A. Al-Halabi, G.-J. Kroes, E.F. van Dishoeck, *J. Chem. Phys.* **124**, 064715 (2006)
2. D.R. Bates, L. Spitzer, *Astrophys. J.* **113**, 441 (1951)
3. E.A. Bergin, M. Tafalla, *Ann. Rev. Astron. Ap.* **45**, 339 (2007)
4. E.A. Bergin, Y. Aikawa, G.A. Blake, E.F. van Dishoeck, in *Protostars and Planets V*, ed. B. Reipurth, D. Jewitt, K. Keil (Arizona, Tucson, 2007), 751
5. E.A. Bergin, J. Alves, T. Huard, C.J. Lada, *Astrophys. J.* **570**, L101 (2002)

6. S.E. Bisschop, H.J. Fraser, K.I. Öberg, E.F. van Dishoeck, S. Schlemmer, *Astron. Astrophys.* **449**, 1297 (2006)
7. J.H. Black, A. Dalgarno, *Astrophys. Lett.* **15**, 79 (1973a)
8. J.H. Black, A. Dalgarno, *Astrophys. J.* **184**, L101 (1973b)
9. J.H. Black, A. Dalgarno, *Astrophys. J.* **203**, 132 (1976)
10. J.H. Black, A. Dalgarno, *Astrophys. J. Supp.* **34**, 405 (1977)
11. J.H. Black, E.F. van Dishoeck, *Astrophys. J.* **322**, 412 (1987)
12. S.E. Butler, T.G. Heil, A. Dalgarno, *Astrophys. J.* **241**, 442 (1980)
13. J.S. Carr, J. Najita, *Science* **319**, 1504 (2008)
14. P. Caselli, F.F.S. van der Tak, C. Ceccarelli, A. Bacmann, *Astron. Astrophys.* **403**, L37 (2003)
15. C. Ceccarelli, P. Caselli, E. Herbst, A.G.G.M. Tielens, E. Caux, in *Protostars and Planets V*, ed. B. Reipurth, D. Jewitt, K. Keil (Arizona, Tucson, 2007), 47
16. A.C. Cheung, D.M. Rank, C.H. Townes, D.D. Thornton, W.J. Welch, *Phys. Rev. Lett.* **21**, 1701 (1968)
17. M.P. Collings, M.A. Anderson, R. Chen, J.W. Dever, S. Viti, D.A. Williams, M.R.S. McCoustra, *Mon. Not. Roy. Astron. Soc.* **354**, 1133 (2004)
18. A. Dalgarno, unpublished results (1954)
19. A. Dalgarno, *Q. J. Roy. Astr. Soc.* **27**, 83 (1986)
20. A. Dalgarno, *Faraday Discussions* **133**, 9 (2006a)
21. A. Dalgarno, *Proc. Nat. Ac. Science* **103**, 12269 (2006b)
22. A. Dalgarno, *Ann. Rev. Astron. Ap.* **46**, 1 (2008)
23. A. Dalgarno, S. Lepp, in *Astrochemistry*, IAU Symposium 120 (Reidel, Dordrecht, 1987), 109
24. A. Dalgarno, R.A. McCray, *Ann. Rev. Astron. Ap.* **10**, 375 (1972)
25. A. Dalgarno, R.A. McCray, *Astrophys. J.* **181**, 95 (1973)
26. A. Dalgarno, G. Herzberg, T.L. Stephens, *Astrophys. J.* **162**, L49 (1970)
27. A. Dalgarno, E. Herbst, S. Novick, W. Klemperer, *Astrophys. J.* **183**, L131 (1973)
28. A. Dalgarno, M. Yan, W. Liu, *Astrophys. J. Supp.* **125**, 237 (1999)
29. T.M. Dame, D. Hartmann, P. Thaddeus, *Astrophys. J.* **547**, 792 (2001)
30. B.T. Draine, W.G. Roberge, A. Dalgarno, *Astrophys. J.* **264**, 485 (1983)
31. Y. Gao, C.L. Carilli, P.M. Solomon, P.A. VandenBout, *Astrophys. J.* **660**, L93 (2007)
32. T.R. Geballe, T. Oka, *Nature* **384**, 334 (1996)
33. A.E. Glassgold, J.R. Najita, J. Igea, *Astrophys. J.* **656**, 515 (2007)
34. R. Gredel, A. Dalgarno, *Astrophys. J.* **446**, 852 (1995)
35. R. Gredel, S. Lepp, A. Dalgarno, E. Herbst, *Astrophys. J.* **347**, 289 (1989)
36. T.W. Hartquist, J.H. Black, A. Dalgarno, *Mon. Not. Roy. Astron. Soc.* **185**, 643 (1978)
37. T.W. Hartquist, A. Dalgarno, M. Oppenheimer, *Astrophys. J.* **236**, 182 (1980)
38. M. Harwit, D.A. Neufeld, G.J. Melnick, M.J. Kaufman, *Astrophys. J.* **497**, L105 (1998)
39. E. Herbst, *Nature* **289**, 656 (1981)

40. E. Herbst, W. Klemperer, *Astrophys. J.* **185**, 505 (1973)
41. E. Herbst, Y. Osamura, *Astrophys. J.* **679**, 1670 (2008)
42. E. Herbst, E.F. van Dishoeck, *Ann. Review Astron. Ap.* **47**, in press
43. N. Indriolo, T.R. Geballe, T. Oka, B.J. McCall, *Astrophys. J.* **671**, 1736 (2007)
44. S. Ioppolo, H.M. Cuppen, C. Romanzin, E.F. van Dishoeck, H. Linnartz, *Astrophys. J.* **686**, 1474 (2008)
45. H.A. Kramers, D. ter Haar, *Bull. Astron. Inst. Netherlands* **10**, 137 (1946)
46. S. Lacour, M.K. André, P. Sonnentrucker et al., *Astron. Astrophys.* **430**, 967 (2005)
47. F. Lahuis, E.F. van Dishoeck, G.A. Blake, N.J. Evans, J.E. Kessler-Silacci, K.M. Pontoppidan, *Astrophys. J.* **665**, 492 (2007)
48. S. Lepp, A. Dalgarno, *Astrophys. J.* **324**, 553 (1988)
49. S. Lepp, A. Dalgarno, *Astron. Astrophys.* **306**, L21 (1996)
50. S. Lepp, A. Dalgarno, E.F. van Dishoeck, J.H. Black, *Astrophys. J.* **329**, 418 (1988)
51. J.L. Linsky, B.T. Draine, H.W. Moos et al., *Astrophys. J.* **647**, 1106 (2006)
52. H. Liszt, R. Lucas, J. Pety, in *Astrochemistry: Recent Successes and Current Challenges*, IAU Symposium 231, eds. D.C. Lis, G.A. Blake, E. Herbst (Cambridge University, Cambridge, 2005), 187
53. B.J. McCall, A.J. Huneycutt, R.J. Saykally et al., *Nature* **422**, 500 (2003)
54. M.C. McCarthy, C.A. Gottlieb, H. Gupta, P. Thaddeus, *Astrophys. J.* **652**, L141 (2006)
55. G.J. Melnick, M.L.N. Ashby, R. Plume et al., *Astrophys. J.* **539**, L87 (2000)
56. D.A. Neufeld, A. Dalgarno, *Astrophys. J.* **340**, 869 (1989)
57. D.A. Neufeld, J.D. Green, D.J. Hollenbach et al., *Astrophys. J.* **647**, L33 (2006)
58. K.I. Öberg, G.W. Fuchs, Z. Awad, H.J. Fraser, S. Schlemmer, E.F. van Dishoeck, H. Linnartz, *Astrophys. J.* **662**, L23 (2007)
59. J.R. Pardo, J. Cernicharo, J.R. Goicoechea, M. Guélin, A. Asensio Ramos, *Astrophys. J.* **661**, 250 (2007)
60. S. Petrie, E. Herbst, *Astrophys. J.* **491**, 210 (1997)
61. K.M. Pontoppidan, *Astron. Astrophys.* **453**, L47 (2006)
62. S.S. Prasad, S.P. Tarafdar, *Astrophys. J.* **267**, 603 (1983)
63. W.G. Roberge, D. Jones, S. Lepp, A. Dalgarno, *Astrophys. J. Supp.* **77**, 287 (1991)
64. H. Roberts, E. Herbst, T.J. Millar, *Astrophys. J.* **591**, L41 (2003)
65. C. Salyk, K.M. Pontoppidan, G.A. Blake, F. Lahuis, E.F. van Dishoeck, N.J. Evans, *Astrophys. J.* **676**, L49 (2008)
66. S. Seager, *Space Science Reviews* **135**, 345 (2008)
67. P.M. Solomon, W. Klemperer, *Astrophys. J.* **178**, 389 (1972)
68. P.M. Solomon, N.C. Wickramasinghe, *Astrophys. J.* **158**, 449 (1969)
69. R. Stark, F.F.S. van der Tak, E.F. van Dishoeck, *Astrophys. J.* **521**, L67 (1999)
70. P. Stäuber, S.D. Doty, E.F. van Dishoeck, A.O. Benz, *Astron. Astrophys.* **440**, 949 (2005)

71. P. Stäuber, J.K. Jørgensen, E.F. van Dishoeck, S.D. Doty, A.O. Benz, *Astron. Astrophys.* **453**, 555 (2006)
72. T.P. Stecher, D.A. Williams, *Astrophys. J.* **149**, 29 (1967)
73. T.L. Stephens, A. Dalgarno, *J. Quant. Spectrosc. Rad. Transf.* **12**, 569 (1972)
74. A. Sternberg, A. Dalgarno, S. Lepp, *Astrophys. J.* **320**, 676 (1987)
75. A. Sternberg, A. Dalgarno, *Astrophys. J.* **338**, 197 (1989)
76. A. Sternberg, A. Dalgarno, *Astrophys. J. Supp.* **99**, 565 (1995)
77. A.G.G.M. Tielens, W. Hagen, *Astron. Astrophys.* **114**, 245 (1982)
78. S. Tiné, S. Lepp, R. Gredel, A. Dalgarno, *Astrophys. J.* **481**, 282 (1997)
79. G. Tinetti, A. Vidal-Madjar, M-C. Liang, et al., *Nature* **448**, 169 (2007)
80. M. Tulej, D.A. Kirkwood, M. Pachkov, J.P. Maier, *Astrophys. J.* **506**, L69 (1998)
81. F.A. van Broekhuizen, K.M. Pontoppidan, H.J. Fraser, E.F. van Dishoeck, *Astron. Astrophys.* **441**, 249 (2005)
82. E.F. van Dishoeck, in *Rate Coefficients in Astrochemistry*, eds. T.J. Millar, D.A. Williams (Kluwer, Dordrecht, 1988), 49
83. E.F. van Dishoeck, in *The Molecular Astrophysics of Stars and Galaxies* eds. T.W. Hartquist, D.A. Williams (Oxford University, Oxford, 1998), 53
84. E.F. van Dishoeck, *Ann. Rev. Astron. Ap.* **42**, 119 (2004)
85. E.F. van Dishoeck, *Proc. Nat. Ac. Science* **103**, 12249 (2006)
86. E.F. van Dishoeck, in *Astrophysics in the Next Decade*, ed. H. Thronson, M. Stiavelli, A.G.G.M. Tielens (Springer, Berlin, 2008)
87. E.F. van Dishoeck, J.H. Black, J.H., *Astrophys. J. Supp.* **62**, 109 (1986)
88. E.F. van Dishoeck, M.C. van Hemert, B.J. Jonkheid, *Faraday Discussions* **133**, 231 (2006)
89. C. Vastel, T.G. Phillips, H. Yoshida, *Astrophys. J.* **606**, L127 (2004)
90. F. Walter, F. Bertoldi, C. Carilli, et al., *Nature* **424**, 406 (2003)
91. N. Watanabe, A. Nagaoka, T. Shiraki, A. Kouchi, *Astrophys. J.* **616**, 638 (2004)
92. D.M. Watson, C.J. Bohac, C. Hull et al., *Nature* **448**, 1026 (2007)
93. C.M. Wright, E.F. van Dishoeck, P. Cox, S.D. Sidher, M.F. Kessler, *Astrophys. J.* **515**, L29 (1999)
94. E.L. Wright, J.C. Mather, C.L. Bennett et al., *Astrophys. J.* **381**, 200 (1991)
95. M. Yan, *PhD Thesis*, (Harvard University, 1997)

ATOMIC AND MOLECULAR PROBES FOR ASTRONOMICAL ENVIRONMENTS

STEPHEN H. LEPP

Department of Physics and Astronomy, University of Nevada, Las Vegas
Las Vegas, NV 89154, USA
E-mail: lepp@nevada.edu
http://www.physics.unlv.edu/~lepp

The emission from atoms and molecules provide our best probes into astronomical environments. Observations of atomic and molecular lines provide information on the abundance of species, the gas temperature and density. When combined with astrochemical models one may infer ionization rates along with heating and cooling rates. This paper will review a few examples of using atoms and molecules as probes. Most of the examples are selected from work done in collaboration with Alex Dalgarno.

Keywords: atomic and molecular physics, astrophysics

1. Introduction

Essentially all our information about astronomical environments arrives in the form of photons and we learn about these environments by interpreting these photons. The emission from atoms and molecules is particularly useful because they emit at discrete frequencies determined by the energy levels for a particular species. The interpretation of these emissions has been a large part of astrochemistry research and a detailed introduction can be found in "Spectroscopy of Astrophysical Plasmas" by Dalgarno and Layzer (1987).[1] In this paper I will outline a few examples of using atoms and molecules to infer astrophysical conditions. Most of these examples will be taken from work which I did in collaboration with Alex Dalgarno. Most of what I learned about astrophysical probes came from my long collaboration with Alex, and I would like to thank the organizers for inviting me to contribute to this workshop and volume in his honor.

It was almost exactly twenty-five years ago that I started working with Alex. I remember the year before, just after the American Astronomical Meeting in Boston, when I trekked from my hotel to Alex's office to talk about my research and discuss the possibility of a postdoc position. I met him briefly in the morning but he couldn't talk then and so we made an appointment for the afternoon. I walked back into Harvard Square and returned at the appointed time. We must have talked for about an hour, mostly about my doctoral thesis, which interested him and he seemed to have many ideas as to how it could have been done better. We talked some of the possibility of him having a postdoc position. I remember when I left his office, there was a line of three or four people waiting to see him. And I remember thinking to myself, as I walked back to Harvard Square, that, although he had made a tremendous impression on me, I felt I hadn't made a very good impression on him. In retrospect, I think I felt this mostly because I simply didn't know him very well and as a result didn't feel he was very enthusiastic, perhaps the British reserve which I wasn't used to. I remember in particular thinking to myself, "I don't know if I'll ever get a job in astrophysics, but I know this guy will never hire me." In fact, he did have a postdoc job the next year, I did apply for it and he did hire me. I'm very grateful he did, as I had tremendous fun working with him. I'll write more about Alex at the end, but for now I will continue on using atoms and molecules as astrophysical probes.

The first information from spectra is how many photons have been received and at which frequency. This allows us to infer how many atoms or molecules are in a particular energy state along our line of sight. I will discuss for the most part the interpretation of the level population by chemical models to infer physical conditions. We will discuss using deuterium fractionation and the OH molecule as probes of the environment. I'll finish with the detection of molecular ions in Supernova 1987A.

The interpretation of astronomical spectra involves a mixture of astronomical modeling and laboratory astrophysics. Laboratory astrophysics involves a combination of atomic and molecular theory and experiment to determine atomic and molecular processes needed for astronomical models. These include the energy levels for atoms and molecules, the radiative transition rates and collisional cross sections.

For a simple example, let us consider the two-level atom which has been diagramed in Fig. 1. In our simple model there are two levels, labeled 1 and 2. To find the population of each we consider both collisional transitions and radiative transitions. Collisional transitions, labeled as $K_\uparrow$ and $K_\downarrow$ in Fig. 1, will occur at a rate depending on the density and rate coefficient. For the upward transition the rate, in transitions per cm^3 per second, will be $n_1 n_X K_\uparrow$, where n_1 is the population of level 1 per cm^3, n_X is the population

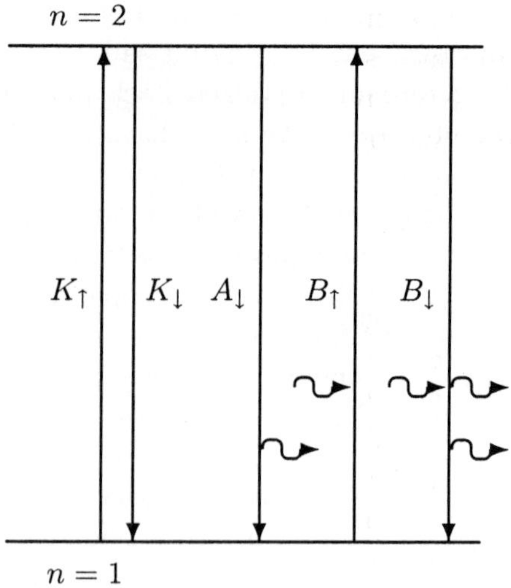

Fig. 1. The Energy Level diagram for a two level atom with transitions show, $K_\uparrow$ and $K_\downarrow$ represent collisional transitions and $A_\downarrow$, $B_\uparrow$ and $B_\downarrow$ radiative.

of the collider causing the transition and $K_\uparrow$ is the rate coefficient, in cm^3 s^{-1}, for the upward transition. The downward transition will have a rate of $n_2 n_X K_\downarrow$, the same form as the upward, except depending on n_2 the population of the upper state, and $K_\downarrow$ the downward rate coefficient. The rates for the radiative transitions are characterized by the Einstein A and B coefficients, where the $A_\downarrow$ is for spontaneous transitions and $B_\uparrow$ and $B_\downarrow$ are for stimulated transitions. The rate for the spontaneous downward transition is $n_2 A_\downarrow$, where $A_\downarrow$ is the Einstein A coefficient for this transition. The rate for the upward and downward stimulated transitions are given by $n_1 I_\nu B_\uparrow$ and $n_2 I_\nu B_\downarrow$ respectively, where I_ν is the intensity of the radiation field at the relevant frequency.

We can now see how level populations might be measured, for example the spontaneously emitted photons might make it to our telescope and be measured. This is common with molecular clouds observed in the radio. Alternatively, a photon from a background continuum source might be missing from the spectra because it was absorbed by an atom or molecule. This is common with clouds observed against a background star, or intergalactic gas observed against a background quasar.

If we can measure both level populations in our simple atom, perhaps both in absorption to a third level for example, then we can use their relative abundance to infer more about the environment. We can solve for the relative level populations in a steady state condition, by equating the total upward rate with the total downward rate and we find the relative level populations are given by:

$$\frac{n_2}{n_1} = \frac{n_X K_\uparrow + I_\nu B_\uparrow}{n_X K_\downarrow + I_\nu B_\downarrow + A_\downarrow} . \tag{1}$$

Thus, the relative level population depends on the density, the radiation field and the rate coefficients. The rate coefficients depend on the temperature, and the upward and downward collision rate coefficients are related to each other by detailed balance as are the Einstein A and B coefficients. This means that the relative populations are primarily dependent on the temperature and density of the collider. At very low densities, the collisions are unimportant and the level populations reflect the temperature of the radiation and at very high densities or very little radiation, the ratio of the level populations reflect the temperature of the collision partner. The density at which the changeover occurs is given by $A_\downarrow / K_\downarrow$ and is called the critical density.

2. Cosmic Background Radiation

This simple model shows how we can use level population measurements as a remote thermometer to measure (depending on conditions) either the temperature, radiation temperature or density remotely. It is interesting to note that in fact the first measurement of the cosmic background radiation was in fact done this way, it just wasn't realized at the time.

The CN molecule was detected in interstellar space by W. S. Adams though it took a few years before it was identified as CN.[2,3] the identification was made by A. McKellar in 1940 and he also noted that it implied a rotational temperature of about 2° K.[3] This was several years before Gamov, Alpher and Herman predicted the Cosmic Microwave Background (CMB) radiation from the big bang[4] and decades before Penzias and Wilson first measured the CMB.[5] We now know it was the first measurement of the temperature of the CMB, even though it was not recognized as such at the time.

This idea of using atoms and molecules as probes into the CMB temperature can be extended to high redshift. This provides us with the ability to measure the CMB temperature at earlier times when the universe hadn't expanded so much. In the standard Big Bang model, the radiation temperature ought to vary as

$$T_z = T_0(1 + z)\,, \tag{2}$$

where T_z is the CMB temperature which should be seen at redshift z and T_0 is the current CMB temperature. This was first suggested shortly after the discovery of CMB radiation.[6] Most of the high redshift CMB temperature measurements have been done with atoms, for example Ge, Bechtold and Black[7] used measurements of neutral carbon against a quasar background source to measure the CMB temperature at $z = 1.9731$. Recent Srianand, Noterdaeme, Ledoux and Petijean[8] made such a measurement with CO in a high redshift Lyman-α system at $z = 2.41837$. Many more measurements can be found in their references. The CO measurement is the highest precision measurement so far and it gives $T_z = 9.15 \pm 0.72$ K which is consistent with the formula in Eq 2, giving 9.315 for for $z = 2.41837$.

3. Deuterated Fractionation

Another way in which atoms and molecules are probes depend on chemical modeling of the species abundances. The first project I did with Dalgarno when I first arrived at the Harvard-Smithsonian Center for Astrophysics involved inferring physical conditions from the relative abundance of molecules and there fractionated forms. In particular we built on the work of Watson,[9] who had modeled the the DCO/HCO ratio.

Since HCO is made primarily by processes starting with H_3^+ we first find the fractionation in this species. In Watson's paper the fractionation of H_3^+ comes from

$$\mathrm{HD} + \mathrm{H_3^+} \leftrightarrow \mathrm{H_2} + \mathrm{H_2D^+} \tag{3}$$

at high temperatures this reaction comes to equilibrium, but at low temperatures the reverse reaction is negligible and it proceeds mostly to the right. This production of H_2D^+ is mediated by dissociative recombination

$$\begin{aligned} \mathrm{H_2D^+} + \mathrm{e} &\leftrightarrow \mathrm{H_2} + \mathrm{D} \\ \mathrm{H_2D^+} + \mathrm{e} &\leftrightarrow \mathrm{HD} + \mathrm{H} \\ \mathrm{H_2D^+} + \mathrm{e} &\leftrightarrow \mathrm{H} + \mathrm{H} + \mathrm{D} \end{aligned} \tag{4}$$

and by reactions with neutrals

$$\begin{aligned} \mathrm{H_2D^+} + \mathrm{X} &\rightarrow H_2 + DX^+ \\ &\rightarrow HD + HX^+\,. \end{aligned} \tag{5}$$

Where X here is used to represent other neutral constituents which can take a proton from H_3^+, such as CO, O, C and so on.

This leads to a fractionation of

$$\frac{n(H_2D^+)}{n(H_3^+)} = \frac{n(HD)}{n(H_2)} f \quad (6)$$

where $n(X)$ represents the abundance of X and f is given by

$$f = k_{\rightarrow}/(k_{\leftarrow} + \frac{\alpha n(e)}{n(H_2)} + \sum k_x \frac{n(x)}{n(H_2)}) \quad (7)$$

where $k_{\rightarrow}$ and $k_{\leftarrow}$ are the reaction rate coefficients for reaction 3, α is the recombination coefficient for reaction 4 and k_x is the rate coefficient for reaction 5.

The H_3^+ makes HCO^+ through

$$H_3^+ + CO \rightarrow H_2 + HCO^+ . \quad (8)$$

This fractionated version H_2D^+ also reacts through the CO version of reaction 5

$$\begin{aligned} H_2D^+ + CO &\rightarrow H_2 + DCO^+ \\ &\rightarrow HD + HCO^+ , \end{aligned} \quad (9)$$

which passes the fractionation from H_3^+ on to HCO^+. If the upper branch of reaction 9 is 1/3 then the fractionation of HCO^+ will be 1/3 the fractionation of H_3^+.

We modified this to take into account that the enhanced fractionation in molecular ions leads to an enhanced abundance of atomic deuterium.[10] This had been noted by Croswell and Dalgarno[11] in their analysis of OH fractionation. In particular atomic deuterium is made from

$$DCO^+ + e \rightarrow CO + D . \quad (10)$$

In regions where the molecules are significantly fractionated this leads to a significant increase in the production of D and thus atomic n(D)/n(H) ratios much larger then the cosmic abundance ratio. This excess deuterium can then cause additional fractionation through reactions such as

$$D + H_3^+ \rightarrow H_2D^+ + H \quad (11)$$

and

$$D + HCO^+ \rightarrow DCO^+ + H . \quad (12)$$

These reactions also convert D back to molecular ions and absorption onto grain surfaces removes D from the gas phase, presumably to go on to make

HD. If grain absorption happens at rate $n(D)n(H_{\mathrm{Total}})k_g$ and reaction n with a rate coefficient k_n, then we can estimate the deuterium abundance by assuming it is primarily made by reaction 10 and primarily removed by grains, reactions 11 and 12. This gives a deuterium abundance of

$$n(D) = \frac{\alpha_{10} n(e) n(DCO^+)}{2k_g n(H_2) + k_{11} n(H_3^+) + k_{12} n(HCO^+)} \,. \tag{13}$$

Using this deuterium abundance we can now estimate the additional fractionation through reactions 11 and 12. This allowed us to find an analytic expression for the fractionation.[10]

In Table 1 we list the limits on the total D to total H ([D]/[H]) for various molecular clouds using these limits. The numbers are taken from Dalgarno and Lepp.[10] The lower limits are fine but the upper limits may be too small as the analysis is still done with a slow dissociative recombination rate for H_3^+. The inferred deuterium ratios are roughly consistent with other measurements. It shows there may be a small cloud to cloud variation in this number. The next generation of radio telescopes such as the Atacama Large Millimeter/submillimeter Array (ALMA).

Table 1. [D]/[H] limits inferred in molecular clouds

Cloud	DCO^+/HCO^+	Lower	Upper
NGC 1333	0.024	1.8×10^{-5}	4.0×10^{-5}
HH 24	0.020	6.0×10^{-6}	1.4×10^{-5}
NGC 2264	0.017	1.2×10^{-5}	2.8×10^{-5}
Ser MC1	0.013	1.0×10^{-5}	2.1×10^{-5}
R CrA	0.017	1.0×10^{-5}	1.3×10^{-5}
NGC 6914	0.028	1.4×10^{-5}	4.0×10^{-5}
NGC 2264	0.017	3.4×10^{-6}	7.5×10^{-6}

4. OH Chemistry

In developing our deuterated chemistry we constructed an astrochemical model which we then expanded to look to include over one hundred species and more than a thousand chemical reactions. The most important of these species, H_3^+, as it initiated most of the chemistry, was not directly observable at the time. We found though a chemical probe in the form of OH which allowed us over a wide range of conditions to infer the abundance of H_3^+[12]

The formation of OH is initiated by a reaction

$$O + H_3^+ \rightarrow OH^+ + H_2 \,, \tag{14}$$

followed by two fast reactions with H_2

$$OH^+ + H_2 \rightarrow H_2O^+ + H \quad (15)$$

$$H_2O^+ + H_2 \rightarrow H_3O^+ + H\,. \quad (16)$$

The H_3O^+ then dissociatively recombines to form

$$\begin{aligned} H_3O^+ + e &\rightarrow H_2O + H \\ &\rightarrow OH + H_2 \\ &\rightarrow O + H_2 + O\,. \end{aligned} \quad (17)$$

There is uncertainty as to the branching ratio for the H_3O^+ dissociative recombination reaction, at the time we assumed that it was mostly to water. This uncertainty still remains with large differences among experiments.[13,14] Observations of interstellar clouds suggest that the H_2O branch is small.[15] Photodissociation by cosmic-ray induced photons will turn some of the H_2O into OH.

The OH is then removed primarily by reactions with O,

$$OH + O \rightarrow O_2 + H\,. \quad (18)$$

Thus we have a chain of reactions which starts with H_3^+ ion reacting with atomic oxygen and ends with OH reaction with atomic oxygen with a number of relatively fast reactions in between. Because the chain starts and ends by a reaction with atomic oxygen, the OH abundance is not sensitive to the abundance of atomic oxygen, unless of course the abundance is so small that the reaction network outlined here is not valid. So the OH abundance is just fixed to the H_3^+ abundance with a ratio determined by the reaction rates of the various reactions. We could now use the H_3^+ abundance to make an estimate of the ionization rate, since H_3^+ is formed following an ionization of H_2.

5. Supernova 1987A

Of course, the most basic probe provided by atoms and molecules is that they are present in the gas. In this section, I would like to describe the one time I was involved in the discovery of a molecular ion in space.

In February 1987, a new supernova was observed in the Large Magellanic Cloud. This was the first supernova visible to the naked eye since the invention of the telescope. It was also the first supernova in which molecules where detected.[16] This led Alex and I to develop a model of the ejecta of supernova in collaboration with my doctoral thesis advisor R. McCray.[17]

The ejecta is initially hot and ionized and dense, but is rapidly expanding and cooling. It is rich in heavy elements produced by the progenitor star and in the explosion. As it expands the ejecta the radioactive decay of unstable nuclei, such as ^{56}Co provide additional heating and ionization. With this model we were able to explain the observed abundances of CO in the Supernova 1987A.

After developing the model Alex suggested I should look through the spectra to see if any of the other abundant molecules in our model were present. One such molecular ion was HeH^+, which was formed by

$$He^+ + H \rightarrow HeH^+ + h\nu \,.$$

It was primarily removed in our models by

$$HeH^+ + H \rightarrow H_2^+ + He$$
$$HeH^+ + e \rightarrow He + H$$

I looked for it in the infrared spectra[18] and found two small lines at the right places for vibrational transitions of HeH^+.

I remember presenting Alex with a graph of the spectra with two small peaks marked right where the band emission from HeH^+ should be. Alex looked at the graph with some scepticism. The two lines I had marked were quite small, and there were two lines which had gone right off scale at either end of the graph. Additionally there were two bands in between which were over ten times bigger.

He asked about the two lines going off the scale and they were both identified as hydrogen recombination lines, which made him a bit happier. Then he asked about the large bands in between and I said they were still unidentified. He said, it was going to be a hard case, because the two small peaks I had found were sitting on the shoulder of those larger peaks and until they were identified, it will be hard to say for certain the smaller peaks aren't part of that structure. He took the graph with him and said he'd see what he could do as he was off to University College London for a meeting.

When he returned, he had talked with Jonathan Tennyson and Steven Millar and they had just had some new calculations of the spectra of H_3^+ and it fit right into those larger peaks. Our model had also produced abundant H_3^+ and so we wrote a paper announced the identification of H_3^+ in the spectra of supernova 1987A.[19] Figure 2 reproduced from this paper shows the model spectra for H_3^+ along with the observed spectra. Also marked in Fig. 2 are the two small lines I had originally identified as HeH^+. This is the first detection of H_3^+ outside the solar system and the first detection of HeH^+ anywhere.

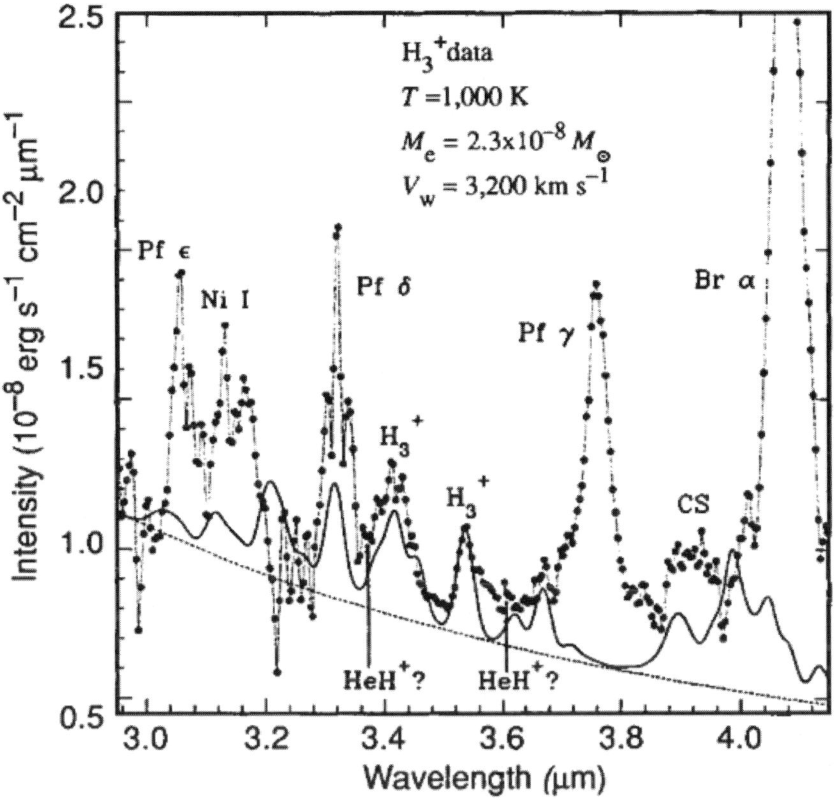

Fig. 2. Fitted H_3^+ spectra (solid) and the spectra from day 192 (filled circles) from Miller, Tennyson, Lepp and Dalgarno (1992). Reproduced from Ref 19 with permission.

6. Conclusion

Atoms and molecules are powerful probes into physical conditions in astronomical environments. The field of astrochemistry, to combine laboratory astrophysics with modeling and astronomical observations has allowed us to learn a great deal about our universe. Molecular probes in particular are likely to grow in importance over the next decade as new radio telescopes, including Atacama Large Millimeter/submillimeter Array (ALMA) and Expanded Very Large Array (EVLA) come online.

Alex Dalgarno has been one of the leaders in developing ways to use atoms and molecules to probe our universe and in this paper I have highlighted a few of the projects I did with him. We have covered only a small fraction of the work he and I collaborated on, which itself is a very small fraction of all of Alex's work. The depth and breadth of Alex's contributions to science are astounding. His contributions to molecular astrophysics alone have had a tremendous impact and earned him the title of the "Father of Molecular Astrophysics".

As I did mention in the introduction, I am very happy that Alex hired me and I had the opportunity to work with him as a postdoc. He was very fun to work with and conversations with him were always interesting. His own insights and understanding of many subjects were almost overwhelming. When I would go in to discuss problems with him, I often would wish I had a tape recording of our conversations to make certain I hadn't missed anything. His memory is absolutely amazing, I remember once suggesting an idea to him, and he said it was a great idea, unfortunately it has already been done. Then he walked over to the shelf in his office and pulled down a 20 year old journal and flipped to the correct page and showed it to me.

Collaboration with Alex was always tremendous fun, I loved bouncing ideas back and forth, and when my suggestion didn't quite work he was very good at not just rejecting it outright, but instead teasing the little bit of good idea hidden in there and sending it back in a more workable form. Doing science with Alex is always a lot of fun. He is the Master Juggler feeding out ideas and taking any ideas back, no matter how badly thrown, fixing them back into the pattern and throwing them back with spot on timing and precision, thrown so perfect they are hard to miss.

Alex was also clean and concise in his writing, and I learned a lot by how my rough drafts came back marked and edited. I learned a lot about English working with Alex, partly after an intensive course of self-study, so that my manuscripts would come back only marked up, rather then rewritten.

Alex has also been very supportive in my own efforts to build a career in science. His advice and his example, on how to be a professor are invaluable. I was lucky in that my Ph.D. thesis advisor, Richard McCray was also supportive. This was brought home to me this summer as I was having breakfast with a few of McCray's other students and one of them commented how lucky I was to have gotten a postdoc with Alex, because not all postdoc advisors were so supportive.

I also consider Alex a good friend, he was one of the busiest people I've ever met, but he always seemed to have time to stop and talk for a bit. It is tremendous fun have a one to one talk with Alex, as he knows so much of so many things and has a great wit. But he also seems to act as a magnet, attracting nice and interesting people into his world. Alex, his friends, his collaborators, his companion Fern, and all others drawn into his circle always make good dinner or conversation companions. I have met many wonderful people through my association with Alex. I would like to thank Alex and all his friends for many a pleasant evening. Finally, and most of all, I thank Alex for bringing me into his world some twenty-five years ago.

Acknowledgments

I gratefully acknowledge funding from NSF grant AST-0607524 and NASA Grant/Cooperative Agreement No. NNX08AE57A.

References

1. A. Dalgarno and D. Layzer, *Spectroscopy of Astrophysical Plasmas* (Cambridge University Press, Cambridge, 1987).
2. W.S. Adams, *Ap. J.* **93**, 11 (1941).
3. A. McKellar, *Publ. Astron. Soc. Pac.* **52**, 187 (1940).
4. R. Alpher, H. Bethe and G. Gamow, *Phy. Rev.* **73**, 803 (1948).
5. A. Penzias and R. Wilson, *Ap. J.* **142**, 419 (1965).
6. J. Bachall and R. Wolf, *Ap. J.* **152**, 701 (1968).
7. J. Ge, J. Bechtold and J. Black, *Ap. J.* **474**, 67 (1997).
8. R. Srianand, P. Noterdaeme, C. Ledoux and P. Petijean, *A&A* **482**, L39 (2008).
9. W.D. Watson, *Rev. Mod. Phys.* **48**, 513 (1976).
10. A. Dalgarno and S. Lepp, *Ap. J.* **287**, L47 (1984).
11. K. Croswell and A. Dalgarno, *Ap. J.* **289**, 618 (1985).
12. S. Lepp, A. Dalgarno and A. Sternberg, *Ap. J.* **321**, 383 (1987).
13. T.L. Williams, N.G. Adams, L. Babcock, C. Herd and M. Geoghegan, *M.N.R.A.S.* **282**, 413 (1996).
14. L. Vejby-Christensen, L. Andersen, O. Heber, D. Kella, H.B. Pedersen, H.T. Schmidt and D. Zajfman, *Ap. J.* **483**, 531 (1997).
15. M. Spaans, D. Neufeld, S. Lepp, G. Melnick and J. Stauffer, *Ap. J.* **503**, 780 (1998).
16. J. Spyromilio, W.P.S. Meikle, R.C.M. Learner and D.A. Allen, *Nature* **334**, 327 (1988).
17. S. Lepp, A. Dalgarno and R. McCray, *Ap. J.* **358**, 262 (1990).
18. W.P.S. Meikle, D.A. Allen, J. Spyromilio and G.-F. Varani, *M.N.R.A.S.* **238**, 193 (1989).
19. S. Miller, J. Tennyson, S. Lepp and A. Dalgarno, *Nature* **355**, 320 (1992).

H^- PHOTODETACHMENT IN ATOMIC PHYSICS AND ASTROPHYSICS

P. C. STANCIL and S. MIYAKE

Department of Physics and Astronomy
and the Center for Simulational Physics,
University of Georgia,
Athens, GA, 30602-2451, USA
E-mail: stancil@physast.uga.edu,
shinyam@physast.uga.edu

H. R. SADEGHPOUR

ITAMP, Harvard-Smithsonian Center for Astrophysics,
60 Garden St.,
Cambridge, MA, 02138, USA
E-mail: hrs@cfa.harvard.edu

B. M. MCLAUGHLIN

David Bates Building, School of Mathematics and Physics,
Queen's University Belfast,
7 College Park,
Belfast BT7 1NN, UK
E-mail: b.mclaughlin@qub.ac.uk

R. C. FORREY

Department of Physics,
Penn State University, Berks Campus,
Reading, PA 19610, USA
E-mail: rcf6@psu.edu

The negative hydrogen ion H^- plays an important role in the continuum opacity of late-type stars and in the kinetics of low-metallicity gas. We review the H^- photodetachment cross section along with other processes involving H^-. We address H^- in early Universe chemistry, the role of oscillator-strength sum-rules in constraining its continuum cross section, and the influence of autodetaching resonances on the efficiency of H^- photodestruction in the reionization era.

Keywords: atomic processes - early universe - galaxies: formation - galaxieis: high redshift

1. Introduction

While H_3^+ is a key species in the complex chemistry of metal-rich interstellar clouds, another two-electron hydrogen ion, H^-, plays a similar role in low-metallicity gas. H^- has long been known to be an important opacity source in the sun and other late-type stars, but it has yet to be detected through resonant spectroscopic features in any astronomical source. In this article, we review the processes important in the non-equilibrium chemistry of H^- with a particular focus on photodetachment and radiative attachment. We discuss the various astronomical environments where H^- plays a fundamental role, the history of H^- investigations from an atomic physics perspective, and the many contributions that Alex Dalgarno has made to the study of this anion.

2. H^- in Astrophysics

It has often been the case that problems in astronomy have driven advances in atomic physics and the hydrogen negative ion is a prime example. In the 1930s, the continuum absorption in the solar spectrum, as well as that of similar late-type stars, was not understood. It was first proposed by Wildt[1] that H^-, which has a small electron affinity of 0.754 eV, might be responsible for this continuum opacity due to its bound-free absorption. This lead Chandrasekhar[2] and Chandrasekhar and Breen[3] to make the first quantum mechanical calculations of bound-free and free-free absorption coefficients for this two-electron anion, respectively. This early history of H^- has been discussed by Rau[4] and recently summarized in Ross *et al.*[5] who mention that the role of H^- as the dominant opacity source in the solar visible and infrared (IR) spectrum was cemented in 1945 by good agreement between Chandrasekhar's calculations and empirically-derived absorption coefficients derived for the Sun by Münch.[6]

The role of H^- as a continuum opacity does not, however, constitute a direct detection. There is a peak in the photodetachment cross section near 0.8 μm, but it is broad. H^- is interesting in that it has only one bound singlet state, the electronic ground level $1s^2\ {}^1S$, and therefore lacks a bound-bound electric dipole spectrum of electronic transitions. These facts make the direct detection of H^- problematic outside the laboratory. However, it is well known, as will be discussed below, that a series of auto-detaching resonances exist in the ultraviolet (UV). These features result from the simultaneous ejection of one electron to the continuum and the excitation of the remaining electron into excited ${}^2P$ states of the residual neutral atomic hydrogen. The physics of these resonances will be discussed below, but their existence prompted searches for them in the interstellar medium (ISM) in the 1970s with the UV satellite *Copernicus*[7] and more recently with the *Far Ultraviolet Space Explorer (FUSE)*.[5] Unfortunately, these searches have only been able to place an upper limit on the H^- abundance in the ISM.

Nevertheless, H^- is believed to play an important role in a variety of astronomical contexts. We consider a number of cases focusing primarily on photo-processes and non-equilibrium chemistry as opposed to opacities.

2.1. *Primordial Gas*

Prior to the formation of the first star or luminous object, the early Universe consisted of a nearly homogeneous expanding and cooling gas of primordial species dominated by protons and electrons. Hydrogen atoms formed by radiative recombination at the start of the recombination era. It was first proposed by McDowell[8] that the first neutral molecule, H_2, would form via the associative detachment process

$$H^- + H \longrightarrow H_2 + e^- \tag{1}$$

following the creation of H^- via radiative attachment

$$H + e^- \longrightarrow H^- + \nu. \tag{2}$$

Interestingly, McDowell quotes an unpublished estimate of process (1) by Dalgarno. In the same year, Dalgarno and Kingston[9] presented calculated rate coefficients for process (2), to be discussed further below.

However, the abundance of H^- is controlled by two processes that can efficiently destroy it: photodetachment

$$H^- + \nu \longrightarrow H + e^- \tag{3}$$

and mutual neutralization

$$H^- + H^+ \longrightarrow H + H. \tag{4}$$

In the post-recombination era of the early Universe, the photons in process (3) are those due to the cosmic background radiation (CBR) field. The CBR is a black-body corresponding to a radiation temperature of $T_r = 2.7(1 + z)$ K, where z is the redshift. Once the radiation temperature falls below ~ 500 K, photodetachment becomes inefficient and process (1) becomes the dominant formation mechanism of H_2. Studies of the hydrogen molecular chemistry in the early Universe have been carried out by numerous authors.[10–12] Due to the expansion of the Universe, the fractional abundance of H_2 with respect to atomic hydrogen reaches a value of only a few times 10^{-5} in the post-recombination gas for $z < 100$.

2.2. *Radiative Feedback from Primordial Objects*

At some intermediate redshift ($z \sim 20 - 50$), the small perturbations in the primordial gas density will grow and ultimately collapse to form a star or other luminous object. H_2 will form by the same processes as discussed in Section 2.1 and it is radiative cooling of H_2 which provides the dominant means of removing the heat generated by the adiabatic collapse as the gas density grows. The efficiency of H_2 cooling depends on the abundance of H_2 which is dependent on the availability of free electrons, H atoms, and H^-. Since the CBR is the only radiation field present and the temperature of the CBR is small by this redshift, H^- photodetachment is not important. While the efficiency of H_2 ultimately declines as the gas becomes optically thick to the cooling lines ($\sim 10^8$ cm^{-3}), the cloud continues to collapse until a star is formed.

The first stars, known as Population III stars, are believed to have been more massive than contemporary Population I stars. Therefore they were likely to have large effective temperatures and would have produced copious UV, FUV, and x-ray radiation. This radiation would then have propagated into the surrounding primordial gas resulting in two possible feedback effects on the abundance of H_2: i) enhancement[13,14] or ii) suppression.[15,16] In the former case, a so-called positive feedback effect, the UV photons ionize atomic H producing electrons which drive the chemistry enhancing the formation of H_2. Since H_2 is the dominant coolant in primordial gas, it was proposed by Haiman, Rees, and Loeb[13] that this effect would accelerate the gravitational collapse and eventual fragmentation of high redshift halos. In turn, this would lead to increased efficiency in Pop III star formation and acceleration of the UV background radiation field responsible for reionization.

In the alternate scenario, UV photons within the Lyman and Werner bands of H_2, energies between $\sim$11 eV and the H ionization threshold, can penetrate large primordial clouds due to the small optical depths of these lines. These photons can then destroy H_2 by absorption to electronically excited states followed by fluorescent decay into the continuum of the electronic ground-state – the so-called Solomon process:

$$H_2 + \nu \rightarrow H_2^* \rightarrow H + H + \nu'. \tag{5}$$

This results in a negative feedback on the abundance of H_2, n_{H_2}, for which an equilibrium estimate

$$n_{H_2} = \frac{k_1 n_H n_{H^-}}{\beta_5} \tag{6}$$

where k_1 is the rate coefficient for associative detachment, β_5 the rate for process (5), and n_H and n_{H^-}, the number densities of H and H^-. A reduction in the abundance of H_2 will directly suppress its cooling efficiency which indirectly influences the efficiency of subsequent Pop III star formation.

Numerous authors have considered such feedback effects, but it is unclear which scenario will dominate.[13–17] Further, only UV photons in the Lyman and Werner bands have been addressed. It was recently pointed out by Glover[18] and Chuzhoy, Kuhlen, and Shapiro[19] that the negative feedback effect could be further enhanced by considering photodetachment of H^-, process (3). Photons with energies between the H^- photodetachment threshold (0.754 eV) and the Lyman limit could efficiently destroy H^- suppressing the formation of H_2 which occurs through the associative detachment process (1).

Chuzhoy *et al.*[19] considered a number of situations and estimated the effect of H^- photodetachment with the suppression factor

$$F_b = 1 + \frac{\beta_3}{k_1 n_H} \tag{7}$$

where β_3 is the H^- photodetachment rate. The abundance n_{H^-} is divided by F_b which directly reduces the abundance of H_2 as given in Eq. (6) assuming that only the three considered processes are important for H_2 formation. They considered the following radiation fields for photodetaching H^-: i) H recombination lines below 10.25 eV, which result from photoionization from the first luminous sources on the surrounding gas, ii) black-body radiation from massive Pop III stars with a upper energy cut-off at the Lyman limit, iii) a power-law spectrum typical of miniquasars, and iv) a FUV background generated from x-ray sources.

Through semi-empirical arguments, Chuzhoy *et al.* deduced that the abundance of H_2 would be significantly reduced, by a factor of $F_b \sim 1000$, if photodetachment of H^- due to recombination photons (case i) was considered. This reduction was further increased by from 10% to a factor of 10 ($F_b \sim 1100 - 10,000$) if continuum black-body flux from massive Pop III stars was added (case i and case ii). The largest suppression factor occurs for the smaller mass stars which have lower effective temperatures T_{eff}.

Instead of a black-body spectrum, a power-law spectrum can be adopted which has the typical form

$$J = J_{21}\left(\frac{E_{\rm ph}}{E_{\rm H}}\right)^{-\alpha} \tag{8}$$

where J_{21} is the intensity in 10^{-21} ergs cm^{-2} Hz^{-1} sr^{-1}, $E_{\rm ph}$ the photon energy, $E_{\rm H}$ the ionization energy of H, and α the power-law index. For $\alpha = 1.7$, typical of a miniquasar, Chuzhoy *et al.* find a further suppression enhancement, or $F_{\rm b} \sim 5000$ (case i and case iii).

Chuzhoy *et al.* considered only the background H^- photodetachment cross section based on a fit to the calculations of Wishart[20] as shown in Fig. 1. Recently, Miyake *et al.*[22] have shown that the contribution from the auto-detaching resonances, not considered by Chuzhoy *et al.*, could be significant. If we consider the ratio of photodetachment rates due to the background (b) and the sum of the background and resonances (r+b), this is approximately equal to the H_2 suppression factor due to H^- photodetachment through the resonances only

$$\beta_3{}^{\rm r+b}/\beta_3{}^{\rm b} \sim F_{\rm r+b}/F_{\rm b} = F_{\rm r}. \tag{9}$$

Therefore, the total H_2 suppression factor is $F = F_{\rm b} \times F_{\rm r}$. Here and in Miyake *et al.*, the H^- photodetachment cross section of McLaughlin *et al.*[21] is adopted in the computation of $F_{\rm r}$.

For the scenario of black-body radiation from a Pop III star (case ii), Miyake *et al.* found an additional enhancement with the suppression factor typically increasing by ~5% for 25,000 K and asymptotically exceeding $F_{\rm r} \sim 1.2$ as $T_{\rm eff}$ exceeds 150,000 K. This behavior is different from the results of Chuzhoy *et al.* who found the black-body contribution to increase with decreasing $T_{\rm eff}$. Overall the effect of the black-body field becomes less important with $T_{\rm eff}$ since the majority of the intensity falls beyond the 13.6 eV cut-off. Nevertheless, the suppression factor is doubled (case i and ii) for $T_{\rm eff} = 40,000$ K with the resonant contribution increasing this by 20%.

Miyake *et al.*[22] also considered the possible enhancement due to the H^- auto-detaching resonances for power-law spectra. Quasars typically have power-law radiation fields for $\alpha \sim 0.5$ - 0.7, massive black holes $\alpha \sim 1$, and the high redshift intergalactic medium (IGM) $\alpha \sim 0.7$ - 1. Constraining the radiation field to photon energies between 10 and 13.6 eV, a resonant enhancement factor of $F_{\rm r} \sim 1.8$ was found, insensitive to the power-law index over the range $\alpha = 0.1 - 5$. This photon energy bin includes the H Lyman lines, the H^- resonances, and the H_2 Lyman-Werner bands, up to the Lyman limit. If the background radiation field after the formation of

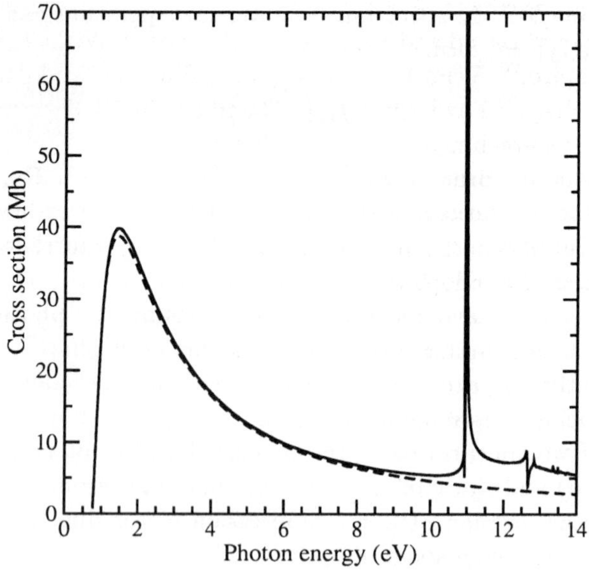

Fig. 1. H^- single photodetachment cross sections. Background from fit to Wishart[20] (dashed line); full cross section from McLaughlin *et al.*[21] (solid line).

the first luminous objects was dominated by miniquasars, or similar UV sources, there would have been a significant negative radiative feedback effect on the creation of H_2 and its cooling efficiency and on the ability of the residual primordial gas to coalesce into another generation of primordial stars[a]. The contribution from H^- auto-detaching resonances are seen to be significant and should be accurately treated in future models. A key component therefore, is the accuracy of the H^- photodetachment cross section including the shape, magnitude, and position of the resonances. In the next section, we describe the status of the atomic physics of H^-.

3. H^- in Atomic Physics

3.1. *Non-resonant Photodetachment*

The background photodetachment cross section has been studied by numerous authors using a variety of techniques following the original calcula-

[a]Note that since the H^- mechanism for forming H_2 is suppressed, other routes such as the charge exchange reaction $H + H_2^+ \rightarrow H_2 + H^+$ would become more important. However, the destruction of H_2^+ due to photodissociation via the FUV radiation field would also contribute to a negative feedback effect.

tions of Chandrasekhar.[2] A chronological, but not comprehensive, list includes Geltman,[23] Broad and Reinhardt,[24] Stewart,[25] Wishart,[20] Abrashkevich and Shapiro,[26] Venuti and Declava,[27] Pindzola and Robicheaux,[28] Kheifets and Bray,[29] Pazdzersky *et al.*,[30] and Frodov.[31] Experimental studies of the photodetachment cross section within a few eV of the threshold were performed by Branscomb and Smith,[32] Smith and Burch,[33,34] and Popp and Kruse.[35] Reasonable agreement has been obtained between the later calculations and the measurements. Typically the cross sections of Wishart[20] have been adopted in most astrophysical modeling applications. McLaughlin *et al.*[21] have performed new calculations using the eigenchannel R-matrix[36] and R-matrix plus pseudo-state[37] method. A combination of the new calculations, previously published results, and measurements were merged to obtain a photodetachment cross section that satisfies a number of oscillator strength sum rules, similar to the approach adopted by Yan, Sadeghpour, and Dalgarno[38] for photoionization of He and H_2. This recommended cross section of photodetachment of H^- was adopted for the astrophysical environments presented in the previous section.

3.2. *Resonant Photodetachment*

As mentioned in the Introduction, while H^- contains only one electronic singlet state, it possesses a rich set of doubly excited states embedded in the one-electron continuum. These resonant states auto-detach to H(n) + e^-, with $n \geq 2$. These resonant structures were first studied theoretically by Macek[39] and observed in the elastic scattering experiments of McGowan, Williams, and Carley.[40] Later experiments summarized in Cohen and Bryant[41] and Balling *et al.*[42] have mapped out the resonances up to and including $n = 8$. Extensive theoretical calculations by Broad and Reinhardt,[24] Sadeghpour *et al.*,[36] Tang and Shimamura,[43] and Kuan *et al.*,[44] among others, have quantitatively reproduced the measurements. In the H($n = 2$) + e^- channel, there are three infinite series of Rydberg Feshbach resonances converging to the H($n = 2$) threshold, including a sharp Feshbach resonance just below the threshold at 10.924 eV and a broad shape resonance just above the threshold at 10.972 eV. The resonances for $n \geq 5$ occur at photon energies greater than the Lyman limit and are therefore not relevant in most astrophysical applications. Nevertheless, they contribute to the oscillator strength sum rules, and to ensuring the accuracy of the total cross section, as discussed below. New R-matrix calculations of the resonances up to $n = 8$ are presented in McLaughlin *et al.*[21]

3.3. *Oscillator Strength Sum Rules*

It was pointed out by Dalgarno and Ewart,[45] that the accuracy of the H^- photodetachment cross section could be checked by comparing the continuum oscillator strength moments $S(k)$ given by

$$S(k) = \int \frac{df}{dE}(E_0 + E)^k dE \tag{10}$$

to other values obtained from initial state properties. The latter, which usually involve initial state matrix elements, can be computed to high accuracy (cf. Dalgarno and Lynn[46]). In Eq. 10, E is the photoelectron energy, E_0 the electron affinity of H^-, and $\frac{df}{dE}$ is the differential oscillator strength for absorption into the continuum. This oscillator strength is related to the photodetachment cross section as $\sigma(E) = 4.03 \times 10^{-18} \frac{df}{dE}$ cm^2, and the photon energy is $E_{\rm ph} = E_0 + E$. In Table 1, "exact" values of the sum rules for $k = -3$ to 2 are summarized from the most recent calculations of Pipin and Bishop[47] and Bhatia and Drachman[48] and using the matrix elements tabulated by Drake.[49] These are compared to the sum rules computed using Eq. (10) and the cross section deduced in McLaughlin *et al.*[21] for single, double, and total detachment. The agreement is seen to be very good for $k = -3$ to -1, but begins to deteriorate for larger values of k. For the astrophysical applications discussed here, this is not a significant issue as the larger values of k are dominated by the higher-energy tail of the cross section which falls-off as $E_{\rm ph}^{-7/2}$. Note the sum rules for k larger than 3 are not defined as the integral in Eq. (10) diverges.

Table 1. Computed Sum Rules for H^-.

Sum Rule	"Exact"	Single	Double	Total
$S(2)$	1.37855	1.261	0.051	1.311
$S(1)$	0.747508	0.596	0.027	0.623
$S(0)$	2	1.693	0.025	1.718
$S(-1)$	14.9685	14.351	0.028	14.379
$S(-2)$	206.165	206.27	0.035	206.30
$S(-3)$	3773.40	3807.01	0.046	3807.05

3.4. *Radiative Attachment*

With an accurate form of the photodetachment cross section, the radiative attachment (process 2) rate coefficient is readily obtained via detailed balance. Using the photodetachment cross sections of Geltman[23] with extrapolations to higher energies by Dalgarno and Ewart,[45] Dalgarno and Kingston[9] computed the radiative attachment rate coefficients from ~1000 to 20,000 K. Thirty-five years later, Stancil and Dalgarno[50] repeated this calculation using more recent theoretical and experimental cross sections and extended the rate coefficients down to 20 K. Both calculations are shown in Fig. 2 and are seen to be in very good agreement for the range of overlapping temperatures. Radiative attachment rate coefficients computed with the recommended cross sections of McLaughlin *et al.*[21] are found to be in excellent agreement with the results of Stancil and Dalgarno.[50] In terms of early Universe postrecombination models, Stancil *et al.*[11] adopted the Stancil and Dalgarno[50] values, while Galli and Palla[10] adopted a fit to the rate coefficients of de Jong.[51] The two fits are in reasonable agreement, with the Galli and Palla values being typically 5% smaller for $T < 2000$ K. The fits do start to significantly diverge above 5000 K. These differences in H^- radiative attachment rate coefficients should not, however, significantly impact the predictions of the abundances of H^- and H_2.

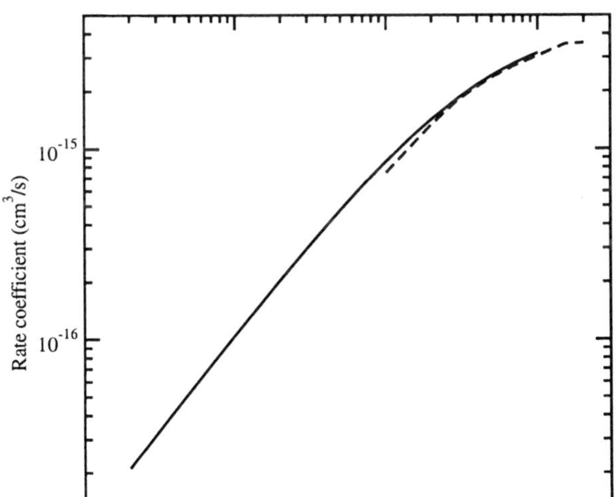

Fig. 2. H^- spontaneous radiative attachment rate coefficients as a function of collision temperature. Stancil and Dalgarno[50] (solid line); Dalgarno and Kingston[9] (dashed line).

3.5. *Stimulated Radiative Attachment*

As intense radiation fields are generally a significant feature of most astrophysical environments, stimulated formation processes may play a role in the chemistry. Stancil and Dalgarno[52] investigated stimulated radiative association of LiH due to the high redshift CBR field. They later extended this work to the formation of H^- by[50]

$$H + e^- + \nu \longrightarrow H^- + \nu + \nu. \tag{11}$$

Fig. 3 displays the rate coefficients for spontaneous plus stimulated radiative attachment for various black-body radiation temperatures T_r. The effect of stimulated attachment was found to be small for $T_r < 3000$ K, because the H^- electron affinity is 0.754 eV ($\sim$9000 K), but the enhancement grows rapidly with T_r for $T_r > 5000$ K. However, Stancil and Dalgarno[50] found that the process had at negligible effect on H^- formation in the early Universe. Stimulated radiative effects may play a role when the residual primordial gas is exposed to radiation from the first stars, as discussed above, providing a positive feedback effect, though this has yet to be studied.

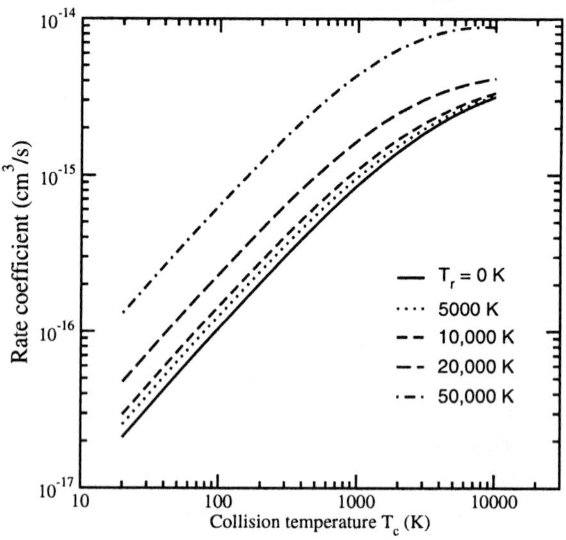

Fig. 3. H^- total (spontaneous plus stimulated) radiative attachment rate coefficients as a function of collision temperature T for various blackbody radiation temperatures T_r. From Stancil and Dalgarno.[50]

3.6. *Other H^- Destroying Processes*

In addition to photodetachment, associative detachment (1) and mutual neutralization (4) are the primary destruction mechanisms for H^-. Calculations for the former process have been made by Dalgarno and Browne,[54,55] Bieniek and Dalgarno,[56] Launay *et al.*,[57] and Cízek *et al.*,[58] while only one experimental investigation has been performed by Schmeltekopf *et al.*[59] Glover, Savin, and Jappsen[53] have pointed out that while there is generally good agreement between the calculations and the measurement, the most recent calculation by Cízek *et al.*, based on a more accurate H_2^- potential energy surface, is a factor of three times larger.

Mutual neutralization of H^- with H^+ has been studied since the early work of Bates and Lewis,[60] but results relevant to astrophysics are primarily limited to the measurement of Moseley *et al.*[61] and the quantum calculation of Fussen and Kubach.[62] Based on the available data, Dalgarno and Lepp[63] deduced rate coefficients which have been adopted in many astrophysical models. However, Glover *et al.*[53] found that the available rate coefficients actually have a scatter of an order of magnitude. Glover *et al.* then investigated the effects of uncertainties in these processes on H_2 formation and cooling in primordial halos. While the uncertainties in processes (1) and (4) led to small variations in protogalaxies forming from cold primordial gas, they found a significant impact if the objects formed from a hot, highly ionized gas. Such a situation is typical of fossil HII regions which are believed to be sites for second generation primordial star formation. More work is therefore needed on both processes to improve models of such environments.

4. Summary

Throughout his career Alex Dalgarno has made, and continues to make, seminal contributions to our understanding of the structure and dynamics of H^-, and other two-electron atomic and molecular species. His remarkable physical insight has shown that even an intractable problem, such as the few-electron Hamiltonian, can be fruitfully attacked with increasing precision, once empirical and numerically accurate facts are employed in a self-consistent manner. Such an approach has allowed us to postulate the importance of H^- photodetachment for the epoch of reionization, where we have been aided by advances in the experimental and theoretical understanding of the absorption continuum and auto-detaching resonant structure of H^-. This knowledge should allow for an accurate treatment of the photodetachment process in astrophysical simulations, once the resonant structure is correctly incorporated. On the other hand, heavy-particle collision processes such as associative detachment and mutual neutralization, important H^- destruction mechanisms, are not known with any sufficient degree of confidence and require further study. All of these processes play key roles in a variety of astrophysical environments. In particular, H^- is a vital, cohesive species that controls the efficiency of Population III star formation, and may influence the time scale of the reionization era of the early Universe.

Acknowledgments

PCS and SM acknowledge support from NSF Grant AST-0607733. The work of BMMcL and RCF was supported by a research development grant from Pennsylvania State University. PCS, BMMcL, and RCF would like to thank the Institute for Theoretical Atomic, Molecular, and Optical Physics at the Harvard-Smithsonian Center for Astrophysics for travel support. The authors would like to dedicate this article to Professor Alex Dalgarno, FRS, on the occasion of his 80th birthday, who has and continues to be an inspiration to us all.

References

1. R. Wildt, *Astrophys. J.* **90**, 611 (1939).
2. S. Chandrasekhar, *Astrophys. J.* **100**, 176 (1944).
3. S. Chandrasekhar and F. H. Breen, *Astrophys. J.* **104**, 430 (1946).
4. A. R. P. Rau, *Astronomy-inspired Atomic and Molecular Physics* Astrophysics and Space Science Library, Volume 271, (Kluwer Academic Publishers, Dordrecht, 2002).
5. T. Ross, E. J. Baker, T. P. Snow, J. D. Destree, B. L. Rachford, B. L. Rachford, M. M. Drosback, and A. G. Jensen, *Astrophys. J.* **684**, 358 (2008).
6. D. Münch, *Astrophys. J.* **102**, 385 (1945).
7. T. P. Snow, *Astrophys. J.* **198**, 361 (1975).
8. M. R. C. McDowell, *The Observatory* **81**, 240 (1962).
9. A. Dalgarno and A. E. Kingston, *The Observatory* **83**, 39 (1963).
10. D. Galli and F. Palla, *Mon. Not. R. Astron. Soc.* **335**, 403 (1998).
11. P. C. Stancil, S. Lepp, and A. Dalgarno, *Astrophys. J.* **405**, 1 (1998).
12. S. Lepp, P. C. Stancil, and A. Dalgarno, *J. Phys. B* **35**, R57 (2002).
13. Z. Haiman, M. J. Rees, and A. Loeb, *Astrophys. J.* **467**, 522 (1996).
14. T. Kitayama, H. Susa, M. Umemura, and S. Ikeuchi, *Mon. Not. Roy. Astron. Soc.* **326** 1353 (2001).
15. Z. Haiman, M. J. Rees, and A. Loeb, *Astrophys. J.* **476**, 458 (1997).
16. Z. Haiman, T. Abel, and M. J. Rees, *Astrophys. J.* **534**, 11 (2000).

17. N. Yoshida, T. Abel, L. Hernquist, and N. Sugiyama, *Astrophys. J.* **592**, 645 (2003).
18. S. C. O. Glover, *Mon. Not. R. Astron. Soc.* **379**, 1352 (2007).
19. L. Chuzhoy, M. Kuhlen, and P. R. Shapiro, *Astrophys. J. Lett.* **665**, L85 (2007).
20. A. W. Wishart, *Mon. Not. R. Astron. Soc.* **187**, 59 (1979).
21. B. M. McLaughlin, H. R. Sadeghpour, P. C. Stancil, A. Dalgarno, and R. C. Forrey, in preparation (2009).
22. S. Miyake, P. C. Stancil, H. R. Sadeghpour, A. Dalgarno, B. M. McLaughlin, and R. C. Forrey, *Astrophys. J.*, submitted (2009).
23. S. Geltman, *Astrophys. J.* **136**, 935 (1962).
24. J. T. Broad and W. P. Reinhardt, *Phys. Rev.* A **14**, 2159 (1976).
25. A. L. Steward, *J. Phys. B* **11**, 3851 (1978).
26. A. G. Abrashkevich and M. Shapiro, *Phys. Rev.* A **50**, 1205 (1994).
27. M. Venuti and P. Decleva, *J. Phys. B* **30**, 4839 (1997).
28. M. S. Pindzola and F. Robicheaux, *Phys. Rev.* A **58**, 4229 (1998).
29. A. S. Kheifets and I. Bray, *Phys. Rev.* A **58**, 4501 (1998).
30. V. A Pazdzersky, V. I. Usachenko, and A. V. Ushnurtsev, *J. Phys. B* **33**, 1135 (2000).
31. A. M. Frolov, *J. Phys. B* **37**, 853 (2004).
32. L. M. Branscomb and S. J. Smith, *Phys. Rev.* **98**, 1028 (1995).
33. S. J. Smith and D. S. Burch, *Phys. Rev.* **116**, 1125 (1959).
34. S. J. Smith and D. S. Burch, *Phys. Rev. Lett.* **2**, 165 (1959).
35. H. P. Popp and S. Kruse, *J. Quant. Spectrosc. Radiat. Transfer* **16**, 683 (1976).
36. H. R. Sadeghpour, C. H. Greene, and Cavagnero, *Phys. Rev.* A **45**, 1587 (1992).
37. D. M. Mitnik, M. S. Pindzola, D. C. Griffin, and N. R. Badnell, *J. Phys. B* **32**, L479 (1999).
38. M. Yan, H. R. Sadeghpour, and A. Dalgarno, *Astrophys. J.* **496**, 1044 (1998).
39. J. Macek, *Proc. Phys. Soc.* **92**, 365 (1967).
40. J. W. McGowan, J. F. Williams, and E. K. Carley, *Phys. Rev.* **180**, 132 (1969).
41. S. Cohen and H. C. Bryant, *Revista Mexicana de Astronomía y Astrofísica* **9**, 148 (2000).
42. P. Balling, *et al.*, *Phys. Rev.* A **61**, 022702 (2000).
43. J. Z. Tang and I. Shimamura, *Phys. Rev.* A **51**, R1738 (1995).
44. W. H. Kuan, T. F. Jiang, and K. T. Chung, *Phys. Rev.* A **60**, 364 (1999).
45. A. Dalgarno and R. W. Ewart, *Proc. Phys. Soc.* **80**, 616 (1962).
46. A. Dalgarno and N. Lynn, *Proc. Phys. Soc.* A **70**, 802 (1957).
47. J. Pipin and D. M. Bishop, *J. Phys. B* **25**, 17 (1992).
48. A. K. Bhatia and R. J. Drachman, *J. Phys. B* **27**, 1299 (1994).
49. G. W. F. Drake, in *Handbook of Atomic, Molecular, and Optical Physics* (Amer. Inst. Phys., New York, 1995), p.161.
50. P. C. Stancil and A. Dalgarno, *Faraday Disc.* **109**, 61 (1998).
51. T. de Jong, *Astron. Astrophys.* **20**, 263 (1972).

52. P. C. Stancil and A. Dalgarno, *Astrophys. J.* **479**, 543 (1997).
53. S. C. Glover, D. W. Savin, and A.-K. Jappsen, *Astrophys. J.* **640**, 553 (2006).
54. A. Dalgarno and J. C. Browne, *Astrophys. J.* **149**, 231 (1967).
55. J. C. Browne and A. Dalgarno, *J. Phys. B* **2**, 885 (1969).
56. R. J. Bieniek and A. Dalgarno, *Astrophys. J.* **228** 635 (1979).
57. J. M. Launay, M. Le Dourneuf, and C. J. Zeippen, *Astron. Astrophys.* **252**, 842 (1991).
58. M. Cízek, J. Horácek, and W. Domcke, *J. Phys. B* **31**, 2571 (1998).
59. A. L. Schmeltekopf, F. C. Fehsenfeld, and E. F. Ferguson, *Astrophys. J.* **118**, L155 (1967).
60. D. R. Bates and J. T. Lewis, *Proc. Phys. Soc. A* **68**, 173 (1955).
61. J. Moseley, W. Aberth, and J. R. Peterson, *Phys. Rev. Lett.* **24**, 435 (1970).
62. D. Fussen and C. Kubach, *J. Phys. B* **18**, L31 (1986).
63. A. Dalgarno and S. Lepp, in *Astrochemistry*, ed. M. S. Vardya and S. P. Tarafdar (Dordrecht, Reidel, 1987), p. 109.

ASTROCHEMISTRY: AN UNFINISHED SYMPHONY

JOHN H. BLACK
Department of Radio & Space Science, Chalmers University of Technology, Onsala Space Observatory, SE-439 92 Onsala, Sweden
E-mail: John.Black@chalmers.se
www.chalmers.se/rss/oso-en/contact/staff/black-john

Alex Dalgarno has introduced many of the themes that define current research in astrochemistry. Variations on these themes will continue to be played out well into the future. Clear and unambiguous tests of theories of formation of interstellar molecules remain elusive; however, the understanding of fluorine chemistry appears to offer one success story. The predicted chemical activity in X-ray dominated regions and in shock waves is being studied with increasingly sensitive observations. Molecules whose chemistry and internal excitation are strongly coupled are especially valuable. In some cases, observations of these reactive species can provide probes of specific chemical rates in interstellar gas. Superthermal OH observed in a shock front appears to offer special diagnostic value. Even cosmologists have recognized that chemistry plays an important role in the early evolution of matter in the Universe around the epoch of recombination. The hydrogen molecule is being used to test cosmological theories of dark matter.

Keywords: Astrochemistry; Interstellar matter; Cosmology

1. Introduction

Alex Dalgarno has enriched the study of astrochemistry by extending it far beyond the realm of the nearby interstellar medium. Thanks to him and his many collaborators and students, molecules and molecular processes are now known to be important in extremely energetic phenomena like supernovae and in the early Universe. I predict that more of his non-astrophysical investigations will find unexpected applications in astrophysics.

The value of a retrospective volume lies in its lessons for the future. Because the field of astrochemistry is littered with the detritus of false starts and wishful thinking, it is instructive to look for examples of testable predictions that have actually been verified. The chemistry of the early Universe is literally hidden in the dark ages of cosmic time; however, it is interesting to ask whether the theory can be tested by observation. In the following, I present three themes with variations: (1) prediction and verification in astrochemistry, (2) spectroscopic probes of molecular shocks, and (3) astrochemistry of the early Universe. A textual analysis of Alex Dalgarno's writing reveals an unusually high incidence of usage of the word "harmony" (J. L. Fox, private communication). This should remind us that our task as scientists is to search out the recurring themes and harmonies from the seeming cacophony of nature.

2. First Movement: A Testable Prediction of Astrochemistry

The development of astrochemistry has seen many instances of wishful thinking encouraged by accidental agreements between theory and observation. In such a complex research field, it is perhaps not surprising that "failures are often more instructive than successes", as Alex once observed in an article entitled *Is interstellar chemistry useful?*[1] Thus it is interesting to ask whether there have been any unqualified successes in astrochemical theory.

The minor chemical element fluorine provides a fascinating case study in interstellar chemistry. A few years ago, Zhu, Krems, Dalgarno, and Balakrishnan[2] investigated the reaction

$$F + H_2 \rightarrow HF + H\,. \qquad (1)$$

This reaction is of fundamental importance in theoretical chemistry, where its behavior in the limit of zero temperature is of interest.[3,4] A result of Ref. 2 was to strengthen the earlier suggestion of Neufeld et al.[5] that gaseous fluorine in interstellar molecular clouds should exist mainly in the form of the HF molecule. Even though the elemental abundance of fluorine in the sun is only $F/H = 3.6 \times 10^{-8}$, a polar molecule like HF can be readily detected at that level through its rotational spectrum. Neufeld, Wolfire, and Schilke[6] incorporated the low-temperature behavior of reaction (1) into an interstellar chemical model to predict the abundances of HF and related molecules in molecular clouds. They proposed that HF is destroyed mainly by reactions with C^+ in photon-dominated regions of molecular clouds (PDR) where C^+ is the predominant ion. This reaction leads to a detectable amount of CF^+, a quantitative prediction that was rapidly confirmed by observation of three rotational transitions of this ion in the Orion Bar PDR.[7] The column density derived from the observations was in good agreement with the abundance predicted by chemical models of

PDRs. In the very near future, HF will be more extensively observed in the interstellar medium at submm wavelengths with the Herschel Space Observatory and at infrared wavelengths through high-resolution absorption spectroscopy. In a broader context, the abundance of fluorine in space is of interest because this element is thought to be a rather specific tracer of nucleosynthetic processes that are accompanied by rapid mass-loss from stars.[8,9] It is still debated whether fluorine in the Galaxy comes mainly from stars on the asymptotic giant branch (low-mass precursors) or from Wolf-Rayet stars and supernovae (massive precursors). Thus it is important that HF can be used as a tracer of the fluorine abundance in the molecular interstellar medium.

3. Spectroscopic Signatures of Processes in Shocks

A defining characteristic of interstellar matter is that the physical and chemical state is usually far from thermodynamic equilibrium. This means that the competition among many microscopic atomic and molecular processes must be analyzed in detail in order to interpret observations. Consequently, it can be difficult to identify the effect of any single process. This difficulty has continued to hamper the application of models of interstellar molecular shocks to the interpretation of astronomical spectra. For example, OH and H_2O are expected to be produced efficiently in shock-heated molecular gas.[10] These same molecules may have high abundances and produce strong emission lines in photon-dominated regions of molecular clouds, too. The relationship between abundance and line intensity is usually complicated and non-linear. Thus it may be difficult to distinguish the predominant kind of chemistry that is responsible for an interstellar spectrum. At first glance, there is a very clear difference in the infrared spectrum between hydrogen molecules excited by ultraviolet fluorescence and H_2 excited by collisions: quadrupole vibration-rotation lines in the highly excited bands, $v = 2 \rightarrow 1$ and $v = 3 \rightarrow 2$ tend to have intensities greater than 1/10 those of the corresponding transitions in the fundamental band $v = 1 \rightarrow 0$ in fluorescence, while the higher transitions are relatively weaker in a collisionally excited gas.[11,12] Unfortunately, the distinction is not so clear in PDRs at high density, $n(H_2) > 10^4$ cm^{-3}, where collisional de-excitation of vibrationally excited H_2 becomes a significant heating mechanism and the intensity ratios of the vibration-rotation lines approach more closely the thermal values often attributed to shock-wave heating.[13]

It is thus gratifying to find a situation where the spectroscopic signature of a particular process is unambiguous. A recent example is provided by the discovery of infrared line emission from superthermal OH molecules in a fast molecular shock.[14] The infrared spectrum of the Herbig-Haro object HH 211 was observed with the Infrared Spectrograph (IRS) on the Spitzer Space Telescope at wavelengths $\lambda = 5$ to $37\ \mu$m. Part of this spectrum is shown in Fig. 1: the resolving power is $\lambda/\delta\lambda = 600$. The spectrum

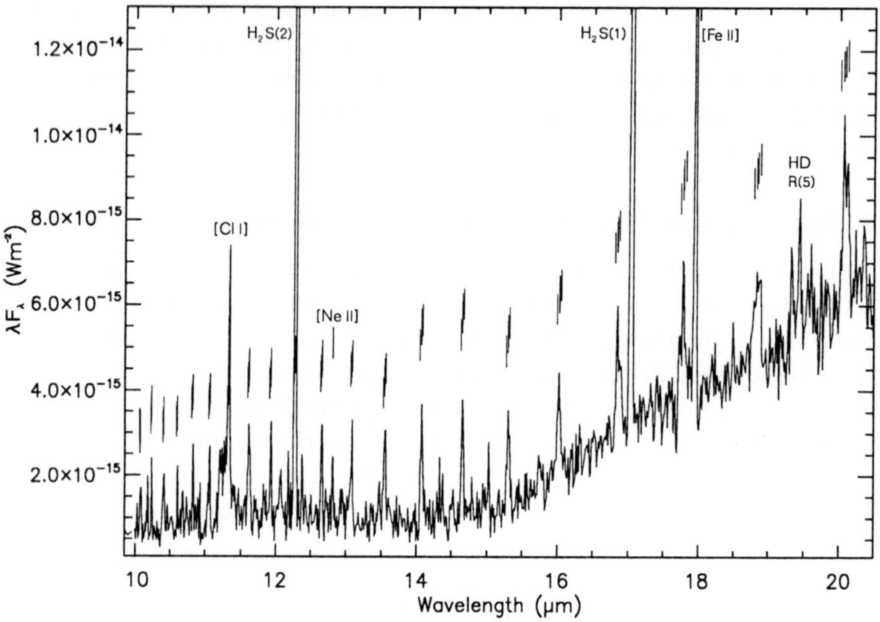

Fig. 1. A portion of the infrared spectrum of HH 211 obtained with Spitzer Space Telescope and based on the data published by Tappe et al.[14] Unlabelled vertical tick-marks indicate the OH lines.

exhibits an extraordinary sequence of rotational lines of OH arising in levels of rotational quantum number up to $J = 69/2$, which has an excitation energy $E/k \approx 28200$ K above the ground state. The intensities remain comparable for lines with upper states $J' = 20/2$ to $69/2$, which implies that the emission cannot arise from a thermal distribution with any physically meaningful temperature. David Neufeld reminded us of very interesting laboratory studies of the photodissociation of H_2O in which the product OH emerges in highly excited rotational levels with very little vibrational excitation.[15,16] *Ab initio* quantum chemical calculations by van Harrevelt and van Hemert[17] confirmed that photodissociation of H_2O through the $\tilde{B}$–$\tilde{X}$ transition at wavelengths $\lambda < 140$ nm leaves the product OH initially in the A $^2\Sigma^+$ state, which subsequently populates highly excited rotational levels of the X $^2\Pi$ state via non-adiabatic crossings between intersecting potential energy surfaces. The fractional distribution of OH states upon photodissociation of H_2O at λ121.6 nm is shown in Fig. 2. In order to test

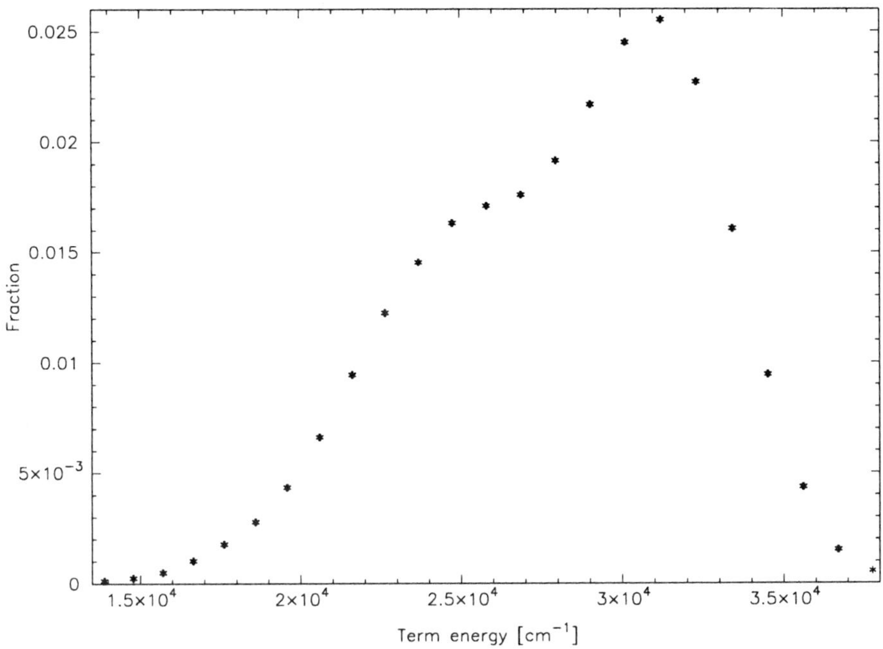

Fig. 2. The fraction of OH in excited states produced by photodissociation of H_2O at λ121.6 nm as a function of term energy (based on Ref 15–17).

this explanation of the infrared spectrum of HH 211, we computed a model of the excitation and radiative transfer of OH that includes:

- a source term for each rotational state proportional to the distribution function in Fig. 2
- a sink term (destruction rate) that balances the total source rate at the required abundance
- 4900 radiative transitions among all bound levels of vibrational states $v = 0$, 1, and 2
- inelastic collisions with H_2
- radiative excitation induced by the observed continuous spectrum of HH 211.

The freely available RADEX code has been crafted to carry out such calculations with a simplified treatment of radiative transfer.[18] This code has been used to generate the synthetic spectrum in Fig. 3, which shows very good agreement with the observed spectrum.

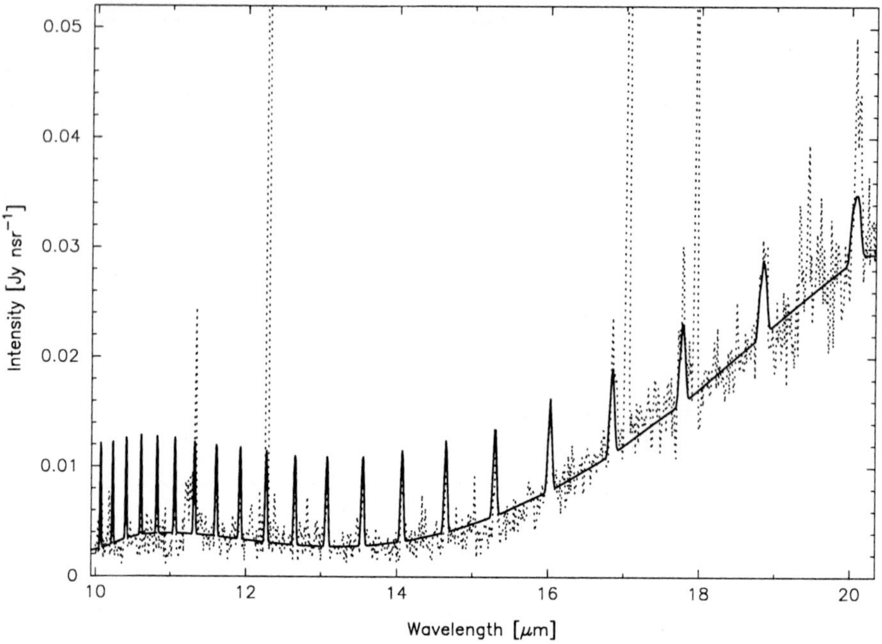

Fig. 3. A comparison between the observed spectrum (dotted) and a theoretical model (solid curve) of HH 211. The intensity in Jansky per nanosteradian refers to the observed flux averaged over the spectrometer apertures.

The photodissociation of H_2O is entirely responsible for the highly excited OH emission. Moreover, the OH source rate required to match the observed emission-line fluxes yields an estimate of the far-ultraviolet intensity incident on the molecular gas. This ultraviolet radiation is likely to be concentrated mainly in the H I Lα line at 121.6 nm wavelength and thus to be identified with the radiative precursor of a dissociating shock wave. The interesting astrophysical result is that the superthermal OH emission provides a diagnostic of the structure of a strong molecular shock, its radiative precursor, and an important part of its cooling. Details of this analysis will be published soon.[19] Perhaps similar OH emission will appear in the infrared spectra of comets, where the the H_2O abundance is high and its photodissociation has a large contribution from the solar H I Lα line.

4. Atomic and Molecular Processes in Cosmology

Dalgarno and collaborators have explored molecular processes that are important in the early Universe.[20,21] Molecules first appeared around the epoch of recombination when the normal matter of the Universe was transformed by radiative recombination from almost fully ionized to almost fully neutral. The small, residual fractions of electrons and protons catalyzed the formation of H_2, which constitutes an important coolant. The presence of a small abundance of molecules, $H_2/H \sim 10^{-5}$, had profound consequences for the gravitational collapse that formed the first stars. It has been startling to observe that some cosmologists have taken a serious interest in atomic and molecular physics (see, for example Ref. 22,23).

There is another way in which molecular physics has invaded cosmology: through spectroscopic tests of exotic effects. One example of great current interest is the search for evidence that fundamental constants vary with cosmic time. Some theories of dark energy invoke rolling scalar fields, which in turn imply that the value of the electron-proton mass ratio, $\mu = m_e/m_p$, should vary. Because the vibrational and rotational energies of light molecules like H_2 are very sensitive to μ, limits on $d\mu/dt$ can be established by comparing laboratory spectra now with spectra of H_2 molecules as they were some 10^{10} years ago. The expanding Universe itself is a time machine, in which the light-travel time to a distant quasi-stellar object (QSO) is measurable as a large redshift. Now that H_2 has been observed in a number of QSO absorption spectra at redshifts $z > 2$, the constancy of the mass-ratio can be tested at the level of $\Delta\mu/\mu \sim 10^{-5}$ over 10^{10} years. Until very recently, the precision of such tests was limited by the laboratory spectroscopy of H_2, but a series of laser spectroscopic measurements by Wim Ubachs and collaborators has largely removed the rest wavelengths from the error budget. The remaining spectroscopic problems have been reviewed recently by Thompson et al.[24,25]

5. Perspective

With a few examples, I have suggested that research areas originally stimulated by Alex Dalgarno and his collaborators continue to grow and to develop even broader significance.

References

1. Dalgarno, A., *Quarterly J. R. A. S.* **27**, 83 (1986).
2. Zhu, C., Krems, R., Dalgarno, A., Balakrishnan, N., *ApJ* **577**, 795 (2002).

3. Balakrishnan, N., Dalgarno, A., *Chem. Phys. Letters* **341**, 652 (2001). Erratum: **351**, 159.
4. Bodo, E., Gianturco, F. A., Balakrishnan, N., Dalgarno, A. *J. Phys. B* **37**, 3641 (2004).
5. Neufeld, D. A., Zmuidzinas, J., Schilke, P., Phillips, T. G., *ApJ*, **488** L141 (1997).
6. Neufeld, D. A., Wolfire, M. G., Schilke, P., *ApJ* **628**, 260 (2005).
7. Neufeld, D. A., Schilke, P., Menten, K. M., Wolfire, M. G., Black, J. H., Schuller, F., Müller, H. S. P., Thorwirth, S., Güsten, R., Philipp, S., *A&A* **454**, L37 (2006).
8. Federman, S. R., Sheffer, Y., Lambert, D. L., Smith, V. V., *ApJ* **619**, 884 (2005).
9. Snow, T. P., Destree, J. D., Jensen, A. G., *ApJ* **655**, 285 (2007).
10. Draine, B. T., Roberge, W. G., Dalgarno, A., *ApJ* **264**, 485 (1983).
11. Black, J. H., Dalgarno, A. 1976, *ApJ* **203**, 132 (1976).
12. Black, J. H., van Dishoeck, E. F., *ApJ* **322**, 412 (1987).
13. Sternberg, A., Dalgarno, A., , *ApJ* **338**, 197 (1989).
14. Tappe, A., Lada, C. J., Black, J. H., Muench, A. A., *ApJ* **680**, L117 (2008).
15. Mordaunt, D. H., Ashfold, M. N. R., Dixon, R. N., *J. Chem. Phys.* **100**, 7360 (1994).
16. Harich, S. A., Hwang, D. W. H., Yang, X., Lin, J. J., Yang, X., Dixon, R. N., *J. Chem. Phys.* **113**, 10073 (2000).
17. van Harrevelt, R., van Hemert, M. C., *J. Chem. Phys.* bf 112, 5787 (2000).
18. van der Tak, F. F. S., Black, J. H., Schöier, F. L., Jansen, D. J., van Dishoeck, E. F., *A&A* **468**, 627 (2007).
19. Tappe, A., Neufeld, D. A., Black, J. H., et al., in preparation (2009).
20. Dalgarno, A., *J. Phys. Conf. Ser.* **4**, 10 (2005).
21. Lepp, S., Stancil, P. C., Dalgarno, A., *J. Phys. B* **35**, R57 (2002).
22. Abel, T., Bryan, G. L., Norman, M. L., *ApJ* **540**, 39 (2000).
23. Abel, T., Bryan, G. L., Norman, M. L., *Science* **295**, issue 5552, 93 (2002).
24. Thompson, R. I., Bechtold, J., Eisenstein, D., Fan, X., Arnett, D., Martins, C., Kennicutt, R., Black, J., *Adv. Space Res.* **42**, 596 (2008).
25. Thompson, R. I., Bechtold, J., Black, J. H., Martins, C. J. A. P., *New Astron.* **14**, 379 (2009).

Session III

Interstellar Medium and Molecular Astrophysics

Tom Hartquist, James Moran, Eric Herbst, Peter Beiersdorfer, David Neufeld, Irwin Shapiro

STILL MORE THOUGHTS ON INTERSTELLAR CHEMISTRY

PATRICK THADDEUS

Harvard-Smithsonian Center for Astrophysics,
60 Garden St., Cambridge, MA 02138

In a cartoon in a well known French mountain climbing series, an Alpine guide is climbing an impossibly sheer granite needle. He has his hand on the tiny platform at the top of the needle, and says to his roped-up companion below, "Ah, the joy of touching virgin rock." The climber cannot see—but the reader can—looking down from above, that the climber's hand is almost touching a sardine can.

Dalgarno was there first.

Again and again, Alex has thought about things which nobody has thought much about before. A good example is provided by our recent discovery of negatively charged molecular ions in the interstellar gas. This came as a surprise to many, because there was a widespread belief that the ionization equilibrium in interstellar space is between neutral molecules and positive ions, and indeed many positive ions like HCO^+ and HNN^+ have been found. It was assumed that negative ions, readily dissociated by starlight, would be exceedingly hard to observe.

But there was a paper by Lepp and Dalgarno twenty years ago which showed that under certain conditions the charge equilibrium was between negative ions and neutrals, and that the abundance of the negative ions could be significant. That work was very much on my mind several years ago when we attempted to assign an enigmatic series of unidentified lines observed by Japanese radio astronomers with the Nobeyama telescope in one of the rich molecular concentrations in our Galaxy. The lines were apparently produced by a carbon chain with six carbon atoms, but no fit to the data could be achieved with a neutral molecule or a positive ion. It was only when we considered the implausible possibility that we were observing a negative ion that we could make sense of the observations. Conclusive proof that we were dealing with a negative ion came when we finally succeeded in finding the Japanese lines in the laboratory. The negative ion turns out to be C_6H^-, and others followed.

Even earlier, Dalgarno and MacCray in 1973 had considered in detail the role of negative ions in the production of the many neutral molecules that were being observed in apace.

It is his far reaching imagination at the service of a penetrating intellect, a good memory, and vast experience that makes Alex such a valuable colleague and guide. It is routine when we start some new enterprise to pause and say, "it is probably time to talk to Alex."

HOW INTERSTELLAR CHEMISTRY (AND ASTROCHEMISTRY MORE GENERALLY) BECAME USEFUL

T. W. HARTQUIST* and S. VAN LOO

School of Physics and Astronomy, University of Leeds, Leeds LS2 9JT, United Kingdom

** E-mail: twh@ast.leeds.ac.uk*

S. A. E. G. FALLE

Department of Applied Mathematics, University of Leeds, Leeds LS2 9JT, United Kingdom

In 1986 Alex Dalgarno published a paper entitled *Is Interstellar Chemistry Useful?*[1] By the middle 1970s, and perhaps even earlier, Alex had hoped that astronomical molecules would prove to: possess significant diagnostic utility; control many of the environments in which they exist; stimulate a wide variety of physicists and chemists who are at least as fascinated by the mechanisms forming and removing the molecules as by astronomy. His own research efforts have contributed greatly to the realization of that hope. This paper contains a few examples of: how molecules are used to diagnose large-scale dynamics in astronomical sources including star forming regions and supernovae; the ways in which molecular processes control the evolution of astronomical objects such as dense cores destined to become stars and very evolved giant stars; theoretical and laboratory investigations that elucidate the processes producing and removing astronomical molecules and allow their detection.

Keywords: active galactic nuclei, AGB stars, airglow, astrochemistry, astronomical spectroscopy, aurorae, brown dwarves, chemical kinetics, cosmic rays, cosmology, exoplanets, interstellar dust, interstellar medium, magnetohydrodynamics, molecular processes, nucleation, planetary nebulae, protoplanetary discs, shocks, star formation, supernovae, surface chemistry

1. Introduction (by T. W. Hartquist)

Jane Fox's eloquent comments about Alex Dalgarno's qualities as a friend, made at the September 2008 symposium honoring him, provided a fitting tribute to Alex's generosity, kindness and thoughtfulness. I am privileged to have Alex as a friend and also to have benefitted from his professional and intellectual support.

Many of Alex's former students and postdocs, as well as other colleagues, have stories about him similar to those that I will relate. We should all wonder how many letters of recommendation Alex has written. That activity must have consumed a tremendous amount of time. However, Alex's professional support of others has often gone far beyond letter writing. He has visited many of us shortly after our moves to new positions and has helped us make good impressions on our new colleagues. In Leeds, as elsewhere, he served on a visiting panel providing advice on how the local physics and astronomy research effort might be developed. The other two panelists visited one day each. In contrast, over a three day period, he spoke one-on-one with every available permanent member of our Physics and Astronomy academic staff to gain a complete and thorough overview of our activity. He produced an insightful report, which due to its concise, reasoned and incisive nature, carried considerable impact. It contributed very positively to our subsequent, successful efforts to establish a new group, with four permanent academic posts, conducting theoretical research on fundamental quantum processes. Alex continued to help Leeds after his 2003 stay. In 2007 he served as an external member of the committee that appointed Paola Caselli as our Professor of Astronomy.

I developed an interest in hydromagnetics and plasma kinetics while working with Alex as a postgraduate student in 1975. Rather than encourage me to focus only on molecular processes, Alex supported my other interests. One problem that I wished to pursue concerns the screening of molecular clouds from ionizing cosmic rays by scattering on Alfvén waves. Alex kindly arranged for me to spend substantial fractions of the summers of 1976 and 1977 in Cambridge, England. While there in 1976, I received a letter from him drawing my attention to a recently published paper by Skilling and Strong[2] on that topic. Alex has often astounded others with his encyclopedic knowledge of a tremendous range of literature. His awareness of so much has often been of great help to others. Alex, Holly Doyle and I made use of results in Ref. 2 in a paper on ionization rates inferred for diffuse molecular clouds.[3]

The title of the present paper echoes that of a paper by Alex[1] to which John Black and Ewine van Dishoeck also referred during their talks at the September 2008 symposium. The remainder of this paper is divided into six sections, half of which mention selected contributions of Alex primarily in molecular astrophysics. My coauthors and I have divided his contributions into those concerning molecular diagnosis, those showing how molecular processes control astrophysical environments and those on the investigation of quantum processes relevant to astrophysics. Sections addressing selected recent works on these themes by other researchers interleave with those summarizing Alex's efforts. I apologize to Alex and to others for the fact that the selections cannot be comprehensive. We have had to neglect a great deal of excellent work done by Alex and by others. Ewine gave a talk at the symposium in which she summarized some of Alex's work in astrochemistry and some recent work of others. The overlaps and the differences between what she said and what I said demonstrate the strength and breadth of Alex's work and of the current state of molecular astrophysics.

2. Alex's Work on Diagnostics

The Copernicus satellite, launched in August 1972, enabled far ultraviolet spectroscopy of the nearest O and B stars and of diffuse interstellar matter along the lines of sight to them. Alex and his students and other collaborators developed the framework and tools to use Copernicus data to probe the natures of diffuse molecular clouds, those molecular clouds having optical depths at visual wavelengths less than about unity. Black and Dalgarno[4] interpreted absorption measurements of the column densities of atomic hydrogen, the ground and next lowest six rotational levels of H_2, HD and OH for the line of sight to ζ Oph. They inferred the intensity of the far ultraviolet radiation impinging on the intervening cloud, the thermal and density structure, the elemental deuterium abundance and the rate at which cosmic rays induce ionization in the cloud. A large body of work on the relevant quantum processes, begun in the late 1960s by Alex and collaborators, underpins the construction of models like that of Black and Dalgarno.[4] Some of the work on processes is described later. Alex had to possess a profound long term vision to construct the foundations upon which the models were built and then to develop and apply the models.

Knowledge of the elemental deuterium abundances in different regions constrains cosmological models and provides insight into the effect of astration on elemental abundances. Studies of the spatial variation of the cosmic ray induced ionization rate are relevant to the understanding of the acceleration and propagation of cosmic rays. Some of Alex's additional findings on deuterium and cosmic rays in diffuse clouds are described in papers by Black and Dalgarno[5] and Hartquist, Black and Dalgarno.[6]

The abundance ratios of some deuterated species, including DCO^+, relative to their protonated counterparts, serve as diagnostics of the fractional ionization in dark molecular clouds. As described more fully later, the fractional ionization governs the role of the magnetic field in star formation in a dark cloud. Dalgarno and Lepp[7] showed that in dark clouds sufficient deuterium is in atomic form to affect the deuterium fractionation. The inference of reliable constraints on fractional ionizations in dark clouds from observations of deuterated and protonated isomers requires consideration of key reactions involving atomic deuterium.

The interaction of cosmic rays with H_2 in interstellar clouds deposits energy and induces excitation. In the Jovian atmosphere, the energy deposition and excitation due to particles accelerated in the solar wind - magnetosphere interaction results in observable H_2 ultraviolet emission. Alex's work on the interpretation of Jovian auroral and airglow emissions[8,9] demonstrated the presence of substantial temperature gradients. We refrain from mentioning other Solar System related studies conducted by Alex, as others writing articles for this volume are addressing those topics. However, a mention of the Jovian work here is appropriate because the approach Alex took in it bears a relationship to that he adopted in some investigations of deposition in extra-Solar System objects, including SN1987A.

Monitoring of the SN1987A supernova ejecta revealed the presence of CO infrared line emission at 112 days. Liu, Dalgarno and Lepp[10] developed a model of the thermal balance and chemistry of the ejecta. Heating is due to radioactive decays which produce γ-rays, and their interaction with matter generates energetic electrons. Radiative association forms CO. The very significant conclusion drawn about the ejecta's dynamics is that microscopic mixing of helium-rich layers with layers rich in carbon and oxygen had to be at most very limited, despite the full development of the Rayleigh-Taylor instability during the explosion. Otherwise, microscopic mixing would have led to the destruction of CO by He^+ at a rate incompatible with the observations.

3. Others' Work on Diagnostics

Studies of cosmic ray ionization rates continue. Observations of H_3^+ infrared absorption now provide constraints on the rates which imply they are an order of magnitude higher in some diffuse interstellar clouds than inferred by Alex and his collaborators in the 1970s.[11] The differences are due to the adoption in the 1970s of a value for the H_3^+ dissociative recombination rate coefficient that is small compared to that now accepted. (See section 7 for a mention of the relevant experimental measurements.) The analysis by Caselli *et al.*[12,13] of millimeter observations of deuterated species and their protonated counterparts has advanced our knowledge of the variation of the fractional ionization in dense molecular cores which are evolving to form stars.

The inference of the dynamics of dense core collapse from millimeter and submillimeter molecular line observations is a very challenging problem. Given that the relative roles of hydromagnetic wave processes and gravity at different stages of collapse continue to be debated, that problem is important. Its solution requires the construction of appropriate dynamical models and the development of accurate models for the gas phase and surface chemistries of dense cores, and involved radiative transfer calculations. Data for multiple lines of multiple species are required. For a simple dynamical description, Tsamis *et al.*[14] have performed relevant chemical and radiative transfer calculations. Much work remains, but their study gives a good indication of what is required.

The angular resolution and sensitivity of ALMA will enable the mapping of molecular distributions in protoplanetary discs. The study of the chemistry of the discs is at an early stage, and a full understanding of what can be learned about disc dynamics from molecular observations does not exist. Ilgner *et al.*[15] are amongst those who have begun to address this issue by considering chemistry for an α disc model. The use of more complicated models of disc dynamics, including gravitational instability and the effects that the magnetic rotational instability has on viscosity, would be interesting. Thi, van Zadelhoff and van Dishoeck[16] performed a notable observational study of simple organic molecules in protoplanetary discs around T-Tauri and Herbig Ae stars with single dish millimeter and submillimeter telescopes. Lahuis *et al.*[17] detected emission from hot organic molecules in a protoplanetary disc with the Spitzer Space Telescope. From data obtained for the disc of TW Hydrae with the Submillimeter Array, Qi *et al.*[18] have concluded that disc chemical models should include active deuterium fractionation. Existing results point to an interesting future for the diagnosis of protoplanetary discs with molecular observations.

4. Alex's Work on Chemical and Quantum Control

Though this article primarily concerns molecular astrophysics, some of Alex's contributions to astrophysics that did not address only molecules are too significant to exclude. The Dalgarno and McCray[19] paper is known to astrophysicists working on a wide variety of nonmolecular sources including supernova remnants. Its title *Heating and Ionization of HI Regions* is a bit deceptive. HI regions are the main topic of the paper, but the cooling of gas at temperatures far higher than those of HI regions is also addressed in it. Its famous Figure 2 shows the cooling rate coefficient for gas from temperatures of about 20K up to 10^8K. Dalgarno and McCray[19] considered molecular coolants but atomic processes received more space than molecular processes. They also summarized research initiated by themselves and C. Bottcher and M. Jura. Bottcher *et al.*[20] showed that supernovae and temporal variations in the population of hot stars are frequent enough that they prevent the interstellar medium from reaching a state in which gas is in two coexisting phases in pressure, ionization and thermal equilibrium. Later in the 1970s, the role of supernovae in establishing the global properties of the interstellar medium became a major topic in interstellar research. Alex and his collaborators pioneered the investigation of this important aspect of galactic astronomy.

Alex's work on heating and cooling has extended to the molecular cooling of the Early Universe. Stancil, Lepp and Dalgarno[21] examined the deuterium chemistry of the pregalactic and protogalactic eras. Though much less abundant than H_2, HD was possibly an important coolant then, due in part to its possession of a dipole moment and H_2's lack of one. This difference between HD and H_2, and the greater mass that D has than H has, leads to the lowest excited level of HD that can be populated by collisions with H having a roughly four times lower energy than the lowest excited level of H_2 that can be populated by collisions with H. Thus, as the temperature drops, HD becomes an increasingly more effective coolant than H_2. HD's dipole moment causes the radiative decay rates of its excited levels to be much larger than those of the corresponding H_2 levels. Consequently, the cooling rate per HD molecule continues to increase with increasing H number density to a higher density than the cooling rate per H_2 molecule does.

Lepp and Dalgarno[22] performed early work on the role that photoabsorption by large molecules or large molecular ions (e.g. free flying polycyclic aromatic hydrocarbons and their negative ions) play in heating interstellar clouds. At the time of their work grain photoelectric heating had been advocated as the primary heating mechanism in diffuse clouds, but it was appearing to be insufficient in at least some sources. Lepp and Dalgarno[22] identified photodetachment from large negative molecular ions as an important previously unconsidered heating process.

Section 2 contains a mention of the relevance of the fractional ionization to the part that magnetic fields have in molecular cloud dynamical evolution and star formation. Oppenheimer and Dalgarno[23] developed a model of the chemistry governing the fractional ionization in dark molecular regions. One of their key realizations is that charge transfer of molecular ions with neutral metal atoms causes a significant reduction in the gas phase recombination rate.

Star forming regions contain shocks driven by the winds and jets of young stars. In many cases these shocks may greatly influence subsequent star formation. In the second half of the 1970s and during the 1980s, the study of the shocks through the observation of infrared H_2 line emission at about 2 microns and of millimeter line emission from other molecules became a major industry. The theoretical models of the shocks had to increase in sophistication to match observational progress. Multifluid hydromagnetic models were developed. Draine, Roberge and Dalgarno[24] contributed significantly in this area by constructing such models and by critically evaluating and compiling data for the processes controlling the thermal balance. The chemistry controlling the ionization balance has a major impact on the shock structure, because the size of the dissipation region depends on the number density of ions. In the paper by Pineau des Forets *et al.*[25] Alex identified the chemistry controlling the ionization structure of shocks in diffuse clouds.

Alex's studies of how chemistry controls the thermal balance of astrophysical sources has included the work in which he explained the existence of large inhomogeneities in the thermal structure of the SN1987A ejecta as the consequence of the variation of the chemical structure throughout the oxygen core.[26]

5. Others' Work on Chemical and Quantum Control

During the 1980s and 1990s T. Ch. Mouschovias and his collaborators, F. H. Shu and his collaborators and others developed a picture of the birth of solar-mass stars in dense molecular cores undergoing gravitationally induced, quasistatic collapse regulated by the magnetic field and ambipolar diffusion. Ambipolar diffusion is the motion of charged species relative to neutrals driven by the magnetic force. In this picture the core forms with a magnetic flux to mass ratio that is too large for gravitationally induced collapse to proceed to higher densities if the flux to mass ratio does not alter. The core is said to be magnetically subcritical. However, due to the low fractional ionization, ambipolar diffusion occurs and reduces the magnetic flux, allowing low Alfvénic Mach number collapse. The ambipolar diffusion timescale depends on the number density of ions, which is the prime reason for the earlier mentions in this article of the fractional ionization in dense clouds.

In the late 1990s, the picture described in the previous paragraph was challenged. For instance, Myers and Lazarian[27] introduced a "turbulent cooling flow" description of the formation and evolution of dense cores. They wished to account for the observed linewidths of features arising in the envelopes of cores being broader than expected from some magnetic and ambipolar diffusion regulated gravitationally induced collapse models.

Ward-Thompson *et al.*[28] have summarized the issues in the debate about the roles of waves and of ambipolar diffusion operating with gravity in the birth of solar-mass stars. (We shall refer to waves rather than turbulence, which is a misnomer because the energy input scale and dissipation scale do not differ sufficiently for turbulence to develop fully.) Good evidence for wave induced evolution exists. However, good evidence for phases during which ambipolar diffusion and gravity operate together also exists (e.g. Chiang *et al.*[29]).

The simulations of Tassis and Mouschovias[30] are representative of work by the Illinois group on dynamics showing the important role that the fractional ionization and charged grains play in the gravitationally induced, ambipolar diffusion regulated collapse phase. The simulations are of thin discs in which the structures vertical to the symmetry planes are in magnetostatic equilibrium.

Van Loo, Falle and Hartquist[31] have demonstrated the importance of ambipolar diffusion and, hence, of the chemistry controlling the fractional ionization during the initial formation of dense cores by the nonlinear steepening of hydromagnetic waves. The relative roles of wave processes and gravitationally induced, magnetically regulated collapse at different phases of core evolution may be worthy of debate. In contrast, the fact that the chemistry controlling the fractional ionization is of great importance in core evolution is very clear.

Multifluid models of shocks in dark regions having hydrogen number densities exceeding about $10^5\,\mathrm{cm}^{-3}$ advanced in complexity during the 1990s, and only now are we on the verge of constructing reliable models of such shocks. The very important Draine *et al.*[24] work concerned shocks that are propagating perpendicularly to the magnetic fields. Of course, real shocks propagate obliquely to the magnetic fields. In dense regions, the collisions of neutrals with charged grains rather than the collisions of neutrals with ions dominate the coupling of neutral flows to magnetic fields. Draine *et al.*[24] treated the effects due to charged grains in an approximate fashion that is suitable for a wide variety of parameters but is not reliable for regions with number densities above about $10^5\,\mathrm{cm}^{-3}$. Pilipp and

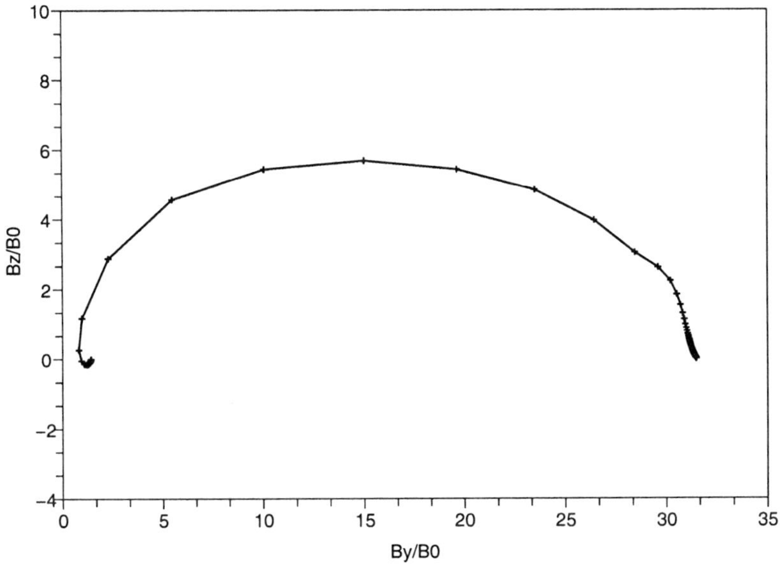

Fig. 1. The rotation of the magnetic field within the shock front of C-type shock. The shock is propagating through a quiescent medium of $n_{\rm H} = 10^6$ cm^{-3} and $B_0 = 10^{-3}$G at 25 km s^{-1}. The angle between the shock propagation and the upstream magnetic field is 45°. The crosses represent the grid spacing of the numerical model.

Hartquist[32] and Wardle[33] initiated efforts to include charged grains more rigorously in models of oblique shocks.

Wardle[33] showed that an attempt to solve the time-independent coupled ordinary differential equations governing steady plane-parallel shock structures through the integration in a downstream direction from upstream boundary conditions cannot yield solutions for fast-mode shocks. Rather integration has to proceed in the upstream direction from downstream conditions. This is not possible unless equilibrium conditions obtain at all points in the flow. Unfortunately, they do not. For instance, the abundance of H_2O, which is an important coolant, does not attain its equilibrium value as gas cools from several hundred degrees until it has been at about 10 K for of the order of 10^5 years or more. This is much longer than the flow time through a shock in a dark dense core. Consequently, a time-dependent approach is necessary. Such an approach including all electric current components in a proper fashion[34] has been combined only recently with a self-consistent treatment of the chemistry controlling the fractional ionization and grain charges and, hence, the coupling between neutral flow and the magnetic field.[35] Figure 1 shows the rotation of the magnetic field within the shock front of a steady C-type shock. The spiral node upstream is one of the reasons why Wardle[33] needed to integrate the steady-state equations in an upstream direction to find a steady fast-mode shock.

Chemistry plays a huge role in controlling the mass loss from highly evolved stars. Ferrarotti and Gail[36] conducted a comprehensive theoretical study of the formation of dust in Asymptotic Giant Branch (AGB) stars. Radiation pressure on the dust greatly influences the mass loss rates and terminal speeds of the outflows. Mass loss rates affect the stellar evolution and the nucleosynthesis products.

6. Alex's Work on Quantum Processes Relevant to Astrophysics

Alex's contributions to the study of processes important for the diagnosis and evolution of astrophysical sources are vast.

The volume of just his work on radiative processes important for modelling diffuse clouds is impressive. This includes the calculation of radiative probabilities for the Lyman and Werner bands of H_2.[37,38] These are important for photodissociation and the pumping of excited rovibrational levels of ground electronic state of H_2. Turner, Kirby-Docken and Dalgarno[39] computed H_2 ground electronic state rovibrational transition probabilities that help determine the level populations resulting from the cascade following pumping. They are also important for radiative cooling in shocked molecular gas in star forming regions. Other work on radiative mechanisms important in diffuse clouds includes that of van Dishoeck, van Hemert and Dalgarno[40] who calculated the OH photodissociation rate.

Collisional processes are also important for H_2 level populations in diffuse clouds. Dalgarno, Black and Weisheit[41] identified the role of H_2 collisions with H^+ in establishing the H_2 ortho-para ratio in diffuse clouds. They also identified H_2 collisions with D^+ as the dominant mechanism for forming HD in diffuse clouds.

The work of Chu and Dalgarno[42] on collisional excitation of CO is relevant to the cooling and observational diagnosis of most detectable diffuse molecular material in the Universe. The adoption by Roberge and Dalgarno[43] of a master equation approach for the calculation of the populations of rovibrational levels of H_2 in shocked gas led to a major advance in the understanding of the collisional dissociation of astrophysical molecules.

Alex has contributed many important studies of charge transfer processes of which that by Butler, Heil and Dalgarno[44] is a good example. The efforts of this trio at the end of the 1970s and the start of the 1980s had a huge impact on models of photoionized regions from planetary nebulae to the broad emission line regions of quasistellar objects. Their work led to the understanding of how charge transfer greatly influences the emission spectra of such plasmas. Of course, Alex's more recent investigations, described elsewhere in this volume, of charge transfer have had major significance for research on the heliosphere, planetary atmospheres, comets and the interstellar soft X-ray background.

Supernova ejecta belong to the long list of different classes of astrophysical sources for which his investigation of quantum processes are relevant. For instance, Dalgarno, Du and You[45] calculated the rates of the radiative association of O with C and C^+ as part of Alex's program on molecules in the ejecta of SN1987A.

7. Others' Work on Quantum Processes Relevant to Astrophysics

The advance of molecular astrophysics has relied on many laboratory efforts involving a wide variety of techniques to obtain reliable reaction rate coefficients for temperatures ranging from about 10K to over 10^3K. For over two decades CRESU devices, in which expansion in supersonic jets cools gas, have facilitated low temperature measurements relevant for the chemistry of star forming regions. Though presented in the context of work on Titan's atmosphere, the results obtained by Berteloite *et al.*[46] for C_4H reactions with hydrocarbons provide recent examples of the type of data relevant for star forming regions that can be obtained with CRESU devices.

Storage ring experiments have produced a smaller volume of data relevant to astrophysics, but in some cases the storage ring results have had considerable impact. The measurement of the H_3^+ dissociative recombination rate by McCall *et al.*[47] has been important for the inference of cosmic ray ionization rates in diffuse clouds from observations of H_3^+, which we mentioned earlier.

The original masthead of *The Astrophysical Journal* described it as *An International Review of Spectroscopy and Astronomical Physics.* Spectroscopy plays a key role in molecular astrophysics just as it does in other areas of astrophysics. An excellent example of the importance of laboratory molecular spectroscopy for astrophysics is provided by the work done by McCarthy *et al.*[48] which led to the first detection of an astrophysical negative molecular ion. C_6H^- exists in the stellar envelope of IRC +10216 and in the dense molecular cloud TMC-1.

Theoretical calculations of line wavelengths and radiative transition rates for H_2O[49] made possible the first detection of water in the atmosphere of an extrasolar planet.[50] Theoretical line list calculations (e.g. Harris *et al.*[51]) are important for work on the atmospheres of cool stars and low metallicity stars and distinguishing brown dwarves from planets.

To this point we have mentioned only gas phase processes. In the past decade the effort to understand grain surface processes affecting astrophysical chemistry has intensified. A key problem concerning the kinetics of interstellar grain surface reactions arises because the number of reactant atoms and molecules on a surface is often too small for a standard rate equation treatment to be appropriate. Caselli, Hasegawa and Herbst[52] introduced a modified rate equation approach, which they have subsequently improved. Their 1998 paper triggered a considerable amount of research on how to treat interstellar grain surface chemistry. Green *et al.*[53] adopted a master equation approach with which they calculated the probabilities that a grain contains specific numbers of atoms and molecules of the different reactant species. The method is suitable for some simple problems but is too computationally expensive for many of interest. Barzel and Biham[54] have developed a technique based on the use of equations they obtained by taking moments of the master equation. It shows considerable promise because it has proven accurate in test calculations and computationally tractable for problems involving many reactant species.

Laboratory studies of surface processes important for astrochemistry have included investigations of H_2 formation and desorption of ices. The short but informative review by Williams *et al.*[55] provides a good introduction to some of the research. While they made a good attempt to note work done elsewhere, the main focus of the Williams *et al.*[55] review is research carried out by several UK groups. Several excellent surface chemistry groups exist throughout the world. The Leiden group is one. Recent work in that group has included a study of the photodesorption of CO ice.[56]

The formation of astrophysical dust in carbon-rich stellar envelopes is better understood than its production in oxygen-rich envelopes. Uncertainties in the field are sufficient that Nuth and Ferguson[57] felt moved to proclaim, in the title of a paper on their relevant experimental results, that *Silicates Do Nucleate in Oxygen-Rich Circumstellar Outflows ...* Astronomical observations had already established this fact, but the problem of explaining it has been so challenging that Nuth and Ferguson[57] had some grounds for announcing their progress in the manner they did. Theoretical efforts to identify the species important in the nucleation of astrophysical dust involve the development and application of methods to evaluate the energies of multiple configurations of various numbers of particles of the nucleating species. Bhatt and Ford[58] have recently published theoretical results on the investigation of MgO as a possible nucleating species around M stars.

8. Closing Remarks

The unusually lengthy list of keywords at the start of this article reflects the broad range of the work that Alex has done in astronomy and the breadth of its impact.

One does not need to possess great prognostic powers to know that molecular astrophysics has a bright future. Herschel is due to be launched in 2009 and ALMA will be operational well before Alex turns 90. When we meet to celebrate that occasion, we will have learned a lot more about astrophysical molecular sources through the use of those two facilities.

References

1. A. Dalgarno, *QJRAS*, **27**, 83 (1986).
2. J. Skilling and A. W. Strong, *A&A*, **53**, 253 (1976).
3. T. W. Hartquist, H. T. Doyle and A. Dalgarno, *A&A*, **68**, 65 (1978).
4. J. H. Black and A. Dalgarno, *ApJ Suppl*, **34**, 405 (1977).
5. J. H. Black and A. Dalgarno, *ApJ*, **184**, L101 (1973).
6. T. W. Hartquist, J. H. Black and A. Dalgarno, *MNRAS*, **185**, 643 (1978).
7. A. Dalgarno and S. Lepp, *ApJ*, **287**, L47 (1984).
8. W. Liu and A. Dalgarno, *ApJ*, **462**, 502 (1996a).
9. W. Liu and A. Dalgarno, *ApJ*, **467**, 446 (1996b).
10. W. Liu, A. Dalgarno and S. Lepp, *ApJ*, **396**, 679 (1992).
11. N. Indriolo, T. R. Geballe, T. Oka and B. J. McCall, *ApJ*, **671**, 1736 (2007).
12. P. Caselli, C. M. Walmsley, R. Terzieva and E. Herbst, *ApJ*, **499**, 234 (1998).
13. P. Caselli, C. M. Walmsley, A. Zucconi, M. Taffalla, L. Dore and P. C. Myers, *ApJ*, **565**, 344 (2002).
14. Y. G. Tsamis, J. M. C. Rawlings, J. A. Yates and S. Viti, *MNRAS*, **388**, 898 (2008).
15. M. Ilgner, Th. Henning, A. J. Markwick and T. J. Millar, *A&A*, **415**, 643 (2004).
16. W.-F. Thi, G.-J. Zadelhoff and E. F. van Dishoeck, *A&A*, **425**, 955 (2004).
17. F. Lahuis, E. F. van Dishoeck, A. C. A. Boogert, K. M. Pontoppidan, G. A. Blake, C. P. Dullemond, N. J. Evans II, M. R. Hogerheijde, J. K. Jorgensen, J. E. Kessler-Silacci and C. Knez, *ApJ*, **636**, L145 (2006).
18. C. Qi, D. J. Wilner, Y. Aikawa, G. A. Blake and M. R. Hogerheijde, *ApJ*, **681**, 1396 (2008).
19. A. Dalgarno and R. A. McCray, *ARA&A*, **10**, 375 (1972).

20. C. Bottcher, R. A. McCray, M. Jura and A. Dalgarno, *ApL* , **6**, 237 (1970).
21. P. C. Stancil, S. Lepp and A. Dalgarno, *ApJ*, **509**, 1 (1998).
22. S. Lepp and A. Dalgarno, *ApJ*, **335**, 769 (1988).
23. M. Oppenheimer and A. Dalgarno, *ApJ*, **192**, 29 (1974).
24. B. T. Draine, W. G. Roberge and A. Dalgarno, *ApJ*, **264**, 485 (1983).
25. G. Pineau des Forets, D. R. Flower, T. W. Hartquist and A. Dalgarno, *MNRAS*, **220**, 801 (1986).
26. W. Liu and A. Dalgarno, *ApJ*, **454**, 472 (1995).
27. P. C. Myers and A. Lazarian, *ApJ*, **507**, L157 (1998).
28. D. Ward-Thompson, P. Andre, R. Crutcher, D. Johnstone, T. Onishi and C. Wilson, *An Observational Perspective of Low-Mass Dense Cores II: Evolution toward the Initial Mass Function*, in *Protostars and Protoplanets V*, eds. B. Reiputh, D. Jewitt and K. Keil, (University of Arizona Press, 2007), pp. 33-46.
29. H.-F. Chiang, L. W. Looney, K. Tassis, L. G. Mundy and T. Ch. Mouschovias, *ApJ* , **680**, 474 (2008).
30. K. Tassis and T. Ch. Mouschovias, *ApJ*, **618**, 769 (2005).
31. S. Van Loo, S. A. E. G. Falle, T. W. Hartquist and A. J. Barker, *A&A*, **484**, 275 (2008).
32. W. Pilipp and T. W. Hartquist, *MNRAS*, **267**, 801 (1994).
33. M. Wardle, *MNRAS*, **298**, 507 (1998).
34. S. A. E. G. Falle, *MNRAS*, **344**, 1210 (2003).
35. S. Van Loo, I. Ashmore, P. Caselli, S. A. E. G. Falle and T. W. Hartquist, *MNRAS*, submitted (2009).
36. A. S. Ferrarotti and H.-P. Gail, *A&A*, **447**, 553 (2006).
37. A. C. Allison and A. Dalgarno, *JQSRT*, **9**, 1543 (1969).
38. A. Dalgarno and T. L. Stephens, *ApJ*, **160**, L107 (1970).
39. J. Turner, K. Kirby-Docken and A. Dalgarno, *ApJ Suppl*, **35**, 281 (1976).
40. E. F. van Dishoeck, M. C. van Hemert, A. C. Allision and A. Dalgarno, *J Chem Phys*, **81**, 570 (1984).
41. A. Dalgarno, J. H. Black and J. C. Weisheit, *ApL* , **14**, 77 (1973).
42. S.-I. Chu and A. Dalgarno, *Proc Roy Soc*, **A343**, 191 (1975).
43. W. Roberge and A. Dalgarno, *ApJ*, **255**, 176 (1982).
44. S. E. Butler, T. G. Heil and A. Dalgarno, *ApJ*, **241**, 442 (1980).
45. A. Dalgarno, M. L. Du and J. H. You, *ApJ*, **349**, 675 (1990).
46. C. Berteloite, S. D. Le Picard, P. Birza, M.-C. Gazeau, A. Canosa, Y. Benila, and I. R. Sims, *Icarus*, **194**, 746 (2008).
47. B. J. McCall, A. J. Huneycutt, R. J. Saykally, T. R. Geballe, N. Djuric, G. H. Dunn, J. Semaniak, O. Novotny, A. Al-Khalill, A. Ehlerding, F. Hellberg, S. Kalliori, A. Neau, R. Thomas, F. Osterdahl and M. Larsson, *Nature*, **422**, 500 (2003).
48. M. C. McCarthy, C. A. Gottlieb, H. Gupta and P. Thaddeus, *ApJ*, **652**, L141 (2006).
49. R. J. Barber, J. Tennyson, G. J. Harris and R. N. Tolchenov, *MNRAS*, **368**, 1087 (2006).

50. G. Tinetti, A. Vidal-Madjar, M.-C. Liang, J.-P. Beaulieu, Y. Yuk, S. Carey, R. J. Barber, J. Tennyson, I. Ribas, N. Allard, G. E. Ballester, D. K. Sing and F. Selsis, *Nature*, **448**, 169 (2007).
51. G. J. Harris, F. C. Larner, J. Tennyson, B. M. Kaminsky, Ya. V. Pavlenko and H. R. A. Jones, *MNRAS*, **390**, 143 (2008).
52. P. Caselli, T. I. Hasegawa and E. Herbst, *ApJ*, **495**, 309 (1998).
53. N. J. B. Green, T. Toniazzo, M. J. Pilling, D. P. Ruffle, N. Bell and T. W. Hartquist, *A&A*, **375**, 1111 (2001).
54. B. Barzel and O. Biham, *J Chem Phys*, **127**, 114703-1 (2007).
55. D. A. Williams, W. A. Brown, S. D. Price, J. M. C. Rawlings and S. Viti, *A&G*, **48**, 1.25 (2007).
56. K. I. Oberg, G. W. Fuchs, Z. Awad, H. J. Fraser, S. Schemmer, E. F. van Dishoeck and H. Linnartz, *ApJ*, **662**, L23 (2007).
57. J. A. Nuth III and F. T. Ferguson, *ApJ*, **649**, 1178 (2006).
58. J. Bhatt and I. J. Ford, *MNRAS*, **382**, 291 (2007).

ATOMS, MOLECULES, AND RADIATION: FROM STAR-FORMING REGIONS TO QUASARS

A. STERNBERG

Sackler School of Physics & Astronomy
Raymond and Beverly Sackler Faculty of Exact Sciences
Tel Aviv University
Ramat Aviv 69978, Israel
E-mail: amiel@astro.tau.ac.il

Atomic, molecular, and optical processes play fundamental roles in regulating the evolution of star-forming regions in galaxies, and of accreting black holes in galaxy nuclei. I present a discussion of the essential physics and chemistry of interstellar photon-dominated regions (PDRs) and X-ray dominated regions (XDRs). In such regions, intense radiation fields determine the atomic and molecular compositions of dense interstellar gas, and control the resulting observable spectroscopic emissions and absorptions.

Keywords: ISM, Atomic and Molecular Processes, PDRs and XDRs

1. Introduction

The Milky Way is a typical spiral galaxy consisting of a few 100 billion stars confined to an extended (15-20 kpc) and thin (100-300 pc) differentially rotating disk, and an inner (3 kpc) stellar spheroid. The total mass of the disk is around 6×10^{10} $M_\odot$. At the Galactic Center there exists a supermassive (4×10^6 $\mathrm{M}_\odot$) black hole, consistent with the notion that most or all galaxy spheroids contain central massive black holes. The luminous portions of the Galaxy (disk and bulge) are embedded in a much more extended ($\sim$ 200 kpc) $\sim 10^{12}$ $\mathrm{M}_\odot$ halo of "dark-matter". The composition of the invisible (cosmological) dark-matter is unknown - one idea is that it consists of weakly-interacting massive particles. Galaxies such as the Milky Way were formed as atomic (hydrogen-helium) gas fell into the gravitational potential-wells of (virialized) dark-matter halos that grew from small gravitationally unstable density fluctutations. As the infalling gas cooled, it fragmented into clouds, and a fraction settled into the disk as cold gas, some of which condensed into stars. Gas cooling and gravitational collapse are regulated by radiative losses due to atomic and molecular line emissions. The collapse times are also controlled by the chemical compositions and ionization fractions, especially in the presence of magnetic fields. Galaxy evolution and star-formation are intimately linked to the basic atomic and molecular processes that determine the chemistry and thermal evolution of the interstellar gas.

The Milky Way, like all disk galaxies is actively forming stars (at a present-day rate of 5 $M_\odot$ yr^{-1}) as is indicated by the presence of the short-lived OB stars. Star-formation occurs within a dynamic and inhomogeneous interstellar medium (ISM) which is a predominantly hydrogen-helium gas with a small admixture of heavy elements ("metals") and sub-micron solid-state "dust-grains". The heavy elements were formed by stellar nucleosynthesis and subsequent ejections and explosions. The ISM consists of several distinct phases, usually referred to as the (coronal) hot ionized medium (HIM: $T \sim 10^6$ K), the warm ionized medium (WIM; $T \gtrsim 10^4$ K), the warm neutral atomic medium (WNM; $\sim 8 \times 10^3$ K), and the cold neutral medium (CNM; ~ 100 K). The HIM, WIM/WNM, and CNM are in approximate pressure equilibrium. The total ISM mass is around 5×10^9 $M_\odot$, and most ($\sim 95\%$) of this mass is in the cold phase. The hot phase occupies most of the volume ($\sim 95\%$). A substantial fraction of the gas is present in dense cold (~ 20 K) self-gravitating molecular (H_2) clouds, with high internal pressures, and with masses up to $\sim 10^5$ $M_\odot$. Crucially, star-formation occurs in the cold dense molecular gas. Stellar EUV ($h\nu > 13.6$ eV) and FUV ($h\nu < 13.6$ eV) radiation from the hot young stars, interacts with the molecular gas, and produces hot photoionized (blister) "H II regions", and adjacent neutral "photon-dominated regions" (PDRs).

Alex[1] has contributed, more than anyone, to our understanding of the myriad atomic, molecular, and optical processes that occur in the multi-phased ISM, especially the cool neutral components.[2] The processes regulate the evolution of interstellar clouds and their collapse into stars. Alex helped to lay the foundation for our understanding of interstellar chemistry,[3–5] including the key role of molecular hydrogen formation in initiating the chemical sequences.[6,7]

Interstellar chemistry is an interplay between bond formation processes including radiative association, associative detachment, and grain-surface formation,

$$\mathrm{X} + \mathrm{Y} \rightarrow \mathrm{XY} + \mathrm{photon} \tag{1}$$

$$\mathrm{X}^- + \mathrm{Y} \rightarrow \mathrm{XY} + \mathrm{e} \tag{2}$$

$$X + Y : g \rightarrow XY + g \quad , \tag{3}$$

bond destruction processes, via photodissociation, dissociative recombination, and collisional dissociation,

$$XY + \text{photon} \rightarrow X + Y \tag{4}$$

$$XY^{+} + e \rightarrow X + Y \tag{5}$$

$$XY + M \rightarrow X + Y + M \quad , \tag{6}$$

and bond rearrangement, via ion-molecule exchange, charge-transfer and neutral-neutral reactions

$$X^{+} + YZ \rightarrow XY^{+} + Z \tag{7}$$

$$X^{+} + YZ \rightarrow X + YZ^{+} \tag{8}$$

$$X + YZ \rightarrow XY + Z \quad . \tag{9}$$

In addition to working out the detailed quantum dynamics for many such interactions, Alex led the way in incorporating the microphysical processes into interstellar cloud models, including treatments of the radiative transfer, and the heating and cooling. The models enabled quantitative interpretations and also predictions for the abundances of observed atomic and molecular species. Alex and his students and postdocs developed models for the study of many interstellar environments, including the classical diffuse clouds[7–9] (with John Black and Ewine van Dishoeck) dense clouds,[10,11] (with Michael Oppenheimer and Stephen Lepp) shocked gas[12,14,15] (with Tom Hartquist, Wayne Roberge and David Neufeld) photon-dominated regions[16,17] (with me), X-ray dominated regions[18–21] (with Min Yan), supernovae[22,23] (with Weihong Liu), and also primordial gas clouds[13,24] (with Stephen Lepp and Phil Stancil).

2. Charge Transfer

I would like to say a few words about charge transfer - it brings back memories. During my senior year in college (1979-80), my classmate Alain Porter and I decided to take an intriguing sounding course entitled "Astronomy 236: Atomic and Molecular Processes in the Interstellar Medium", to be taught by Alex Dalgarno and George Victor.[a] "Who are these people?" we

[a]Sadly, Alain died many years ago. George passed away this past year.

wondered. As it turned out, we were the only students to enroll - none of the graduate students signed-up. I recall finding the material quite challenging, and also being most impressed by how the action of specific microphysical processes could be isolated, identified observationally, and understood quantitatively in terms of fundamental physics - mostly quantum physics and spectroscopy in this course. At around that time, Alex and his PhD student Scott Butler were busy calculating rate coefficients for charge-transfer (CT) reactions, $X^{i+} + H \rightarrow X^{(i-1)+} + H^{+}$, for multiply charged species colliding with hydrogen (and also helium) atoms.[25] For my Astronomy 236 term paper, Alex suggested that I explore the effects of charge-transfer as a neutralization process. I constructed simple models (using a hand-held calculator if I recall correctly) for the abundances of sulfur ions, S^{2+} (slow CT) and S^{3+} (fast CT) in planetary nebulae and H II regions. The goal was to demonstrate the major effect that charge-transfer has on the relative abundances of these ions. (It is strange to recall that it was only in the late-70's that CT was incorporated into model nebulae, with the recognition - as demonstrated in my term paper and elsewhere - that such reactions are often competitive with radiative recombinations, especially in partially ionized zones.) This was my first research project with Alex, and I remember it well. I think of it especially when I try to assign projects to beginning students.

I was unsure about graduate school, and Alex kindly offered me a job as a research assistant for a year. I continued to think about charge-transfer, also as an excitation mechanism for the product ions. One day, it occurred to me that charge transfer production of O^{2+}, with channels into excited singlet $2p3p\ ^1P$ and $2p3s\ ^1P^o$ states, would then necessarily be followed by allowed transitions to the metastable ^1S and ^1D levels. The intensities of the famous oxygen forbidden lines at 436.3 nm (emitted from ^1S) and at 495.9 and 500.7 nm (emitted from ^1D), which serve as classical nebular temperature diagnostics - under the assumption that the metastable levels are populated by electron impact excitation - would be modified by the charge transfer excitations. I mentioned this to Alex, although I was only vaguely aware that this might be a good idea. There were computations to be done, but I think Alex basically had the paper[28] written in his head in the short time it took him to walk from my desk back to his office. Well, as it turned out charge transfer excitation has only a small effect on nebular temperature determinations, but it is there in the allowed O^{2+} emission-line spectra, and it is competitive with the classical Bowen (He^{+}-O^{2+}) resonance fluoresence excitation mechanism.[29–32]

I decided to move to New York City, and to enroll at Columbia University for graduate school. However, I had some unfortunate experiences there including, I recall, with some laboratory equipment, and I changed my mind. I wanted Alex as my PhD thesis advisor. Alex kindly agreed, and I returned to Cambridge as a Columbia U. CfA pre-doc.

3. Photon-Dominated Regions

Back at the CfA, in Alex's group, my interest in interstellar atomic and molecular excitation mechanisms broadened considerably.

In the late 70's and early 80's, advances in detector technology had opened up the submillimeter and far-infrared spectral windows for astronomical observations. Fine-structure line emissions from atoms and ions, such as the $^3P_2$-$^3P_1$ 63 μm line of atomic oxygen, and the $^2P_{1/2}$–$^2P_{3/2}$ 158 μm line of C^+ were detected for the first time.

C^+ was of particular interest. Long before it was first detected in Orion[33] the C^+ line was recognized theoretically as a likely major cooling transition for the neutral atomic hydrogen (H I) components of the ISM.[34] Alex pointed out the crucial and dominating role of neutral impact excitation of C^+ by H atoms[2,35] (see also[36]). C^+ cooling plays a central role in the theory of the two-phase CNM/WNM medium,[37,38] dominating the energy losses of the cool phase. However, the early theoretical considerations focused primarily on the relatively low density gas ($n_{\rm H} \lesssim 50$ cm^{-3}), where the excited level is subthermally populated and the C^+ emissivities are low (see Fig. 1). For fixed density, the cooling efficiency decreases sharply for temperatures T below the transition energy of 92 K, and levels off at higher T, where the Boltzmann factor in the collisional excitation rate coefficient approaches unity. At fixed gas temperature the efficiency increases linearly with density up to the critical hydrogen density of 10^4 cm^{-3}, above which collisional deexcitations quench the radiative emissions. For conditions in the classical WNM ($n_{\rm H} \sim 0.4$ cm^{-3} $T \approx 8000$ K) and CNM ($n_{\rm H} \approx 60$ cm^{-3} $T \approx 50$ K), the C^+ emission efficiencies are well below the maximum possible. However, for densities greater than 10^2 and up to at least 10^4 cm^{-3}, and for T above $\sim 10^2$ K, the cooling efficiencies are significantly enhanced. This range of elevated temperature and density corresponds to conditions characteristic of so-called "dense PDRs", which are dense neutral gas clouds exposed to intense FUV fields - the surface layers of star-forming molecular clouds. For example, a major goal of the upcoming *Herschel* mission will be to map and detect the 158 μm C^+ cooling line in nearby and distant

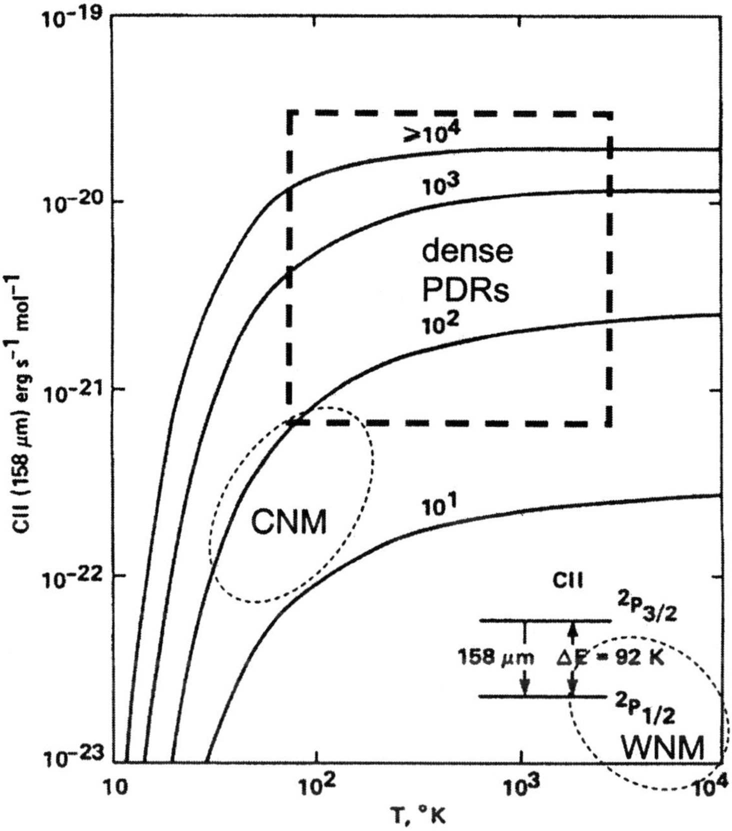

Fig. 1. C^+ line cooling per particle as a function of the gas temperature and hydrogen density, for "neutral impact" excitation with hydrogen atoms (adapted from[39] which was based on[35]).

galaxies. The interpretation will be based on the H-C^+ collisional excitation rate coefficient - first computed by Alex, decades ago.

The stellar FUV photodissociates molecules (e.g., H_2 and CO) and Tielens and Hollenbach[39] introduced the term "photo-dissociation regions" (PDRs) (see also[40]). However, molecules are also selectively produced in PDRs, (e.g. in reactions initiated by the association of C^+ and H_2) and the chemistry differs from cosmic-ray driven ion-molecule sequences that occur in the shielded interiors of the clouds. In view of this, Alex coined the alternative and better term "photon dominated regions" (still PDRs, as Alex wryly pointed out to me).

The starburst galaxy M82 (3.3 Mpc) provides a nice example of a global, optically obscured, PDR. Optical imaging shows a central dusty starburst with large scale $H\alpha$ filaments extending out of the galaxy in a galactic superwind. Infrared, submillimeter, and millimeter-wave spectroscopy of this source reveals a thermal dust continuum, and a wealth of PDR emission lines. The observed fine-structure line strengths have typical line-to-continuum ratios of 0.1–1%[41]

The fine-structure line intensities may be understood with the recognition that most of the incident FUV energy is absorbed by dust particles and reradiated as quasi-thermal IR. A small fraction of the FUV heats the gas via the photoelectric ejection of electrons from the grain surfaces.[34,42–44] In this process energetic electrons are ejected from the grains at a rate that depends on the FUV intensity and the grain charge, set by the balance between photoemission and recombination.

The typical "work-function", for interstellar grains is $W = 6$ eV, and the expected photoelectic yield, or the fraction of electrons that are actually dislodged off the grains, is typically $Y = 0.1$. Therefore, for a mean incident FUV photon energy of 10 eV, a fraction $([10 - 6]/10) \times 0.1 = 0.04$ of the original photon energy is available to heat the gas. This fraction, or "heating efficiency", decreases if the grains become positively charged, as they do when the photoelectron emission rates become large relative to the grain-electron recombination rates. This argument predicts cooling emission lines with intensities of order 1% of the thermal IR dust continua, consistent with observations.

As Alex and I showed, radiative excitation of H_2 followed by collisional deexcitations is an additional important heating mechanism in dense PDRs[16] (see §5). Predicted, and observed, gas temperatures then range from a few 100 K up to 1000 K in the outer cloud layers, down to 10-20 K at large depths where the FUV is fully attenuated and where other processes such as cosmic-ray heating and cosmic-ray induced photodissociation[26,27] dominate.

The chemical structure may be understood by considering theoretical models,[16,17,46] in the simplest picture (see Fig. 2) as a plane-parallel slab of gas that absorbs FUV radiation incident from one side. Any EUV radiation capable of ionizing hydrogen is assumed to be absorbed in a thin ionization front adjacent to the PDR. The penetrating FUV radiation dissociates molecules, and ionizes atoms, as it penetrates the cloud. The photorates are attenuated with increasing cloud depth (or visual extinction A_V) as the FUV radiation is absorbed by the dust. PDRs consist of several distinct zones and transition layers, from outer photodissociated gas to inner fully molecular gas. The initiating H/H_2 transition occurs closest to the cloud surface where the combined effects of dust opacity and H_2 absorption-line "self-shielding" reduce the photodissociation rate. The H/H_2 transition point depends on the efficiency with which H_2 forms on the grain surfaces. The C^+ to C to CO transition, as first considered by Alex and Michael Oppenheimer,[11] occurs deeper in the cloud. (The C^+ layer is a prominent source of near-red metastable emission lines of atomic carbon. The excitation mechanism is recombination followed by cascading.[48]) A similar S^+ to S to SO,SO_2,CS transition zone occurs at still larger A_V. The sizes of the various transition layers are controlled by several parameters, including most importantly the cloud density (or pressure), and the intensity of the illuminating radiation field. Additional controlling parameters are the grain FUV scattering properties, the cloud geometry and clumpiness, and the elemental gas phase abundances.

The ionization state decreases with cloud depth, from high values $\sim 10^{-4}$ due to the photoionization of carbon, to much lower values $\lesssim 10^{-7}$ in the shielded cores, where the positive charge is carried by molecular ions (see e.g.,[10]). Cloud collapse may be regulated in part by the loss of magnetic support. The flux loss occurs via ambipolar diffusion which becomes slow as the ionization fraction becomes large and the field is frozen-in, as in the outer C^+ layer. This may lead to self-regulated star-formation in which clouds expand - and become more transparent and photoionized and hence magnetically stable - in response to energy injection by stellar outflows and radiation fields produced by contracting and star-forming inner cores.[49]

4. Molecular Hydrogen

With the (somewhat belated) recognition that dense PDRs produce significant cooling radiation in galaxies, it was most natural for us to consider, as well, the problem of molecular hydrogen (H_2) line emissions from PDRs.

Though very abundant, H_2, is usually difficult to observe directly because its small moment of interia leads to widely spaced energy levels which are inefficiently excited at the cold temperatures of quiescent molecular clouds. Also, because H_2 is symmetric, the rotational and vibrational transitions are dipole forbidden, and emissions occur only via weak quadrupole transitions. However, in PDRs H_2 is efficiently excited, leading to widespread and detectable emissions.

Herzberg[50] was the first to suggest that it should be possible to observe H_2 via quadrupole rotational and vibrational transitions. At the time,

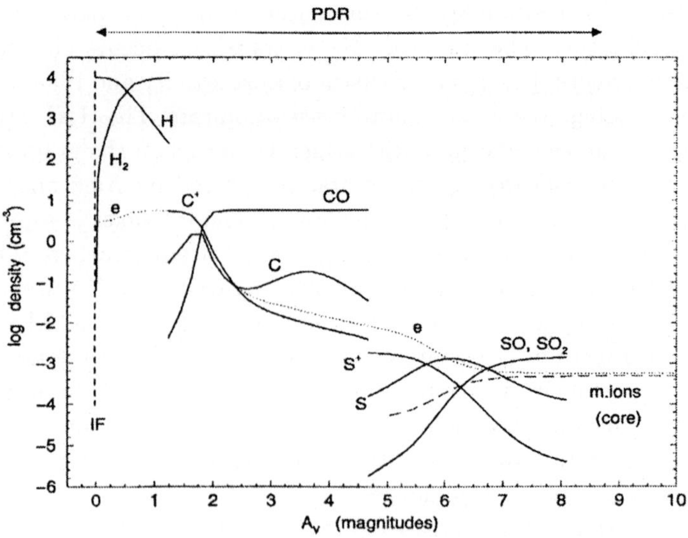

Fig. 2. Schematic structure of a PDR. FUV radiation is incident from the left.

Herzberg was trying to find a way to detect H_2 in the atmospheres of the giant planets. This was ultimately successful.[51–54] It took many more years for H_2 to be detected in the ISM. The first detections were via (allowed) far-ultraviolet absorption lines in the electronic Lyman and Werner bands, in sounding-rocket observations.[55] Such observations are limited to optically thin diffuse clouds with suitable background stars serving as UV continuum sources. With the advent of sensitive near-infrared detectors (again in the 70's), quadrupole H_2 ro-vibrational emissions were finally detected in the Becklin-Neugebauer object in the Orion nebula[56] A prominent transition is the near-infrared 1-0 S(1) line at 2.12 μm. Since then, such observations have become routine, and H_2 has been detected in circumstellar shells and planetary nebulae around evolved stars, Herbig-Haro objects and protostellar jets, shocked gas and PDRs in star-forming regions, and in external starburst galaxies and active galactic nuclei.

Prior to the interstellar detections Alex had been working steadily to compute the necessary molecular data that would be required for quantitative analysis of observations that he realized were sure to come. This included computations of the (X-B) Lyman and (X-C) Werner band oscillator strengths[57] the spontaneous radiative dissociation probablities from the B and C states,[58,59] and the quadrupole transition probabilities between all

301 ro-vibrational levels of the ground X-state.[60,61] The computations were of high precision, as beautifully demonstrated in the astonishing match between theory and laboratory measurements of the spontaneous radiative dissociation continuum from the B and C states.[62] Alex and John Black[8] incorporated the basic H_2 data into diffuse cloud models, and made predictions for the H_2 populations in FUV excited ro-vibrational states and the resulting quadrupole infrared emission line cascade. This was the first computation of the interstellar "fluorescent" FUV-pumped H_2 emission-line spectrum. Such emissions were finally detected, in reflection nebulae[63,64] almost a decade later.

The diffuse cloud models focussed on low density gas exposed to weak interstellar FUV fields. With the recognition of the important role of PDRs, Alex and I constructed models for much denser and optically thicker gas clouds, exposed to very intense radiation fields. Luminous H_2 emissions would be expected, in addition to the bright fine-structure lines. For this purpose, we gave special emphasis to FUV-pumping in combination with collisional excitations and deexcitations of the ro-vibrational states. We considered a wide range of gas densities (pressures) and illuminating FUV fields. This included a treatment of the thermal balance and depth dependent chemistry, for one of the first full-fledged "PDR codes" (see[65]). The results were published in our major "SD89" paper[16] (by then I was already doing my postdocs in Tel Aviv and MPE/Garching). John and Ewine also considered the radiative excitation of H_2 for various gas densities and FUV field strengths.[66] Alex and I considered thermalization of the level populations by elastic and inelastic H-H_2 and H_2-H_2 collisions, and showed how the H_2 emission line spectrum makes a transition from a "fluorescent" form in low density gas ($n \lesssim 10^4$ cm^{-3}) to a "shock-like" form in dense ($n \gtrsim 10^4$ cm^{-3}) gas (see also[68] and[67]).

A beautiful observational example is the colliding pair of galaxies known as the "Antennae" system. Collisions and merging of galaxies are believed to be important processes in the hierarchical build-up of "large scale structure" in the Universe. Galaxy collisions often lead to "starbursts" as the associated compression of the ISM gas leads to elevated star-formation efficiencies. An optical image (*Hubble Space Telescope*) of the Antennae reveals many blue star-clusters containing large numbers of recently formed massive hot stars. However, there is quite a bit of obscuration by interstellar dust. Near and mid-infrared imaging (first with the *Infrared Space Observatory* and more recently with the *Spitzer Space Telescope*) shows that much of the energy being released by the star-formation is from within the optically obscured region. Near-IR Keck (ground-based) spectroscopy of the embedded star clusters reveals a spectrum containing a wealth of vibrational H_2 emission lines.[69] The luminosity in just the 1-0 S(1) line alone is 9600 $L_\odot$. The relative intensities show that the excitation mechanism is indeed FUV pumping. This is the Black & Dalgarno spectrum in a starburst galaxy!

5. X-ray Dominated Regions

In X-ray dominated regions (XDRs), penetrating (keV) X-rays heat the gas and drive the chemistry, as opposed to FUV photons as in PDRs. XDRs began to receive attention from theorists[18,20,70,71] especially with the recognition that considerable amounts of molecular gas may surround accreting supermassive black holes in active galactic nuclei (AGN). Such systems are prodigious sources of X-rays. Warm XDRs may be the origin of the spectacular H_2O megamasers observed in AGN.[72–74]

XDRs may also be important in envelopes and protoplanetary disks around young stellar objects.[75,76]

There are two important differences between PDRs and XDRs. First, because the X-rays are absorbed in photoionization events, as opposed to the FUV dust absorption that occurs in PDRs, the X-ray intensity falls off much more slowly in XDRs. Second, the radiative heating efficiency, that is, the fraction of the radiation energy converted to heat and then reradiated in cooling emission lines, is larger in XDRs than in PDRs.

A key parameter in XDRs is the ratio ("ionization parameter"), H_X/n_H, of the local X-ray energy deposition rate H_X to hydrogen gas density n_H[20,70] For large H_X/n_H, the gas remains primarily atomic, with large (0.1–0.01) ionization fractions, and high ($\sim 10^4$ K) temperatures. The H/H_2 ratio, ionization fraction, and gas temperature, decrease as H_X/n_H becomes small. Molecules form with high efficiency at intermediate values of H_X/n_H where the gas temperature is maintained at a level of a few hundred K.

It is of interest to consider the X-ray energy deposition in the limit of a fully molecular gas. In this case, energetic electrons are produced by photoionization

$$\mathrm{H_2 + X - ray \rightarrow H_2^+ + e^*} \quad . \tag{10}$$

The energetic electrons, e^*, lose energy via Coulomb scattering with free thermal electrons, by electronic excitations of the H_2 molecules (and trace H atoms), and by collisional ionizations of the H_2. For small fractional ionizations Coulomb scattering is a minor channel. The excitations are followed by emissions of UV photons which are absorbed by, and heat, the dust.

The induced UV photons also mediate the chemistry by photodissociating molecules.[26] Importantly, the electron impact ionizations

$$e^* + H_2 \rightarrow H_2^+ + e + e^* \tag{11}$$

are immediately followed by proton transfer

$$H_2^+ + H_2 \rightarrow H_3^+ + H \quad , \tag{12}$$

and dissociative recombination

$$H_3^+ + e \rightarrow H_2 + H \quad . \tag{13}$$

Most of the ionization energy of 15.4 eV is converted into kinetic energy of the dissociating fragments, i.e., heats the gas.[77] As calculated by Alex, Min Yan and Weihong Liu, the energy lost to each energetic electron per ionization event is on average 37.7 eV.[21] Therefore, a heating efficiency of about 15.4/37.7=0.4 is expected in XDRs, significantly larger than in PDRs. This effect is seen in computations of the fine-structure emission line intensities in XDRs, with large resulting line-to-continuum ratios.

The Seyfert 2 galaxy NGC 1068 (15.5 Mpc) is likely a good source for a quantative study of XDR chemistry. Millimeter-wave interferometry shows that in NGC 1068 molecular emissions are produced in a kpc scale star-forming ring, and in an inner 100 pc scale region surrounding the active nucleus. Molecules have been detected in the nuclear region, including CO, HCN, CN, CS, HCO^+, HOC^+, and SiO (see[78–81]). The mm-wave maps show clear differences between the circumnuclear ring, and the nucleus. In particular the HCN/CO intensity ratio is small in the ring, and strikingly large in the nucleus. Part of this difference can be attributed to a higher gas density in the nucleus which then favors emission from molecules with large dipole moments such as HCN. A radiative transfer analysis[79] shows that the large HCN/CO ratio in the nucleus is due in part to an enhanced HCN/CO *abundance* ratio. In fact, recent observations[81] imply an HCN/CO abundance ratio of $\sim 10^{-3}$ in the nucleus, an extraordinarily large value compared to Galactic molecular clouds.

Alex and Stephen Lepp suggested[18] that the large HCN/CO abundance ratio in the nucleus of NGC 1068 might be due to an enhanced ionization rate due to X-rays from the accreting black hole. They also suggested that CN would persist at high ionization rates. Figure 3 shows recent computations[82] that support these ideas. Figure 3 shows the computed abundances (relative to H_2) for several species including CN, HCN, and CO, as functions of the parameter n_H/ζ_{-17} (cm^{-3}), where ζ_{-17} is the H_2 ionization rate in units of 10^{-17} s^{-1}. The abundances are for a pure gas-phase chemistry in

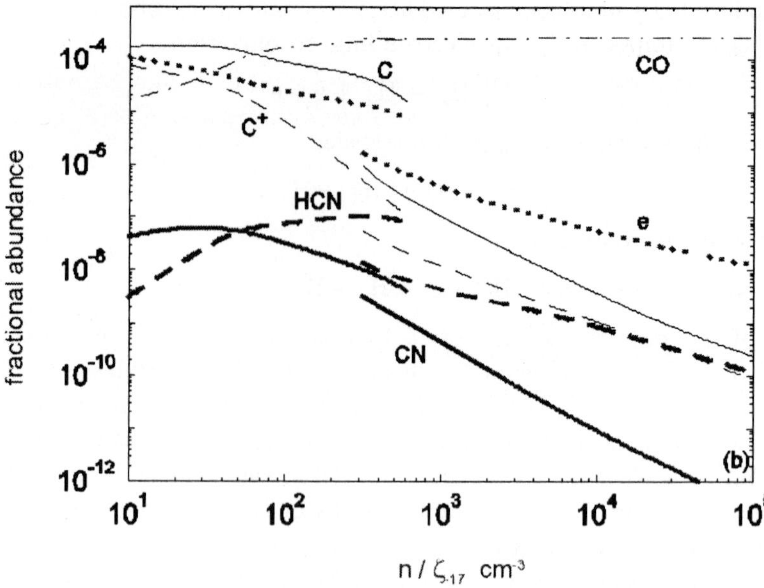

Fig. 3. Carbon-Nitrogen chemistry near black holes. Abundances are shown as functions of the density/ionization-rate (adapted from[82]).

steady state, for Galactic ISM abundances of the heavy elements carbon, nitrogen, and oxygen. At high density or low ionization rate, the gas is in a low-ionization phase (LIP). In the LIP the CN and HCN abundances, and the HCN/CO and CN/HCN density ratios, are small. At low density or high ionization rate the gas is in a high-ionization phase (HIP). In the HIP, the CN and HCN abundances become large. Furthermore, HCN/CO approaches $\sim 10^{-3}$ in the HIP, similar to what is observed in the nucleus of NGC 1068. Also, the CN/HCN ratio increases with the ionization rate. This is nicely consistent with the recent observations that reveal CN/HCN ratios $\gtrsim 1$ in the nucleus of 1068.

Large CN/HCN ratios are also expected in PDRs though for somewhat different reasons. How may PDRs and XDRs be distinguished chemically? This is a vexing question[19,83] and requires further theoretical study. One idea is to make use of the fact that in XDRs multiply charged atomic ions are expected to coexist together with the neutral molecular gas. The multiply charged species are produced by inner-shell X-ray ionization and Auger decay, processes that do not occur in FUV irradiated gas in PDRs. Ions such as C^{2+} and S^{2+} have been put forth as possible candidates, provided they are not rapidly neutralized by charge transfer or gas-grain reactions. Fine structure lines of S^{2+} at 33.5 and 18.7 μm would then be potentially useful probes. As noted by Alex and Min Yan, if these ions react rapidly with H_2 they may lead to enhanced CH^+ and SH^+ in XDRs.

Acknowledgements

Alex has been my mentor, colleague, and friend, for thirty years. My debt to Alex is beyond measure. I thank Jim Babb, Kate Kirby, and Hossein Sadeghpour for organizing this wonderful and inspiring celebration. I thank the Institute for Atomic and Molecular Physics (ITAMP) for hospitality and support.

References

1. Dalgarno, A. 2008, ARAA, 46, 1
2. Dalgarno, A., McCray, R.A. 1972, ARAA, 10, 375
3. Bates, D.R., & Spitzer, L.Jr. 1951, ApJ, 113, 441
4. Herbst, E., & Klemperer, W. 1973, ApJ, 185, 505
5. Dalgarno, A., & Black, J.H. 1976, RPPh, 39, 573
6. Black, J.H., & Dalgarno, A. 1973, ApL, 15, 79
7. Black, J.H., & Dalgarno, A. 1977, ApJS, 34, 405
8. Black, J.H., & Dalgarno, A. 1976, ApJ, 203, 132
9. van Dishoeck, E.F., & Dalgarno, A. 1984, ApJ, 277, 576
10. Oppenheimer, M., & Dalgarno, A. 1974, ApJ, 187, 231
11. Oppenheimer, M., & Dalgarno, A., 1974, ApJ, 192, 29
12. Hartquist, T.W., Dalgarno, A., & Oppenheimer, M. 1980, ApJ. 236, 182
13. Dalgarno, A. & Lepp, S. 1987, in Astrochemistry: IAU Symposium, Goa India, (Dordrecht, Reidel Publishing), p.109
14. Draine, B.T., Roberge, W.G., Dalgarno, A. 1983, ApJ, 264, 485
15. Neufeld, D.A., & Dalgarno, A. 1989, ApJ, 340, 869
16. Sternberg, A., & Dalgarno, A. 1989, ApJ, 338, 197
17. Sternberg, A., & Dalgarno, A. 1995, ApJS, 99, 565
18. Lepp, S., & Dalgarno, A. 1996, AA, 306, L21
19. Sternberg, A., Yan, M. & Dalgarno, A. 1997, IAUS, 178, 141
20. Yan, M. 1997, Phd thesis, Harvard University
21. Dalgarno, A., Yan, M., & Liu, W. 1999, ApJS, 125, 237
22. Liu, W., & Dalgarno, A. 1992, ApJ, 396, 679
23. Liu, W., & Dalgarno, A. 1995, ApJ, 454, 472
24. Stancil, P.C., Lepp, S., & Dalgarno, A. 1996, ApJ, 458, 401
25. Butler, S.E., Heil, T.G., & Dalgarno, A. 1980, ApJ, 241, 442
26. Sternberg, A., Dalgarno, A. & Lepp, S. 1987, ApJ, 321, 383
27. Lepp, S., Dalgarno, A., & Sternberg, A. 1987, ApJ, 321, 383

28. Dalgarno, A., & Sternberg, A. 1982, ApJ, 257, L87
29. Dalgarno, A., & Sternberg, A. 1982, MNRAS, 200, P77
30. Shields, G.A., Dalgarno, A., & Sternberg, A. 1983, PhRvA, 28, 2137
31. Sternberg, A., Dalgarno, A., & Roueff, E. ComAp, 13, 29
32. Selvelli, P., Danziger, J., Bonifacio, P. 2007, AA, 464, 715
33. Russell, R.W., Melnick, G., Gull, G.E., & Harwit, M. 1980, ApJ, 240, L99
34. Spitzer, L.Jr., 1948, ApJ, 107, 6
35. Dalgarno, A., & Rudge, M.R.H. 1964, ApJ, 140, 800
36. Barinovs, G., van Hemert, M.C., Krems, R., & Dalgarno, A. 2005, ApJ, 620, 537
37. Field, G.B., Goldsmith, D.W., & Habing, H.J. 1969, ApJ, 155, L149
38. Wolfire, M.G., McKee, C.F., Hollenbach, D., Tielens, A.G.G.M. 2003, 587, 278
39. Tielens, A.G.G.M., & Hollenbach, D. 1985, 291, 722
40. Crawford, M.K., Genzel, R., Townes, C.H., & Watson, D.M. 1985, ApJ, 291, 755
41. Genzel, R., Harris, A.I., & Stutzki, J. 1989, ESASP, 290, 115
42. Draine, B.T. 1978, ApJS, 36, 395
43. Bakes, E.L.O., & Tielens, A.G.G.M. 1994, ApJ, 427, 822
44. Weingartner, J.C., & Draine, B.T. 2001, ApJS, 134, 263
45. Kaufman, M.J., Wolfire, M.G., Hollenbach, D.J., & Luhman, M.L. 1999, ApJ, 527, 795
46. Jansen, D.J., Spaans, M., Hogerheijde, M.R., & van Dishoeck, E.F. 1995, AA, 303, 541
47. Oppenheimer, M., & Dalgarno, A. 1975, ApJ, 200, 419
48. Escalante, V., Sternberg;, A., & Dalgarno, A. 1991, ApJ, 375, 630
49. McKee, C.F. 1989, ApJ, 345, 782
50. Herzberg, G. 1938, ApJ, 87, 428
51. Kuiper, G.P. 1949, ApJ, 109, 540
52. Herzberg, G. 1952, ApJ, 115, 337
53. Kiess, C.C., Corliss, C.H., & Kiess, H.K. 1960, AJ, 132, 221
54. Spinrad, H. 1963, ApJ, 138, 1242
55. Carruthers, G.R. 1970, ApJ, 161, L81
56. Gautier, T.N., Fink, U., Larson, H.P., & Treffers, R.R. 1976, ApJ, 207, L129
57. Allison, A.C., & Dalgarno, A. 1969, AD, 1, 289
58. Dalgarno, A., & Stephens, T.L. 1970, ApJ, 160, L107
59. Stephens, T.L., & Dalgarno, A. 1972, JQRST, 12, 569
60. Turner, J., Kirby-Docken, K., & Dalgarno, A. 1977, ApJS, 35, 281
61. Wolniewicz, L., Simbotin, I. & Dalgarno A. 1998, ApJS, 115, 293
62. Dalgarno, A., Herzberg, G., & Stephens, T.L. 1970, ApJ, 162, 49D
63. Sellgren, K. 1986, ApJ, 305, 399
64. Gatley et al. 1987, ApJ, 318, L73
65. Rollig et al., 2007, AA, 467, 187 2007
66. Black, J.H., & van Dishoeck, E.F. 1987, ApJ, 322, 412
67. Shaw, G., Ferland, G.J., Abel, N.P., Stancil, P.C., van Hoof, P.A.M. 2005, ApJ, 624, 794

68. Draine, B., & Bertoldi, F. 1996, ApJ, 468, 269
69. Gilbert, A.M. et al. 2000, ApJ, 533, L57
70. Maloney, P.R., Hollenbach, D.J., & Tielens, A.G.G.M. 1996, ApJ, 466, 561
71. Meijerink, R., & Spaans, M. 2005, AA, 436, 397
72. Neufeld, D.A., Maloney, P.R., & Conger, S. 1994, ApJ, 436, L127
73. Nakai, N, Inoue, M., & Miyoshi, M. 1993, Nature, 361, 45
74. Moran, J. 2008, ASPC, 395, 87
75. Glassgold, A.E., Najita, J., & Igea, J. 1997, ApJ, 480, 344
76. Stauber, P., Doty, S.D., van Dishoeck, E.F., & Benz, A.O. 2005, AA, 440, 949
77. Glassgold, A.E., & Langer, W.D. 1973, ApJ, 186, 859
78. Tacconi, L.J., Genzel, R., Blietz, M., Cameron, M., Harris, A.I., & Madden, S. 1994, 426, L77
79. Sternberg, A., Genzel, R., & Tacconi, L. 1994, ApJ, 436, L131
80. Helfer, T.T., & Blitz, L. 1995, ApJ, 450, 90
81. Usero, A., Garca-Burillo, S., Fuente, A., Martn-Pintado, J., & Rodrguez-Fernndez, N.J. 2004, AA, 419, 897
82. Boger, G.I., & Sternberg, A. 2005, ApJ, 632, 302
83. Krips. M., Neri, R., Garcia-Burillo, S., Martin, S., Combes, F., Gracia-Carpio, J., & Eckart, A. 2008, ApJ, 677, 262
84. Lepp, S., & Dalgarno, A. 1988, ApJ, 335, 769

MOLECULAR ASTROPHYSICS WITH THE SPITZER SPACE TELESCOPE: STUDIES OF INTERSTELLAR SHOCKS

DAVID A. NEUFELD

Department of Physics and Astronomy, The Johns Hopkins University, 3400 North Charles Street, Baltimore, MD 21218, USA

The Infrared Spectrograph onboard the *Spitzer Space Telescope* has proven to be a powerful tool for the study of astrophysical molecules, particularly in interstellar regions where the gas has been heated by shock waves. Recent observations – which have provided valuable information about the simple hydrides H_2, HD, H_2O and OH – are reviewed here.

1. Introduction

In this paper, I will summarize the implications of recent Spitzer observations of interstellar hydrides. This is not intended as a general review of the field, but rather a discussion of selected results obtained with the following collaborators: Paule Sonnentrucker and Yuan Yuan (Johns Hopkins), Dan Watson, Judy Pipher, Bill Forrest, Kyoung Hee Kim, and Joel Green (Rochester), Gary Melnick and Volker Tolls (CfA), Ron Snell (UMass), Michael Kaufman (SJSU), David Hollenbach (SETI Institute), and Ted Bergin (Michigan).

When mid-infrared emissions are observed from interstellar hydrides, they inevitably probe gas that is considerably warmer than that present within quiescent molecular clouds. Such gas resides in special regions in which the gas temperature is elevated – by the passage of interstellar shocks, for example. Accordingly, this introductory section will comprise a brief description of the origins and effects of interstellar shock waves, followed by a discussion of the observational capabilities provided by *Spitzer Space Telescope*.

1.1. *The Origin and Effects of Interstellar Shocks*

The possibility of interstellar shock waves is immediately evident from a comparison of the typical thermal and Alfven velocities in a molecular cloud – of order few x 0.1 km/s and ~ few km/s, respectively – with the higher velocities of bulk motion associated with a wide variety of astrophysical phenomena. The most important of these are related to stellar activity. The births of stars are accompanied by supersonic (and super-Alfvenic) gas outflows with velocities that may exceed 100 km/s: because their velocities exceed the maximum signal speed, these create shock waves where they interact with the surrounding molecular gas clouds. And the deaths of massive stars in supernova explosions expel material at velocities of 30,000 km/s into the interstellar medium (ISM), sending shock waves that may propagate many parsecs from the sites of the explosions.

Shock waves heat and compress the interstellar gas through which they propagate. Where they encounter molecular gas clouds, they may substantially modify the chemical composition of the gas. Slower shocks warm the gas, driving reactions with energy barriers that may be negligibly slow in the cold quiescent ISM but rapid in shock-heated regions; they also sputter dust grains and the icy mantles that coat them, releasing material into the gas phase. Faster shocks may result in the dissociation of molecules, and in the ionization of the resultant atoms.

1.2. *Infrared Spectroscopy with Spitzer*

Mid-infrared spectroscopy provides a powerful tool for studying neutral atoms, atomic ions, and molecules within warm, shock-heated regions. The Infrared Spectrograph (IRS) onboard the *Spitzer Space Telescope* provides long-slit spectroscopy at wavelengths in the 5.2 – 37 μm range, with an angular resolution and sensitivity that is greatly improved relative to the *Infrared Space Observatory (ISO)* that preceded *Spitzer*.

The 5.2 – 37 μm spectral region provides access to a diverse selection of emissions from interstellar shock waves: fine structure emissions of Ne^+, Ne^{++}, Si^+, P^+, S, S^{++}, Fe^+, Fe^{++} ; vibrational transitions of CO_2 and C_2H_2 ; and – the focus of this article – pure rotational transitions of the hydrides H_2, HD, OH and H_2O. Figure 1 shows an example of a *Spitzer*/IRS spectrum, obtained toward the supernova remnant 3C391 [1]. The Short-Low modules (top two panels) cover the 5.2 – 14 μm spectral region with a spectral resolving power $\lambda/\Delta\lambda \sim 60$, while the Short-High module (middle two panels; same data with different vertical scales) and Long-High module (bottom two panels) cover the 10 – 20 μm and 19 – 37 μm regions with a spectral resolving power $\lambda/\Delta\lambda \sim 600$.

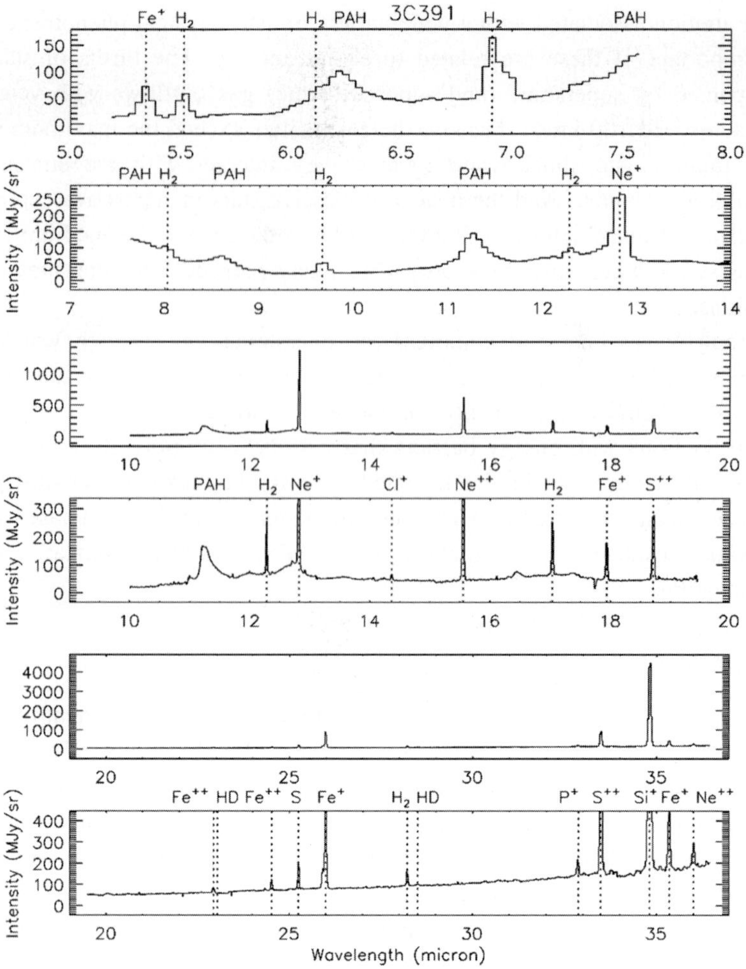

Figure 1: Spitzer/IRS spectrum of the supernova remnant 3C391 (from the paper of Neufeld *et al.* [1]). The spectrum is an average over a 25" (HPBW) diameter circular aperture centered at $\alpha = 18^h49^m21.9^s$, $\delta = -0^0 57'22''$ (J2000.0)

Because of the high sensitivity of the IRS instrument, extended regions can be mapped fairly rapidly, yielding spectral line maps that allow the spatial distribution of the various atomic and molecular emissions to be compared, using – for example – the method of Principal Component Analysis (PCA; see [1]). Figure 2 shows examples of spectral line maps obtained toward the supernova remnant 3C391.

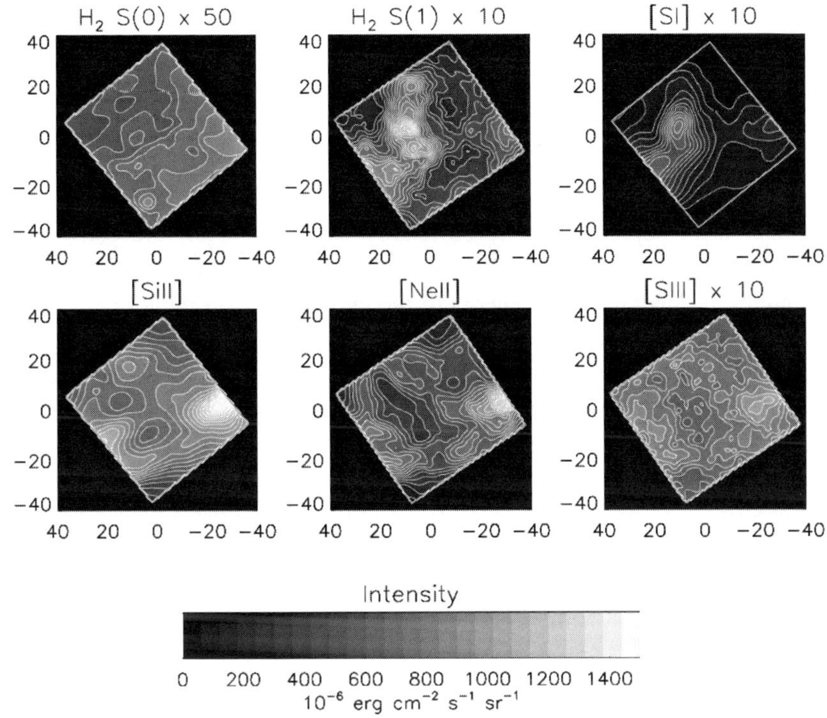

Figure 2: Spectral line maps obtained toward the supernova remnant 3C391 (from the paper of Neufeld *et al.* [1])

In sources where shock-heated gas is prevalent and the 3 – 10 μm spectrum is dominated by H_2 emissions – as in the supernova remnant IC443, for example – photometric observations with the Infrared Array Camera (IRAC) on *Spitzer* provide an additional and extremely efficient tool for determining the distribution and properties of the warm molecular gas, the four IRAC bands – centered at 3.6 μm, 4.5 μm, 5.6 μm, and 8 μm – being dominated by the H_2 v=1–0 Q(5), 0–0 S(9), 0–0 S(7), and 0–0 S(5) transitions, respectively [2].

2. Spitzer Observations of Interstellar Hydrides

2.1. *Molecular Hydrogen*

At temperatures in the range ~ 100 – 4000 K, and H_2 densities smaller than ~ 10^6 cm^{-3}, the cooling of warm molecular gas is dominated by quadrupole-allowed pure rotational emissions from H_2. With access to the lowest 8 rotational transitions of H_2, the *Spitzer*/IRS probes the dominant cooling lines of the gas (and may be supplemented by IRAC photometry in sources where H_2 emissions dominate the 3 – 10 μm spectral region).

Because the H_2 lines are optically-thin, with spontaneous rates that have been reliably determined by Alex Dalgarno and his collaborators [3, 4] the fractional level populations may be readily obtained. They are conveniently represented by rotational diagrams – where the logarithm of the column density is plotted as a function of the energy of the upper state – examples of which are shown in Figure 3 for four supernova remnants. These and other sources typically exhibit rotational diagrams that show: (1) positive curvature; and (2) a "zigzag" behavior indicative of a subthermal ortho-to-para ratio (OPR), the departures from equilibrium being largest for the states of lowest energy. These behaviors indicate, respectively (1) that an admixture of gas temperatures is present along the sight-line to the source; and (2) that the gas has been heated so recently that the H_2 ortho-to-para ratio has not yet had time to reach the equilibrium value appropriate to its current, elevated temperature.

The observed H_2 rotational diagrams are in good agreement with the predictions of models for shocks in the presence of interstellar magnetic fields. For propagation velocities less than 40 – 50 km/s, such shocks are of "C-type" in the designation of Draine [5], with the flow velocity changing continuously within a region where the ions and electrons drift relative to the neutral molecules. The emission properties of such shocks were first investigated by Alex and his collaborators [6], and independently by Chernoff, Hollenbach & McKee [7]; C-type shocks are expected to be luminous sources of H_2 emission.

The rotational diagrams observed by *Spitzer* require the presence of multiple shock velocities within the telescope beam, with an admixture of velocities that is consistent with paraboloidal bow shocks [2]. The observed OPR are consistent with models that invoke reactive collisions with H as the dominant process of equilibration (e.g. [8]). In this picture, the shocks propagate in molecular gas with an initial OPR that approaches the low equilibrium value in the cold interstellar medium. The shocked gas is heated up to temperatures > 300 K, at which the equilibrium OPR would be 3. Because the reactive collisions that convert ortho-H_2 to para-H_2 possess an energy barrier ~ 0.4 eV, and because the gas temperature and the abundance of H atoms both increase with the shock velocity, faster shocks are more efficient in increasing the OPR ratio within the limited time period for which the temperature is elevated; this consideration provides a quantitative explanation for why the low-lying H_2 states exhibit a lower OPR than the higher states [9].

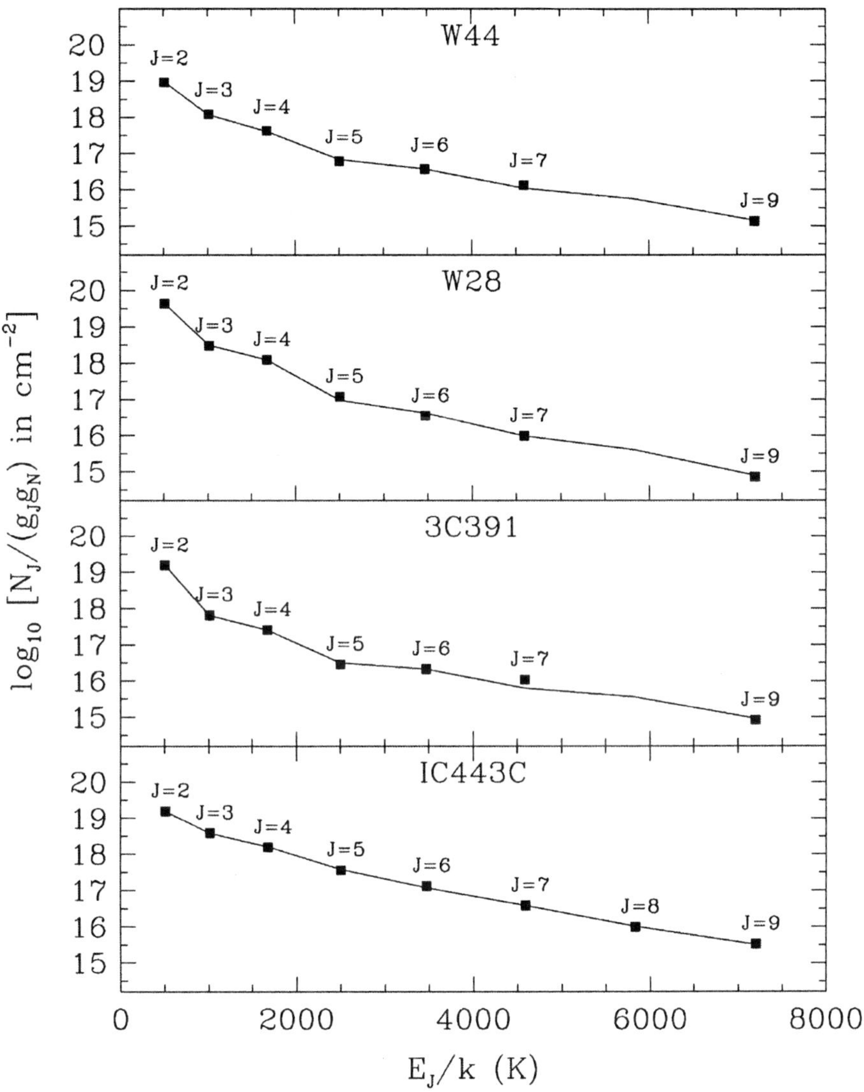

Figure 3: H_2 rotational diagrams for four supernova remnants (from Neufeld *et al.* [1])

2.2. *Hydrogen Deuteride*

In some of the shocked regions where H_2 has been observed with Spitzer, HD rotational emissions have also been detected [10, 11]. The brightest transition accessible to *Spitzer*/IRS is the R(3) transition at 28.502 μm; in some sources, the R(4) and R(6) transitions have also been detected. Figure 4 shows maps of the HD R(3) and R(4) emissions in IC443, together with maps of the H_2 S(2) and S(3) transitions, which originate in states of comparable energy; the morphological similarity of the HD and H_2 maps supports the identification of HD.

Because the column density and the temperature distribution of the warm H_2 has been extremely well-characterized by the observations of the H_2 S(0) to S(7) transitions, an HD/H_2 abundance ratio can be derived. Typical results lie in

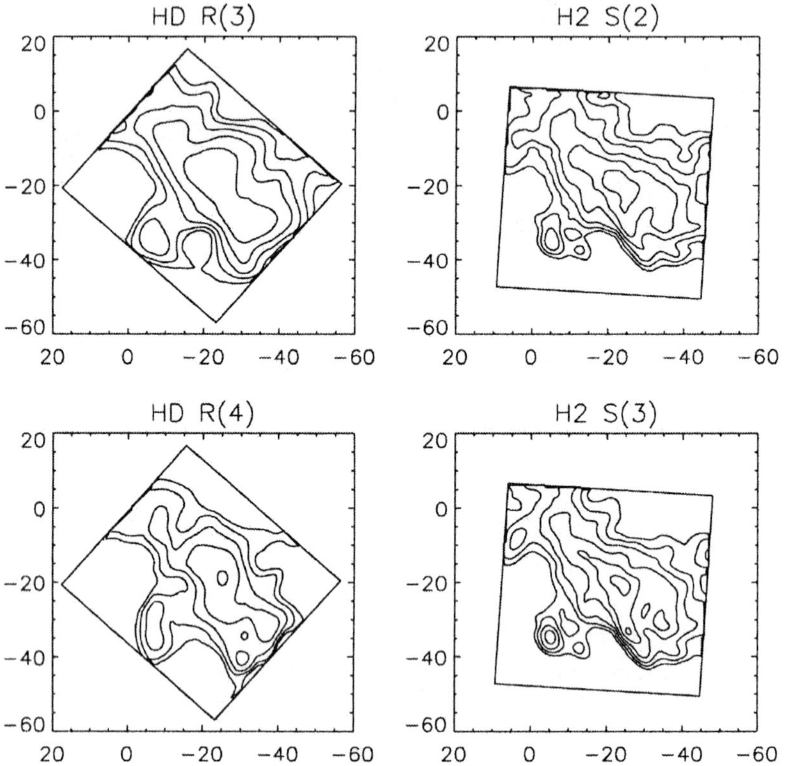

Figure 4: Spectral line maps for the HD R(3), HD R(4), H_2 S(2) and H_2 S(3) transitions in IC443 clump C (from the paper of Neufeld *et al.* [10])

the range 1 – 2 x 10^{-5}, in agreement with values obtained previously from *ISO* observations of the Orion Molecular Cloud. If, as predicted by chemical models for the sources observed by *Spitzer*, HD accounts for almost all the deuterium in the gas-phase, then the implied gas-phase deuterium abundance is $[D/H]_{gas}$ ~ 5 – 10 x 10^{-6}, a factor 3 – 5 below the primordial [D/H] abundance ratio predicted by big-bang nucleosynthesis (e.g. [12]).

Spitzer measurements of the $n(HD)/n(H_2)$ ratio in shocked regions are broadly consistent with previous studies of interstellar n(D)/n(H) in atomic clouds (e.g. [13]), measured by means of ultraviolet observations with the Far-Ultraviolet Spectroscopic Explorer (FUSE). These indicate a relatively constant $[D/H]_{gas}$ ratio ~ 15 x 10^{-6} for sight-lines to stars close to the Sun, with large (factor 5) variations on sight-lines to stars at larger distances, d (i.e. for d > 100 pc, like all the shocked regions observed with *Spitzer*); the observed variability in $[D/H]_{gas}$ has been attributed to the likely effects of variable deuterium depletion [13], resulting from the presence of deuterated polycyclic aromatic hydrocarbons.

2.3. *Water Vapor*

The chemical effects of interstellar shocks are perhaps most apparent from observations of water vapor. Observations with the *Submillimeter Wave Astronomy Satellite* (*SWAS*), which is sensitive to water vapor in both warm and cold interstellar gas – and with *ISO*, which is primarily sensitive to warm water – suggest that water abundances are dramatically enhanced within warm, shock-heated regions. The typical $n(H_2O)/n(H_2)$ ratio is ~ 10^{-10} – 10^{-9} in cold quiescent clouds and is ~ 10^{-6} – few x 10^{-4} in warm shock-heated material.

There are two possible enhancement mechanisms in shocks, and their relative importance remains uncertain: (1) water production via warm gas-phase chemistry, and (2) the release of icy grain mantles due to sputtering. Starting with the earliest models for chemistry in C-type interstellar shocks (e.g. [5]), such models are unanimous in predicting the rapid production of water via the endothermic reaction $O + H_2 \rightarrow OH + H$, followed by $OH + H_2 \rightarrow H_2O + H$. These reactions, which are negligibly slow at the temperatures of cold quiescent molecular clouds, become rapid at temperatures above ~ 400 K and result in the efficient conversion of all gas-phase oxygen not bound in CO into water vapor. The second possible enhancement mechanism results from the presence of large abundances of water ice – frozen to the surfaces of dust grains – that are revealed by infrared absorption spectra [14]. These icy grain mantles can be eroded by sputtering in shocks [15], releasing water vapor into the gas phase.

The relative importance of these two scenarios might be probed by an accurate determination of the H_2O ortho-to-para ratio [16]; chemical production in the gas-phase would lead to an OPR of 3, whilst the vaporization of icy mantles might yield the lower OPR typically observed in comets.

Rotational emissions from warm water vapor are most readily detected in the far-infrared spectral region. Such emissions were widely detected by *ISO*, primarily with the use of the Long Wavelength Spectrometer (LWS), which covered the 45 – 197 μm spectral range with a beam size ~ 75" (HPBW). Shock-heated water vapor also emits in the mid-infrared region accessible to *Spitzer*/IRS, which provides a spatial resolution more than a factor of 10 better than that of *ISO*/LWS. Recent *Spitzer* observations [16] of the protostellar outflow associated with NGC 2071 have revealed several spectral lines of water in the 29 – 37 μm region. The most prominent of these emissions, the $7_{25} - 6_{16}$

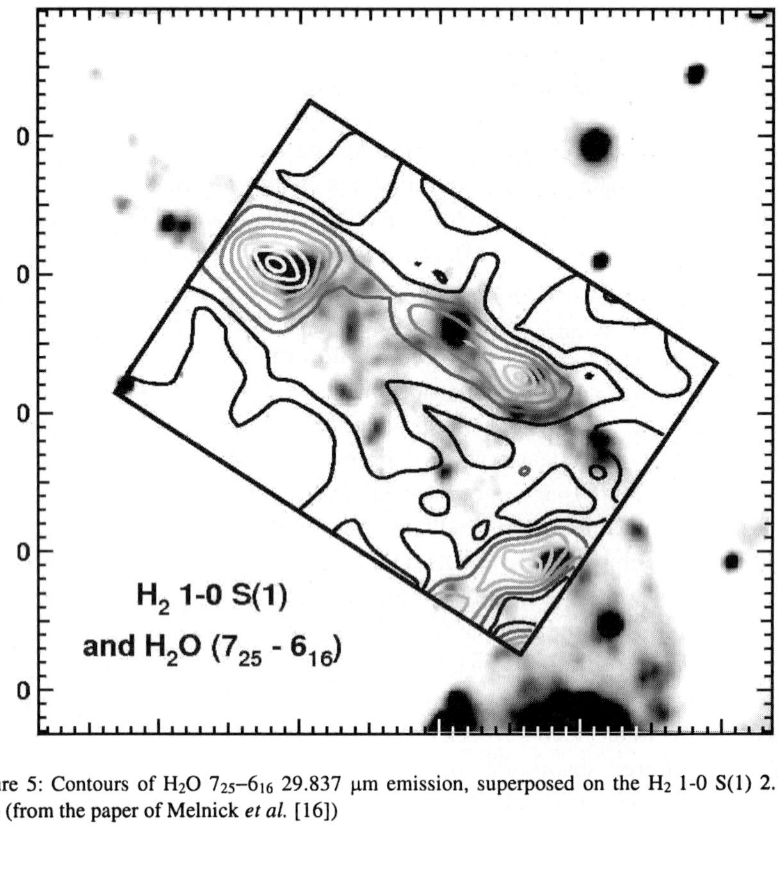

Figure 5: Contours of H_2O 7_{25}–6_{16} 29.837 μm emission, superposed on the H_2 1-0 S(1) 2.12 μm map (from the paper of Melnick *et al.* [16])

and $6_{34} - 5_{05}$ transitions at 29.837 μm and 30.899 μm, can be mapped; these display a spatial distribution that is strikingly similar to that of the warm H_2 (Figure 5). The derived H_2O abundance lies in the range 2×10^{-5} to 6×10^{-4}, the large uncertainty resulting from uncertainties in the total gas density. The best-fit H_2O ortho-to-para ratio is 2 – a value suggestive of grain mantle sputtering as the origin of the observed water vapor – but the accuracy is probably insufficient to rule out definitively an OPR of 3. A definitive study of the H_2O ortho-to-para ratio will have to await the *Herschel Space Observatory*, discussed briefly in Section 3 below.

2.4. *Hydroxyl Radicals*

The hydroxyl radical is a second oxygen-bearing molecule that is observed behind interstellar shock waves. Far-IR emissions of OH, involving low-excitation rotational states, have been widely observed from shocked regions with the use of *ISO*. In addition, maser emissions in the 1720 MHz Λ-doubling transition are frequently associated with supernova remnant shocks [17].

Recent *Spitzer* observations of the protostellar outflow HH211 – reported by Tappe *et al.* [11] – have revealed a remarkable mid-IR spectrum that shows a forest of high-*J* rotational states of OH. Hydroxyl emissions cover the spectral range of the high-resolution spectrometers, the highest transition accessible to Short-High being the $^2\Pi_{3/2}$ $J = 69/2 - 67/2$ line, with an upper state of energy $E_U/k = 28{,}200$ K. In addition to displaying states of remarkably high excitation, the spectrum is noteworthy in two other respects: (1) as the upper state energy, E_U, increases, there is no measurable diminution in the line intensities, and (2) there is no detectable emission from vibrationally-excited states of OH.

These unusual features of the OH emission spectrum in HH211 point to a very specific excitation mechanism: the photodissociation of H_2O by H Lyman alpha radiation at 121.6 nm. While H_2O photodissociation by UV radiation in the 140 – 190 nm region (in the so-called "first absorption band") produces OH with a rotational temperature ~ 500 – 1000 K [18] that is far too small to explain the observed spectrum, photodissociation by UV radiation shortward of 140 nm (in the second absorption band) selectively produces OH in states of high rotational excitation, a result that is well-established by both experimental and theoretical studies [19, 20, 21]. Here, the OH is initially produced in the $A^2\Sigma^+$ state, which subsequently yields excited rotational states (with $J = 35 - 50$) of the $X^2\Pi$ ground electronic state; in addition, the ground vibrational state is strongly favored over excited vibrational states [21].

The observed OH spectrum in HH211 can then be explained by a picture in which water vapor – produced either by gas-phase chemistry or the release of grain mantle material – is photodissociated by shock-emitted UV radiation, which is dominated by the H Lyman alpha line. This photodissociation produces OH in rotational states with $J \sim 35 - 50$; a radiative cascade ensues in a series of mid-infrared transitions with $\Delta J = -1$.

A simple argument shows that this excitation mechanism will likely dominate collisional excitation in populating high-J rotational states of OH behind the shock. If photodissociation dominates the destruction of H_2O, and if the destruction of OH is balanced by its formation via reaction of OH with H_2, then the excitation rate is $k\, n(\mathrm{OH})\, n(\mathrm{H_2})$, where $k \sim 6 \times 10^{-12} \exp(-2{,}000\mathrm{K}/T)$ is the rate coefficient for the reaction $\mathrm{OH} + \mathrm{H_2} \rightarrow \mathrm{H_2O} + \mathrm{H}$. By contrast, the rate of collisional excitation is $q\, n(\mathrm{OH})\, n(\mathrm{H_2})$, where q is the collisional excitation rate to states with $J > 35$. While q is unknown for states of such high excitation, it is unlikely to exceed $10^{-11} \exp(-30{,}000\mathrm{K}/T)$. Thus, at the likely temperature (< 4000 K) of the OH-emitting region, collisional excitation will be a factor of at least 10^3 less efficient than H_2O photodissociation in populating the highest-excitation states observed by *Spitzer*.

3. Future Studies

With the end of *Spitzer*'s cryogenic phase – expected in the first half of 2009 – the IRS instrument will cease operations. However, mid- and far-IR spectroscopy has a bright future, with several new instruments expected to begin operations during the next few years. In particular, the *Herschel Space Observatory* will facilitate high-spectral-resolution heterodyne spectroscopy with its HIFI instrument over the 157 – 625 μm wavelength range, and moderate resolution imaging spectroscopy with its PACS instrument over the 55 – 210 μm range. These instruments promise to provide a superb tool for studying molecular gas in shocked regions, allowing the dominant cooling transitions of OH, H_2O and CO to be mapped. Particularly when combined with earlier *Spitzer* observations of H_2, *Herschel* observations of shocked regions will yield a particularly powerful probe of the physical and chemical conditions.

Along with *Herschel*, the *Stratospheric Observatory for Infrared Astronomy* (*SOFIA*) will provide important spectroscopic capabilities at both mid- and far-IR wavelengths, including access to the HD $J = 1 - 0$ transition through heterodyne observations with the GREAT instrument. In the longer term, the *James Webb Space Telescope* (*JWST*) – with its MIRI instrument -- will observe the 5 – 28 μm spectral region, allowing sub-arcsecond resolution observations of the H_2 rotational transitions that dominate the cooling of warm molecular gas.

4. Concluding Remarks

As the pages in this volume have made clear, Alex Dalgarno's influence on many fields of physics, astrophysics, and chemistry has been immense. I will conclude by adding a few personal remarks about his role as a mentor.

I was a graduate student with Alex from 1983 to 1987. It was a very happy time, and I learned a tremendous amount of science during the process, of course, starting with the famous graduate course on Atomic and Molecular Physics that was taught by Alex and the late George Victor. But looking back on this now, I realize I learned much more than science from Alex. In particular, by observing his example, and without even realizing it at the time, I find that I learned a lot about the right way to conduct oneself as a scientist and a professor. I think this is perhaps the most important thing, and I hope I have carried this with me – however, imperfectly – into my professional life since I graduated. The right way – the moral way – isn't always the easiest way, and we all know that it isn't the path that is always followed.

On the science side, I observed how to act with the utmost scientific integrity and intellectual honesty. I had many questions for Alex during my graduate studies, and of course he usually had the answers at his fingertips or was able to figure them out as we spoke. But very occasionally he said "I'm not sure", or "I don't know" or "let me back off on what I said". Of course, it makes a great impression on an aspiring scientist when someone of Alex's stature says he isn't sure about something, because it allows you to admit that you are unsure about something – and when you start out you're unsure about almost everything. And, of course, those three little words "I don't know" are what make science possible.

On the mentoring side, I observed how Alex treated everybody – even the most junior scientists – with fairness, and respect and generosity and encouragement. At science meetings, I began to notice, for example, that some of the leaders in the field only wanted to interact with other people they considered important, while Alex was always eager to talk to the most junior people and find out what they were working on.

At the conference dinner that took place during the celebratory symposium that this volume documents, Alex said that he considered his former students and postdoctoral fellows to be his real legacy. And while we, his former students and postdoctoral fellows may not really believe that – we may think his legacy is measured by the groundbreaking scientific results and papers and citations – it is clear that Alex really meant what he said and has acted accordingly throughout his career, to our great advantage.

References

1. D. A. Neufeld, D. J. Hollenbach, M. J. Kaufman, R. L. Snell, G. J. Melnick, E. A. Bergin, and P. Sonnentrucker, *Astrophys. J.*, **664**, 890 (2007).
2. D. A. Neufeld and Y. Yuan, *Astrophys. J.*, **678**, 974 (2008)
3. J. Turner, K. Kirby-Docken, and A. Dalgarno, *Astrophys. J. Suppl.*, **35**, 281 (1977)
4. L. Wolniewicz, I. Simbotin, and A. Dalgarno, *Astrophys. J. Suppl.*, **115**, 293 (1998)
5. B. T. Draine, *Astrophys. J.*, **241**, 1021 (1980)
6. B. T. Draine, W. G. Roberge, and A. Dalgarno, *Astrophys. J.*, **264**, 485 (1983)
7. D. F. Chernoff, C. F. McKee, and D. J. Hollenbach, *Astrophys. J.*, **259**, L97 (1982)
8. R. Timmermann, *Astrophys. J.*, **498**, 246 (1998)
9. D. A. Neufeld, G. J. Melnick, P. Sonnentrucker, E. A. Bergin, J. D. Green, K. H. Kim, D. M. Watson, W. J. Forrest, and J. L. Pipher, *Astrophys. J.*, **649**, 816 (2006).
10. D. A. Neufeld, J. D. Green, D. J. Hollenbach, P. Sonnentrucker, G. J. Melnick, E. A. Bergin, R. L. Snell, W. J. Forrest, D. M. Watson, M. J. Kaufman, *Astrophys. J.*, **647**, L33 (2006)
11. A. Tappe, C. J. Lada, J. H. Black, and A. A. Muench, *Astrophys. J.*, **680**, L117 (2008)
12. D. N. Spergel, *et al.*, *Astrophys. J. Suppl.*, **148**, 175 (1993)
13. J. L. Linsky *et al.*, *Astrophys. J.*, **647**, 1106 (2006)
14. D. C. B. Whittet, M. F. Bode, D. W. T. Baines, A. J. Longmore, and A. Evans, *Nature*, **303**, 218 (1983)
15. B. T. Draine, *Astrophys. & Space. Sci*, **233**, 111 (1995)
16. G. J. Melnick, V. Tolls, D. A Neufeld, Y. Yuan, P. Sonnentrucker, D. M. Watson, E. A. Bergin, and M. J. Kaufman, *Astrophys. J.*, **683**, 876 (2008)
17. M. Wardle and F. Yusef-Zadeh, *Science*, **296**, 2350 (2002)
18. P. Andresen, G. S. Ondrey, B. Titze, and E. W. Rothe, *J. Chem. Phys*, **80**, 2548 (1984)
19. D. H. Mordaunt, M. N. R. Ashfold, and R. N. Dixon, *J. Chem. Phys*, **100**, 7360 (1994)
20. R. van Harrevelt, and M. C. van Hemert, *J. Chem. Phys*, **112**, 5787 (2000)
21. S. A. Harich, D. W. H. Hwang, X. Yang, J. J. Lin, R. N. Dixon, *J. Chem. Phys*, **113**, 10073 (2000)

A COMMON ORIGIN FOR INTERSTELLAR CH^+ AND H_3^+

CESARE CECCHI-PESTELLINI

INAF - Osservatorio Astronomico di Cagliari
St. 54 Loc Poggio dei Pini, Capoterra 09012, Italy
E-mail: ccp@ca.astro.it

In this work I propose that the large abundances of CH^+ and H_3^+ observed in the diffuse interstellar medium may have a common origin, and I identify the driving agent in the dissipation of turbulence. Results of the present model are consistent with the observational constraints along the line of sight towards ζ Persei, and may be extended to other diffuse lines of sight.

1. Introduction

Diffuse clouds are detected towards bright background sources in the absorption lines of a wide variety of atoms and molecules [1], typically along relatively low-extinction lines of sight ($A_V \leq 2$ mag). The chemistry of such clouds has been considered for a long time as simpler to understand than that of denser regions, as their chemistry and thermodynamics would be directly linked to the local density of the diffuse interstellar radiation. In the traditional view, such clouds have no embedded heat sources, contain only the simplest diatomic molecules, and the timescales for most chemical processes are short enough to attain chemical equilibrium, so that the physical conditions are supposed to vary slowly at subparsec scales.

Despite the apparent simplicity, the chemistry of diffuse clouds is not well established, and there are remarkable discrepancies between models and observations. A cold chemistry generally reproduces column densities of OH, C_2 and CH, but underestimates CO [2], CN [3], CH^+ (e.g., [4]) and HCO^+ [2] by a large factor. Moreover, the agreement between models and observations might be illusory, as equilibrium excitation analyses provide evidences that, e.g., a large amount of CH originates from CH^+ chemistry [5,6]. In addition, millimeter observations suggest the existence of a rich chemistry even at low visual extinction (cf. [7] and reference therein). Further uncertainty on the nature of diffuse clouds is generated by the discovery along diffuse lines of sight of concentrations of the molecular ion H_3^+ [8,9] as large as those observed in dark clouds [10]. Such abundances appear inconsistent with model predictions based on a high removal rate via electronic dissociative recombination [11,12].

The basic problem for the chemistry is how to incorporate the ambient oxygen and carbon into molecules. Oxygen is ionized by slow charge transfers with H^+, while the abundant ionized carbon does not react rapidly with H_2. Thus, most theoretical models rely on the endothermic reactions $C^+ + H_2 \rightarrow CH^+ + H$ ($\Delta E/\kappa \sim 5000$ K) and $O + H_2 \rightarrow OH + H$ ($\Delta E/\kappa \sim 3000$ K). The need for energy sources in excess of the average energy density of diffuse and translucent clouds has given rise to consideration of models involving localized volumes of warm gas created and substained by dissipation of gas kinetic energy, as within MHD shocks [13], or coherent small-scale vortices in MHD turbulence [14,15]. Indeed, many observational facts, such as line width-size relations, the large Reynolds numbers of the gas motions and cloud clumping, are consistent with the existence of turbulence within interstellar clouds (cf. [16], and reference therein). Thus, in the last decade, the classical picture of two-phase diffuse interstellar medium has been modified incorporating small amounts of warm gas intimately associated with the cold neutral phase [14,17,18]. These tiny warm regions might promote the first steps in the chemistry of cold neutral medium, not only by producing the required concentrations of CH^+, but also inducing the formation of HCO^+, an important chemical precursor of CO. Actually, if the large observed abundance of HCO^+ is inserted into standard chemical models of diffuse clouds, the observed variation of carbon monoxide column density can be explained without resorting to any other assumption [2].

Attempts to reproduce the unexpected large concentrations of H_3^+ in the direction of a variety of diffuse sources have been performed relying on the ion-molecule scheme in a cold cloud initiated by primary ionization of hydrogen by cosmic rays or X-rays (e.g. [8,20,21]). In particular, along the line of sight towards ζ Persei, McCall et al. [12] suggest the presence of a very high cosmic-ray ionization rate $\zeta = 1.2 \times 10^{-15}$ s^{-1}, larger than the standard interstellar value by a factor of 30. Although the H_3^+ abundance can be fit exactly, species such as OH, CO, S, C, and HD result overproduced by relatively large factors [19]. A detailed PDR model, that reproduces the abundances observed towards ζ Persei to within a factor of three, has been put forward by Le Petit et al. [19]. The adopted ionization rate, $\zeta = 2.5 \times 10^{-16}$ s^{-1}, is intermediate between the classical value ($1 \div 5 \times 10^{-17}$ s^{-1}) and that advocated in [12]. The column density of

H_3^+, N_3, was about a factor 2.8 lower than the observed value. Increases in ζ to bring N_3 into agreement are costrained by the column density of OH. To reproduce the CH^+ abundance and those of the excited rotational populations of H_2, Le Petit et al. [19] consider the addition of an MHD shock.

The purpose of this work is to present a viable single mechanism able to provide the observed large concentrations of both H_3^+ and CH^+ in diffuse interstellar clouds. Such a mechanism would alleviate the constraints posed by the column density of OH, whose concentration increases with the increase of both gas temperature (CH^+ formation) and fractional ionization (H_3^+ formation), as well as by the column densities of other molecular diagnostics such as HD [22]. I propose the dissipation of turbulence as driver agent not only of the hydrocarbon chemistry but also of the H_3^+ large abundance. The formation problem for H_3^+ in conventional cold gas models is a problem of supply, a too low formation rate. Since it has been recently found that the rate of the H_3^+ electronic recombination, the main destruction channel in diffuse clouds, decreases significantly with increasing gas temperature [23], a transient non-equilibrium chemistry at high temperature might solve the problem. Dissipative bursts locally and temporarily heat the gas to temperatures up to $T_{\rm h} \sim 2,000$ K, well above that of the ambient diffuse gas, determining favourable conditions to the establishment of large concentrations of H_3^+. Such a chemistry is attractive because it offers the possibility of linking three important interstellar chemical observables: (*i*) it increases the formation rate of CH^+; (*ii*) it inhibits the destruction rate of H_3^+; and (*iii*) it produces pure rotational H_2 excitation [24,25].

Details of the chemical model are presented in the next Section, together with a discussion of results in the light of the observational constraints derived along the line of sight towards ζ Persei. Conclusions are summarized in Section 3.

2. The Formation of H_3^+ in Diffuse Interstellar Clouds

H_3^+ is produced by the reaction of H_2 with the H_2^+ ion, resulting from the cosmic-ray ionization of H_2. In a diffuse gas with visual extinction lower than $A_V = 2$ mag, the fractional ionization is nearly constant and H_3^+ is removed largely by dissociative recombination with electrons. Hence if ζ is the cosmic-ray ionization rate, the steady state abundance of H_3^+, n_3, is approximately

$$n_3 = \frac{\zeta}{\alpha(T)} \frac{1}{n_e/n_2}, \tag{1}$$

where $\alpha(T)$ is the rate of dissociative recombination at the gas temperature T, n_e and n_2 are the numerical densities of electrons and molecular hydrogen, respectively.

The space-time intermittent distribution of the dissipation rate of kinetic energy in turbulent interstellar clouds generates transient and localized warm regions embedded in cooler gas. I summarize the thermal history of a gas pocket trapped in such regions by means of temperature profiles parametrized in lifetime, $\Delta t_{\rm h}$, and peak temperature, $T_{\rm h}$. I adopt a step function, i.e. I neglect both the energy injection and the cooling phases. This choice is not terribly arbitrary (cf. Fig.1 in Joulain et al. [15]).

A parcel of gas with total hydrogen density $n_{\rm H}$ and initial kinetic temperature $T_{\rm c}$ is heated to a temperature characteristic of those turbulent dissipation events which are neither the most powerful (the rarest) nor the weakest, i.e. $T_{\rm h} = 500 - 2000$ K [15], over a time-scale $\Delta t_{\rm h}$ of the order, at most, of a few thousand years. Then the gas kinetic temperature drops to the initial value. Repeated impulsive heating events are not included. The non-equilibrium models I present in this work are calculated using as initial conditions the results of statistical equilibrium calculations at gas temperatures and densities characteristic of the diffuse interstellar medium [26,27]. The initial chemical distribution is obtained exploiting a model similar to the one described in Casu et al. [28]. In the present work a recent version of the UMIST database [29] has been implemented, with new rates of the electronic recombination of both H_3^+ [23] and CH^+ [30] incorporated in the network. The results, computed at a single point using average density and temperature, are scaled through total hydrogen column densities. In Table 1 are reported the reactions rates relevant to the chemistry of H_3^+ and CH^+.

I assume that H_3^+ forms esclusively in turbulent fluctuations and derive the fraction of warm gas necessary to account for the observed column densities for the case of the line of sight towards ζ Persei. Considerable uncertainties attend the values of the cosmic-ray ionization rates in the interstellar medium, but there are indications that the range could be narrow, $\zeta \sim 10^{-16} \div 10^{-15}$ s^{-1}, in going from dense cores to the intercloud medium [22]. The electronic density is not well constrained. Under the assumption that most electrons come from the photoionization of carbon, McCall et al. [12] derived an electron fraction $n_e/n_2 = x_e = 3.8 \times 10^{-4}$ along the line of sight towards ζ Persei. The absorption mesaurements yield the column densities $N_3 = 8 \times 10^{13}$ cm^{-2} [12], $N_{\rm H} = 1.6 \times 10^{21}$ cm^{-2} [12], and $N_1 = 6.3 \times 10^{20}$ cm^{-2} [31], that are, respectively, the integrals of n_3,

Table 1. H_3^+ and CH^+ relevant reaction rates

	rate (cm^3 s^{-1})	temperature range	process	source
α	$1.27 \times 10^{-6} T^{-0.48} - 1.3 \times 10^{-8}$	$T < 4,000$ K	$H_3^+ + e$	Ref.23
k_F	$1.0 \times 10^{-10} e^{-4640/T}$	$154 < T < 3,000$ K	$C^+ + H_2$	Ref.29
k_1	7.5×10^{-10}		$CH^+ + H$	Ref.29
k_2	1.2×10^{-9}		$CH^+ + H_2$	Ref.29
k_e	$2.0 \times 10^{-7} (300/T)^{0.37}$	$T < 1,200$ K	$CH^+ + e$	Ref.30
	$2.3 \times 10^{-7} \sqrt{300/T}$	$T > 1,200$ K		

$n_{\rm H}$, and n_1 along the line of sight. In the calculation, I adopt mean values, derived by the observed column densities, such as $x_1 = N_1/N_{\rm H} = 0.4$, $x_2 = (N_{\rm H} - N_1)/2N_{\rm H} = 0.3$, and x_e. Furthermore, I assume that all gas-phase carbon is ionized, and set its volumne density to $n_{\rm C^+} \sim n_e$, as implyed in the derivation of x_e [12]

If turbulent spots have mean lifetimes $\Delta t_{\rm h} \sim (\alpha n_e)^{-1} \sim 1,000$ yr, where α is the dissociative electronic recombination rate of H_3^+ and $n_{\rm H} > 20$ cm^{-3}, $n_3^\star$ reaches approximately its steady state concentration at temperature $T_{\rm h}$. Thus, the fraction of warm gas along the line of sight giving rise to the observed column density of H_3^+ results

$$f_3^\star \sim \frac{\alpha(T_{\rm h}) x_e N_3 n_{\rm H}}{\zeta N_{\rm H}}. \tag{2}$$

For temperatures $T_{\rm h}$, lying between 800 and 2,000 K, $\alpha \sim 2 \div 3 \times 10^{-8}$ cm^3 s^{-1}, considerably slower than in a cold gas. Thus, Eq.(2) reads as

$$\frac{f_3^\star}{10^{-2}} \sim 0.5 \div 0.8 \frac{n_{\rm H}}{\eta}, \tag{3}$$

where $\eta = \zeta/(10^{-16}\ {\rm s}^{-1})$. If the density of the gas is $n_{\rm H} \leq 10 \times \eta$ cm^{-3}, the aggregate column density of warm gas is lower than 10% of the total hydrogen column density along the line of sight towards ζ Persei.

The existence of warm gas in the diffuse interstellar medium has been considered along many lines of sight to account for the observed abundances of CH^+. Thus, if H_3^+ originates in turbulent dissipative structures, its concentration must be related to the abundance of CH^+. Towards ζ Persei, the column density of CH^+ has been estimated by van Dishoeck and Black [32] and found to be $N_{\rm CH^+} = 3.5 \times 10^{12}$ cm^{-2}. The formation of CH^+ involves the endothermic reaction between C^+ and H_2, while the destruction occurs mainly through reactions with atomic hydrogen, molecular hydrogen, and electrons. During a thermal spike the time evolution of the CH^+ abundance follows strictly the rise and the fall of gas temperature, reaching quickly stationary concentrations even in the warm phase [14]. The fractional abundance of CH^+ w.r.t. H_2 results

$$\frac{n_{CH^+}}{n_2} = \frac{k_F(T_h) n_{C^+}}{k_1 n_1 + k_2 n_2 + k_e n_e}, \tag{4}$$

where $n_1 = x_1 n_H$ is the numerical density of atomic hydrogen, k_F is the formation rate, and k_1, k_2, and k_e are destruction rates via collisions with atomic hydrogen, molecular hydrogen, and electrons, respectively (see Table 1). The fraction $f_+^\star$ of hydrogen column density along the line of sight devoted to the production of CH^+ is

$$f_+^\star = \left(\frac{N_{CH^+}}{x_2 N_H}\right) \frac{k_1 x_1/x_2 + k_2 + k_e x_e}{k_F(T_h) x_e}. \tag{5}$$

For temperatures of turbulent spots larger than 700 K, $f_+^\star$ is few percent of the line of sight.

Results of numerical calculations are presented in Fig.1, where $\eta f_3^\star$ and $f_+^\star$ as functions of temperature of turbulent spots are shown. The gas density reference value is $n_H = 30$ cm^{-3}, in agreement with gas densities inferred along lines of sight showing the presence of warm H_2 [33], and consistent with values exploited by Joulain et al. [15] in their study of non-equilibrium chemistry in the dissipative structures of interstellar turbulence. $\eta f_3^\star$ scales linearly with gas density because the abundance of H_3^+ tends to a constant value independent of density [20]. When $T_h \sim 800$ K the fractions of warm gas needed for the production of H_3^+ and CH^+ coincide. Moroever, setting $\eta = 2.5$, as in Le Petit et al. [19], the warm gas linear filling factor results lower than 8%.

The fraction of warm gas needed to reproduce the observed column density of H_3^+ scales with the ratio $x_3 = N_3/N_H$. For instance, along the line of sight towards Cyg OB2 No.12 this ratio results $x_3 = 3 \times 10^{-8}$ [34], a factor of 2 lower than towards ζ Persei, with a consequently lower fraction of warm gas.

There is not a straight link between the observed column densities and the actual volume density structure of the region. In this model is simply assumed that the density distribution is uniform so that $n_1/n_{C^+} = N_1/N_{C^+}$, $n_2/n_{C^+} = N_2/N_{C^+}$, and $N_{C^+}/N_H \sim x_e x_2$. Consequently, I do not express my results in terms of a specific model for the line of sight towards ζ Persei. Nevertheless, it is important to discuss the plausibility of my somewhat rigid assumptions on molecular concentrations. Indicating with Γ (s^{-1}) the photodissociation rate and with $\mathcal{R}$ (cm^3 s^{-1}) the formation rate of molec-

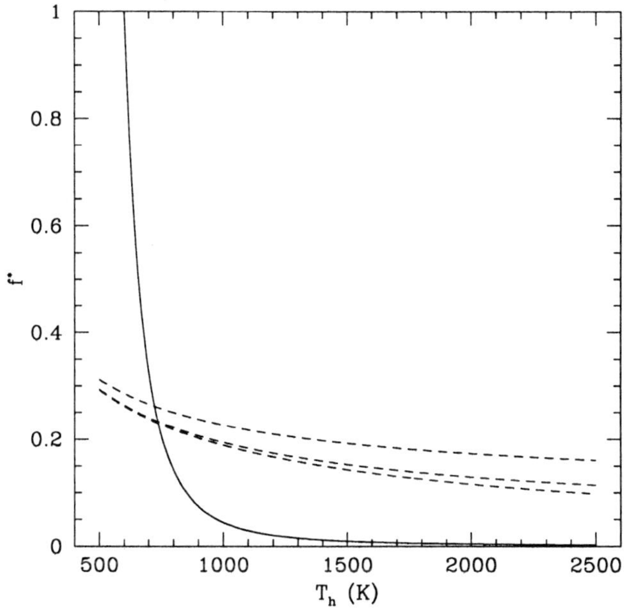

Fig. 1. Fractions (%) of warm gas along the line of sight needed to provide the observed column densities of CH^+ (solid line) and H_3^+ (dashed lines). $f_3^\star$ is normalized to $\zeta = 10^{-16}$ s^{-1}. The lifetimes of thermal spikes are 500 yr, 1,000 yr, and 2,000 yr (top to bottom). The gas density is $n_H = 30$ cm^{-3}.

ular hydrogen on dust grain surfaces, I obtain

$$n_1/n_2 \sim (\Gamma + \zeta)/(\mathcal{R} n_H). \tag{6}$$

Draine and Bertoldi [35] provide a simple approximation to the H_2 photodestruction rate as

$$\Gamma \sim \left(\frac{N_2}{10^{14}/\mathrm{cm}^{-2}}\right)^{-0.75} e^{-\tau_d}\,\Gamma_0, \tag{7}$$

where τ_d is the dust optical depth, and $\Gamma_0 \sim 5 \times 10^{-11}$ s^{-1} the unshielded photodestruction rate. Using $\mathcal{R} n_H = 1 \div 2 \times 10^{-15}$ s^{-1}, values characteristic of lines of sight where warm H_2 has been detected [33], and $n_1/n_2 = x_1/x_2$, I find that the "mean" turbulent spot is located at $\sim 0.2 \div 0.25$ mag from the edge of a cloud whose thickness is approximately $A_V \sim N_H/(2 \times 10^{21}) = 0.8$ mag. The above result is derived for a cold gas. However, the lifetimes of coherent vortex filaments in dissipative turbulence are much shorter than the evolution time of the H_2 level system, and on that timescale no significant modifications of both photodestruction rate and molecular hydrogen density occur [25]. Dissipative bursts may occur anywhere in a cloud, with their properties not much affected by variations in their locations or in the gas density [15]. However, too close to the edge of a cloud ($A_V < 0.1$ mag) the increase in the photodissociation rate limits considerably the abundance of H_2, and thus the production of H_3^+ (and CH^+). On the other hand, in more shielded regions the ratio x_e decreases ($A_V > 0.3$ mag from photodestruction of atomic carbon [29]), increasing the concentration of H_3^+ (e.g. [20]).

Finally, I discuss the rotational population of H_2. Both the molecular fraction x_2 and the column densities of the rotational states $J = 3-5$ detected along the line of sight towards ζ Persei [19] are very similar to the ones observed, with a similar Doppler parameter, along the line of sight towards HD 96675 [33]. Moreover, the column density of CH^+ for HD 96675 is very similar to the value derived towards ζ Persei. In the case of HD 96675, Cecchi-Pestellini et al. [25] estimate that the amount of warm H_2 and the corresponding excitation temperature are well reproduced using vortex temperatures around 800 K, and a warm gas filling factor of few percent.

3. Conclusions

In the cold diffuse medium temperature spikes induced by localized dissipation of turbulence may increase the concentrations of the species H_3^+. Using laboratory and observational data I have constrained the fraction of warm gas along the line of sight towards ζ Persei to be approximately lower than 10 %. The total aggregate size of turbulent gas is consistent with the observed column density of CH^+ if the mean temperature of dissipative structures is approximately 800 K. Temperatures in this range are also able to provide the observed amount of warm H_2 as well as the correspdensityonding excitation temperature. The derived kinetic temperature, $T_h \sim 800$ K, pertains to typical dissipation events [15].

The computed linear filling factor is moderately larger than the value inferred by Falgarone et al. [24] through the observations of five pure rotational lines of H_2 (a few %) along a line of sight through the Galaxy which avoids regions of massive star formation. However, the fraction of warm gas scales with x_3, the mean fractional concentration of H_3^+ along a line of sight. McCall et al. [34] provide a compilation of a number of lines of sight towards quasi-diffuse sources. With the exception of HD 20041, it is found that x_3 is in the range of values $x_3/10^{-7} \sim 0.25 \div 0.5$, in which

ζ Persei resides at the upper end. Provided that the electron concentration derived towards ζ Persei could be extended to all the other lines of sight, the derived fractional aggregate size of dissipative structures appears to be an upper limit.

Unless invoking very high hydrogen ionization rates (see [12]), H_3^+ appears to be asssociated with low density extended gas layers (cf. Eq.2), while heavier species, such as C_2, CN and to some extent CO (see [36]) require the presence of denser regions [19,20]. There are both theoretical and observational evidence that two separate chemical regimes are present in the diffuse interstellar medium: (*i*) the abundance of H_3^+ follows the increase in the fraction of hydrogen that is converted to molecular form with increasing depth inside a cloud and tends to a constant value independent of density [9,20]; (*ii*) H_3^+ has been detected along diffuse lines of sight with and without the presence of heavier species [34]; (*iii*) at least in the case of Cyg OB2 No.12 the velocity profile of H_3^+ appears more similar to that of K I than to C_2, CN, or CO [34]; (*iv*) an essentially unshielded interstellar region can develope detectable amounts of molecular material [36]. Molecular regions should be typically on a Solar system scale [19,20,37]. Indeed, observational evidendence for such microstructure has been accumulating for the last three decade (cf. [37]). If this is the case is still unclear, and data analysis produces ambiguous results. Cecchi-Pestellini and Dalgarno [38] show that C_2 absorption observations, one of the key diagnostics of density in diffuse interstellar clouds, are insensitive to large density fluctuations of small linear filling factor (less than $\sim 10\%$). On the other side, Liszt and Lucas [2] rule out the possibility, showing that profile variations in HCO^+ absorption lines are not the results of AU-sized inclusion of high hydrogen volume density.

The present model is clearly an idealization and, like most of those which have been proposed to explain the unexpected large concentrations of H_3^+, has its limitations. In particular, the physical properties of the background cold gas are assumed not to vary along the line of sight. Nevertheless, the present proposal allows a quantitative prediction of the influence of turbulent thermal spikes on H_3^+ column density and its relation with important diagnostics of warm gas.

References

1. E.F. van Dishoeck, in *The Molecular Astrophysics of Stars and Galaxies*, T.W. Hartquist and D.A. Williams, Eds. (Clarendon Press, Oxford, 1998), p. 53.

2. H.S. Liszt and R. Lucas, *A&A* **355**, 333 (2000).
3. H.S. Liszt and R. Lucas, *A&A* **370**, 576 (2001).
4. J.H. Black, A. Dalgarno and M. Oppenheimer, *ApJ* **199**, 633 (1975).
5. R. Gredel, G. Pineau des Forêts and S.R. Federman, *A&A* **389**, 993 (2002).
6. J. Zsargó and S.R. Federman, *ApJ* **589**, 319 (2003).
7. H.S. Liszt, J.Pety and R. Lucas, *A&A* **486**, 493 (2008).
8. B.J. McCall, T.R. Geballe, K.H. Hinkle and T. Oka, *Science* **279**, 1910 (1998).
9. T.R. Geballe, B.J. McCall, K.H. Hinkle and T. Oka, *ApJ* **510**, 251 (1999).
10. T.R. Geballe and T. Oka, *Nature* **384**, 334 (1996).
11. V. Kokooline, C.H. Greene and B.D. Esry, *Nature* **412**, 891 (2001).
12. B.J. McCall, A.J. Huneycutt, R.J. Saykally et al., *Nature* **422**, 500 (2003).
13. D.R. Flower and G. Pineau de Forêts, *MNRAS* **297**, 1182 (1998).
14. E. Falgarone, G. Pineau de Forêts and E. Roueff, *A&A* **300**, 870 (1995).
15. K. Joulain, E. Falgarone, G. Pineau des Forêts and D.R. Flower, *A&A* **340**, 241 (1998).
16. B.G. Elmegreen and J. Scalo, *ARA&A* **41**, 211 (2004).
17. E. Falgarone, D.C. Lis, T.G. Phillips, A. Pouquet, D.H. Porter and P.R. Woodward, *ApJ* **436**, 728 (1994).
18. J. Pety and E. Falgarone, *A&A* **412**, 417 (2003).
19. F. Le Petit, E. Roueff and E. Herbst, *A&A* **417**, 993 (2004).
20. C. Cecchi-Pestellini and A. Dalgarno, *MNRAS* **313**, L6 (2000).
21. R. Gredel, J.H. Black and M. Yan, *A&A* **375**, 55 (2001).
22. A. Dalgarno, *PNAS* **103**, 12269 (2006).
23. B.J. McCall, A.J. Honeycutt, R.J. Saykally et al., *Phys. Rev. A,* **70**, 052716 (2004).
24. E. Falgarone, L. Verstraete, G. Pineau de Forêts and P. Hily-Blant, *A&A* **433**, 997 (2005).
25. C. Cecchi-Pestellini, S. Casu and A. Dalgarno, *MNRAS* **364**, 1309 (2005).
26. M.J. Shull et al., *ApJ* **538**, L73 (2000).
27. P. Sonnentrucker, S.D. Friedman, D.E. Welty, D.G. York and T.P. Snow, *ApJ* **576**, 241 (2002).
28. S. Casu, C. Cecchi-Pestellini and S. Aiello S., *MNRAS* **325**, 826 (2001).
29. J. Woodall, M. Agúndez, A.J. Markwick-Kemper and T.J. Millar, *A&A* **466**, 1197 (2006).
30. C.H. Sheehan and J.-P. St.-Maurice, *Adv. Spa. Res.* **33**, 216 (2004).
31. R.C. Bohlin, B.D. Savage and J.F. Drake, *ApJ* **224**, 132 (1978).
32. E.F. van Dishoeck and J.H. Black, *ApJS* **62**, 109 (1986).
33. C. Gry, F. Boulanger, C. Nehmé, G. Pineau des Forêts, E. Habart and E. Falgarone, *A&A* **391**, 675 (2002).
34. B.J. McCall, H.K. Hinkle, T.R. Geballe et al., *ApJ* **567**, 391 (2002).
35. B.T. Draine and F. Bertoldi, *ApJ* **468**, 269 (1996).
36. A. Heithausen, *A&A* **393**, L41 (2002).
37. T.A. Bell, S. Viti, D.A. Williams, I.A. Crawford and R.J. Price, *MNRAS* **357**, 961 (2005).
38. C. Cecchi-Pestellini and A. Dalgarno, *MNRAS* **331**, L31 (2002).

Session IV

Atomic and Molecular Processes in the Solar System and Planetary Atmospheres

Ray Flannery, Cecil Laughlin, Maurice Cohen, Alex Dalgarno, Brendan McLaughlin, Michael Jamieson, William McConkey

ALEXANDER DALGARNO

JANE L. FOX

Department of Physics, Wright State University, Dayton, OH 45435

Alex Dalgarno saved my career.

I entered the Graduate School at Harvard in the Chemistry Department in September of 1973 from the University of Michigan, where I had (unfortunately) majored in Chemistry. Chemistry is considered a "laboratory science" but I had no talent in the laboratory. I had merely fallen in love with the ideal gas law in high school, and I had stubbornly allowed this to determine my course of studies, most of which had nothing to do with the ideal gas law, and much of which was focused on laboratory studies.

Although I changed my degree program at Harvard to Chemical Physics, which was a much better fit, there was very little choice in the Chemistry Department for potential advisors, since one of the two theoreticians had recently started doing experimental work, and the other had the reputation of taking on only the best student in any given class.

In the summer of 1974, I was working as a teaching assistant for a summer school course in introductory chemistry, and floundering for a direction in research. The inmates of Prince House, the name of the Harvard building that housed the graduate students in theoretical chemistry, urged me strongly (and, probably would have physically carried me up Garden St.) to look up a Professor of Astronomy named Alexander Dalgarno, who was not only world-renowned scientifically, but also had a reputation for taking on and graduating many students in a reasonable period of time. Other graduate students from the Chemistry Department before me, and after me, had "fled" to the Observatory to work with Alex, including, in the class before me, Debbie Watson, who later became one of my office mates and a good friend.

I first met with Alex in September, 1974, and he agreed to take me on for a reading course. My first assignment was to carry out a $1/Z$ expansion of two excited electronic states of the lithium isoelectronic sequence.

This involved using a canned program that required one to input large (but carefully selected) numbers of "configurations", which "interacted" somehow to predict energies and oscillator strengths for these states of lithium and lithium-like ions. Debbie Watson showed me how to use this program.[1] The number of "configurations" that could be input was constrained by the small amount of shared memory on the Observatory CDC computer, which was of the order of 128 kB.

Running programs on the CDC computer in those days required punching a card for each line of code and data, carrying large boxes of these cards down the hall, and waiting in line to read them into a very large and very loud "card reader". This increased my upper body strength, engendered patience, and probably damaged my hearing. More often than once, I fed a rubber band into the card reader, taking it off line until the rubber band could be removed. (I said that I was not talented in experimental work!) Although I continued with the "configuration interactions" work, and eventually published a manuscript on it,[2] the paper has been referenced only 17 times in 30 years, partly because it was the subject of a comment,[3] which suggested that we had neglected "radiative autoionization".

Nevertheless, I was willing to try to do whatever Alex suggested. In 1975, I spent the entire summer attempting to prove that an unbound state of helium was bound. When Alex returned from travel at the end of August, I told him that the state was bound for "2.1-ium".

Very shortly thereafter, I found myself on the receiving end of a one-hour introductory lecture on "aeronomy", a subject that I didn't know existed. Alex told me that research in this field required a lot of reading, and synthesizing large amounts of information, which, as it turned out, suited me quite well. Thus I went from theoretical investigations of the properties of lithium and helium to modeling the upper atmosphere of Mars, without quite knowing what had hit me. I do not remember agonizing over this transition. It is remarkable that Alex then had such a large group doing so many different types of research, including, in addition to theoretical atomic and molecular physics, theoretical astrophysics, and aeronomy. I am grateful for his broad range of interests and for his insight that I might be better suited to the field of planetary aeronomy than to theoretical atomic physics.

Since the Viking mission to Mars was due to return information on the upper atmosphere in mid-1976, my first aeronomy project was to compute the expected intensities of some ultraviolet emissions of argon from the Martian thermosphere. Russian researchers had suggested, from the failure of a getter-ion pump on the spacecraft Mars 6 (*e.g.* Ref. 4), that the atmosphere of Mars was composed of a large proportion of argon. This project required (among other things) finding cross sections for electron-impact processes for species such as argon and carbon dioxide. In those days, long before the era of online searches, finding publications that contained these data required spending weeks in the library, looking up and cross-referencing relevant publications using a printed source, which I believe was called the Chemical Abstracts Index, and consisted of several hundred printed volumes. When I once told Alex that I was tired of searching for and preparing data, and wanted to do something new, he replied simply that "sometimes research is boring". The memory of this truism has encouraged me to persevere through the routine aspects of research, for which I am grateful. I was also inspired to assign the most boring aspects of research to graduate students.

I went on to predict the argon emission line intensities using a Mars thermosphere model, and found that there would be no detectable emissions even if the atmosphere were pure argon. I am grateful to Alex, however, for guiding me in this unsuccessful study, because it had prepared me for later, more realistic studies of the thermosphere/ionosphere of Mars. To paraphrase Samuel Beckett, I had tried again, and failed again, but this time I had failed better.

After the mass spectrometer measurements were returned from the Viking landers in 1976 (*e.g.* Ref. 5). I used the Martian thermospheric density profiles to study the ionosphere, emission intensities, and heating rates of the the upper atmosphere of Mars. Later I carried out a similar investigation of the upper atmosphere of Venus. Although I had never taken a course in any aspect of atmospheric sciences or numerical methods, I managed to learn just enough from Alex, George Victor, and the few existing aeronomy books to carry me through this process.

Seven papers based on my 1978 thesis on *The Upper Atmospheres of Mars and Venus* were eventually published. Some of them languished for quite a while, gathering only a few citations, until very recently, with the successful European aeronomy missions, Mars Express and Venus Express. Thus thirty years later some of these publications have finally become highly relevant. For example, Ref. 6, the major paper from my thesis on the aeronomy of Mars, now has garnered 119 citations, many of them very recently, despite the necessarily computationally simplistic nature of the study. I am grateful to Alex for his tremendous powers of foresight, and for suggesting these studies, the importance of which I was only vaguely aware at the time.

That I finally produced a reasonably successful thesis is due in large measure to Alex's patience, his broad interests, and his ability as a mentor. I do not know where I would have ended up, if it weren't for Alex and his interest in aeronomy. (I might perhaps have become a computer programmer, or an ornithologist.) I found that aeronomy was a field that I could finally "get my head around", where physical intuition, rather than mathematical complexity, was required, and where the ideal gas law was finally important. It is remarkable that Alex felt I could succeed in such an endeavor. I am grateful for Alex's interest in and ability to teach me aeronomy "on the fly" as I carried out my thesis research.

Alex's expectations were high and gender-neutral. He managed to communicate his expectations without appearing to prod. He tolerated the times when I appeared to accomplish nothing. For example, he did not intervene when, licking my wounds over my failure to finish my PhD in time for June 1978 graduation, I spent a month in a profound funk doing library research about the history of investigations of Mars. This research became the source of the first chapter of my thesis. I was delighted later to find that Alex actually liked it, and I am grateful to him for having tolerated (and continuing to pay me during) my mini "sabbatical".

Since my graduation from Harvard in November 1978, I have continued to study the aeronomy of Mars, Venus, Earth, Jupiter and Titan, and have used numerical techniques and those of synthesis and analysis similar to those I learned as a graduate student, and as a post-doctoral fellow with Mike McElroy in 1978–1979. I am grateful to Alex and to Mike and to members of their research groups for guiding me in applying these techniques.

I have never, however, been quite able to emulate Alex's legendary terse style of prose. One of the seven papers that became part of my thesis[7] was 1.3 pages long, a feat that I would have said was impossible, had I not experienced it. Alex believes in writing only what is necessary, and not one word more. I recall once writing the phrase "rather similar" in a manuscript. When Alex read it, he gently reminded me that "similar" meant "rather similar". So I crossed off the "rather", thus making the manuscript one word shorter. Herzberg,[8] in his text on diatomic molecules, often uses the phrase "entirely similar", and I shudder to think how Alex might react to that wording. (Perhaps "entirely similar" means "the same", thus creating the potential to cut several letters from many parts of the text.) Although I am grateful to Alex for exposure to this style of publishing, when I try to write succinct papers, I usually incur excess page charges.

In 1981 Alex again saved me again from a disastrous appointment as an assistant professor at the University of Illinois, where a pompous faculty member had tried to force me to align my research program with his interests, and to carry out experimental (!) research on the terrestrial stratosphere; as support for this demand, he once told me "Lady modelers are a dime a dozen". I became intent on proving him wrong.

I fled back to Harvard to Alex's group in July 1981. Over the next three years, I tried to work, but gradually seemed not to accomplish much. Again, Alex didn't give up on me. I later found out that I had been severely ill for many years. The neurologist who diagnosed my illness in early 1985 was surprised that I could work at all in my condition. I am grateful for Alex's continuing support during these three years, and during the last 24+ years, as I have become increasingly debilitated. Indeed, I consider my election as a fellow of the American Geophysical Union a minor miracle that would not have happened without Alex there to make the case.

I was privileged to spend seven years of my academic career working directly with Alex. For this, for all the reasons enumerated above, and finally, I am most grateful to Alex for being one of the ten (or perhaps even five) most important people in my life.

Thank you, Alex.

References

1. D. K. Watson and and S. V. ONeil, A $1/Z$ expansion study of the $1s^2 2s^2\,{}^1S$, $1s^2 2s2p\,{}^1P$, and $1s^2 2p^2\,{}^1S$ states of the beryllium isoelectronic sequence, Phys. Rev. A **12** (1975), p. 729.
2. J. L. Fox and A. Dalgarno, Radiative transition probabilities of the $1s2p^2\,{}^2P$ and $1s2p^2\,{}^2D$ states of the lithium isoelectronic sequence, Phys. Rev. A **16** (1977), p. 283.
3. C. A. Nicolaides and D R. Beck, Comment on lifetime of Li $1s^2 2p\,{}^2P$ state–How probable is radiative autoionization?, Phys. Rev. A **17** (1978), p. 2116.
4. V. I. Moroz, Argon in the Martian atmosphere: Do the results of Mars 6 agree with the optical and radio occultation measurements?, Pis'ma Astron. Zh. **1** (1975), p. 36; translated in Icarus **28** (1976), p. 159.
5. A. O. Nier and M. B. McElroy, Structure of the neutral upper atmosphere of Mars: Results from Viking 1 and Viking 2, Science **194** (1976), p. 1298.
6. J. L. Fox and A. Dalgarno, Ionization, luminosity and heating of the upper atmosphere of Mars, J. Geophys. Res. **84** (1979), p. 7315.
7. J. L. Fox, A. Dalgarno, E. R. Constantinides, and G. A. Victor, The nitrogen dayglow on Mars, J. Geophys. Res. **82** (1977), p. 1615.
8. G. Herzberg, *Molecular Spectra and Molecular Spectra, I. Spectra of Diatomic Molecules*, Second Edition, Van Nostrand Reinhold, New York, 1950.

ATOMIC AND MOLECULAR PROCESSES IN THE SOLAR SYSTEM AND PLANETARY ATMOSPHERES: AN OVERVIEW AND SOME EXAMPLES

T. E. CRAVENS

Dept. of Physics and Astronomy, Malott Hall, 1251 Wescoe Hall Dr.
University of Kansas, Lawrence, KS 66045, USA

This chapter provides a brief summary of the overall role of atomic and molecular (ATM) processes in solar system environments and a few examples of Alex Dalgarno's contributions. The bulk of the chapter is devoted to the role of ion-neutral collisional processes at comet Halley and at Saturn's satellites Enceladus and Titan.

1. Atomic and Molecular Processes in the Solar System and Contributions of Alex Dalgarno

Atomic and molecular (ATM) processes help determine the structure, composition, and dynamics of atmospheres throughout the solar system including the solar atmosphere, and planetary, cometary, and satellite atmospheres and ionospheres. For decades, Alex Dalgarno has been the pivotal figure in the application of ATM processes to both the interstellar medium and to planetary atmospheres. Alex's contributions come not just from his own work and publications, but also from the mentoring and instruction of a remarkably large number of students, post-doctoral associates, and colleagues [1]. In this section, a brief synopsis of the important atomic and molecular processes in the solar system is given.

First, consider the solar corona, which is a powerful source of extreme ultraviolet (EUV) and x-ray radiation. A balance between electron-atom ionizing collisions and dielectronic recombination collisions determines the charge-states of ion species in the million degree coronal gas (cf. [2]). Most of the coronal radiation is in the form of line radiation due to electron-ion collisions and the subsequent radiative cascades.

Ionizing radiation from the solar corona is largely responsible for the formation of planetary ionospheres in the solar system. Neutral atoms and molecules in the atmospheres are ionized and dissociated by this radiation. The associated deposition of energy (via dissociation products or photoelectrons from ionization) contributes to the heat (or energy) balance of the atmospheres and ionospheres (e.g., [3]).

Ionospheric plasma is also created via the precipitation of fast particles (e.g., electrons or protons) from an external plasma environment (e.g., a magnetosphere or the solar wind) into a planetary atmosphere, followed by ionizing or dissociating collisions. Such "auroral" precipitation also results in heating and photon emission (i.e., airglow) via collisional processes. The x-rays produced at Jupiter via the precipitation of energetic oxygen and sulfur ions provide one example of auroral photon emission [4, 5].

Following the creation of ions, electrons, and atoms within the atmosphere via external inputs of photons or fast particles, other atomic and molecular processes take over in controlling the composition and structure. For example, ion-neutral chemistry and recombination help determine electron-ion ionospheric composition and density (i.e., the electron density). Alex has made important contributions to this area, including important papers on ionization, heating, airglow, and chemistry of the ionospheres of Earth, Venus, Mars, and comets [6, 7]. The efficiency with which fast neutral atoms resulting from dissociation processes deposit their energy in an atmosphere requires a detailed consideration of collision processes [8, 9].

Atomic and molecular processes also determine how interstellar neutrals are distributed within the solar system and how the solar wind interacts with these neutrals. The solar wind is the extension of the solar corona out into interplanetary space [10]. As the hot ionized gas flows outward from the sun, thermal energy is converted to bulk kinetic energy and the flow becomes supersonic at a few solar radii. The solar wind runs into the interstellar plasma at the heliopause, located near 150 AU [11]. A shock (the "termination shock") forms upstream of the heliopause and has been observed near 95 AU [12] . The region of space within the heliopause is called the heliosphere. Interstellar plasma cannot penetrate the heliopause; however, due to the long collisional mean free path, interstellar atomic hydrogen and helium flows relatively unimpeded into the solar system, where it interacts with the solar wind. Charge exchange collisions between the incoming neutral H atoms and stagnated plasma located near the heliopause create a modest build-up of the H density (called the "hydrogen wall") [13]. As the H atoms travel into the inner solar system they undergo attrition due to photoionization by solar radiation [14]. The relatively small photoionization cross section for helium explains why the amount of He gas does not become attenuated until very close to the sun [15].

Charge exchange reactions between solar wind protons and interstellar neutrals, or photoionization of the interstellar neutrals, create protons that are initially almost at rest with respect to the solar wind flow. These ions are then accelerated by the Lorentz force associated with the motional electric field in the solar wind (and the interplanetary magnetic field) such that they are "picked-up" and move with the flow. These pick-up ions are a lot hotter than the ambient protons and increase the thermal pressure, thus altering the dynamics as well as the energetics of the flow [15]. The mechanism is discussed in more detail in the next section.

A particularly interesting application of ATM physics to the heliosphere is the production of x-rays from charge exchange of heavy solar wind ions with neutrals (see the chapter in this volume by Kharchenko). Approximately 0.1% of the solar wind is composed of highly-charged ion species (e.g., C^{6+}, O^{7+}, O^{6+}, Fe^{13+},....) that can undergo charge exchange collisions with neutrals. The product ions are left in excited states and thus emit EUV or soft x-ray photons. These ions originate in the million-degree solar corona, and the charge states are "frozen in" as the solar wind flows outward (cf. [16]). This solar wind charge exchange (SWCX) x-ray production mechanism was invoked to explain the x-rays observed from comet Hyakutake [17]; see reviews by Lisse *et al.*, Cravens, and Krasnopolsky *et al.* [18, 19, 20]. As discussed later in this chapter, comets are copious producers of neutral gas, making them suitable candidates for this x-ray source. Solar wind ions also undergo charge exchange collisions with any neutral they encounter in the solar system, including interstellar neutrals. Alex Dalgarno, with students and colleagues, has made a significant contribution to the study of heliospheric x-ray emission and the spectrum of these x-rays [21, 22].

2. Ion-Neutral Collisions in the Inner Coma of Comet Halley

The nucleus of a comet is made of ice and dust and is typically a few kilometers across [23]. When the comet approaches within a few AU of the sun, the ice sublimates and produces an atmosphere (or coma) of water vapor and dust that extends hundreds of thousands of kilometers into space [24]. Other neutral species also present in the gas include carbon monoxide, ammonia, and formaldehyde [25]. Comet Halley was especially active during its 1986 apparition with a gas production rate Q of about 10^{30} s^{-1}. Many *in situ* measurements were made near Halley by the several spacecraft that encountered it [26].

In the outer reaches of the cometary coma, the mean free path for an incident solar wind ion colliding with a neutral is very large. Neutral atoms and molecules are ionized by solar EUV and x-ray radiation and the resulting ions (i.e., H_2O^+, OH^+, O^+, H^+....) respond to the solar wind electric and magnetic fields via the Lorentz force. The spatial trajectories of these ions are cycloids, combining gyration (with gyrofrequency given by $\Omega = qB / m$, where q is the charge, B is the magnetic field strength, and m is the ion mass) and **E** x **B** drift motion. The **E** x **B** drift direction is approximately in the solar wind direction, and the new ions are said to be "picked up" by the flow [27]. The new cometary ions are more fully assimilated into the solar wind flow via wave-particle interactions associated with low frequency plasma waves [28]. Cometary ion pick-up slows down the flow due to the mass addition and due to the increase of the thermal pressure associated with the hot pick-up ions. Magnetic lines of force in interplanetary space are "frozen into" the highly electrically conducting flow (cf. [29, 10]) so that they "pile up" on the sunward side of the cometary coma and drape around the denser inner coma forming a magnetic tail.

In the inner coma of an active comet like Halley (i.e., cometocentric distances, r, less than about 10^4 km), where the neutral density starts to become larger, collisional processes, including chemistry, become important (see reviews by Cravens [30] and Flammer [31]). The total neutral density is described by:

$$n_n(r) = Q / (4\pi\, u_n\, r^2) \qquad (1)$$

where Q is the total cometary gas production rate and u_n is the neutral outflow speed ($u_n \approx 1$ km/s).

Cold ionospheric-type plasma was detected in the inner coma of comet Halley [32] and found to contain H_2O^+, H_3O^+, NH_4^+, and other ion species [33, 32, 24]. For cometocentric distances less than a few thousand km, H_3O^+ is the main ion species (Figure 1). The ion production rate due to photoionization (in the optically thin region of the coma) is given by $P = I\, n_n$, where the ionization frequency ($I \approx 10^{-6}$ s^{-1} at 1 AU) can be found by integrating the product of the photoionization cross section and the solar flux over wavelength. The H_2O^+ ions created this way then react with water to produce H_3O^+ions:

$$H_2O^+ + H_2O \rightarrow H_3O^+ + OH. \qquad (2)$$

The rate of this important ion-neutral reaction is high at low temperatures ($k = 2.08 \times 10^{-9}$ cm^3 s^{-1}) for the reasons discussed by McDaniel *et al.* [6] (also see the discussions by Cravens and Korosmezey [34] and by Lishawa *et al.* [35]).

All the major ions species (H_3O^+, H_2O^+, NH_4+...), once produced, undergo dissociative recombination (DR). The DR rate coefficient for H_3O^+ is $\alpha = 3.39 \times 10^{-4}/T_e^{0.9}$ [36] and this coefficient depends on the electron temperature, which varies with cometocentric distance. ATM processes (not discussed) are also essential for determining this temperature [37, 38, 39]. The total ion loss rate is given by $L = \alpha n_e^2$ (assuming that $n_i = n_e$). In most of the inner coma, chemical processes are faster than transport processes, and photochemical equilibrium (i.e., $I n_n = \alpha n_e^2$) can be used to obtain the following approximate expression for the density:

$$n_e \approx [I Q / (4\pi\alpha u_n)]^{1/2} (1 / r). \tag{3}$$

Figure 1 shows the density profiles for the most important ion species in the inner coma of comet Halley as measured in 1986 by the mass spectrometer onboard the Giotto spacecraft. Within a cometocentric distance of 10^4 km, equation (3) does a rather good job of predicting the ionospheric density for comet Halley.

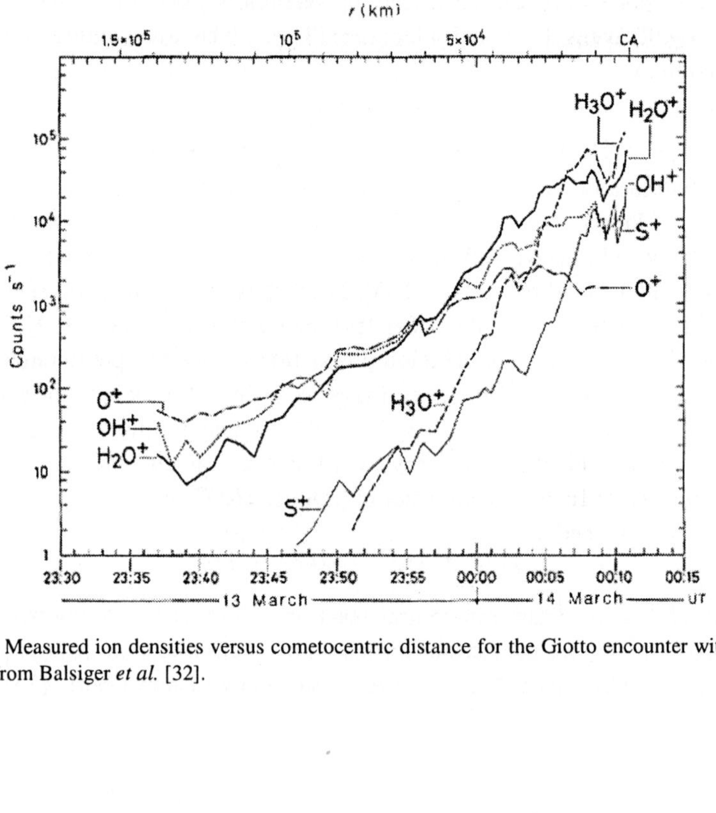

Figure 1. Measured ion densities versus cometocentric distance for the Giotto encounter with comet Halley. From Balsiger *et al.* [32].

Ion-neutral collisions also play an important role in the dynamics and energetics of cometary ionospheric plasma, particularly near the diamagnetic cavity. The Giotto magnetometer observed this field-free cavity for distances of about 5000 km from the nucleus of comet Halley [40]. Figure 2 shows the magnetic field strength measured in the inner coma. The dominant process determining the location and structure of this cavity are ion-neutral collisions [41, 42, 43]. Consider the plasma momentum equation from a single-fluid magnetohydrodyamics (MHD) description [10]:

$$\rho\left(\frac{\partial \mathbf{u}}{\partial t}+\mathbf{u}\cdot\nabla\mathbf{u}\right)=-\nabla(p_e+p_i)-\rho\nu_{in}(\mathbf{u}-\mathbf{u}_n)+\mathbf{J}\times\mathbf{B}+\rho\mathbf{g} \tag{4}$$

where $\rho = n_e m_i$ is the total plasma mass density, m_i is the average ion mass ($m \approx 19$), **u** is the bulk flow velocity, p_e and p_i are electron and ion thermal pressures, respectively, and **g** is the gravitational acceleration ($\mathbf{g} \approx 0$ for comets). The ion-neutral momentum transfer collision frequency is $\nu_{in} \approx k_{in}\, n_n$ (see [6, 34]). Equation (2) is Newton's second law for an ionized fluid parcel immersed in a neutral gas. The parcel accelerates in response to the net force that includes the pressure gradient force, the Lorentz force per unit volume (given in terms of the current density, **J**, and the magnetic field, **B**), and the ion-neutral friction term. In the inner coma, the friction and the **J** x **B** terms dominate. The **J** x **B** term can be rewritten as $-\nabla(B^2/2\mu_0)+(1/\mu_0)\mathbf{B}\cdot\nabla\mathbf{B}$, using Ampere's law, $\mu_0\mathbf{J}=\nabla\times\mathbf{B}$.

The ion and neutral gases flow together in the field-free diamagnetic cavity, but in the magnetized region outside the cavity ions gyrate and are "stuck in" the field, such that the **J** x **B** force applies. Just outside the cavity boundary, the ions are stagnated ($\mathbf{u} \approx 0$) and neutrals flow past them, exerting an outward collisional force. This outward force on the plasma is balanced by an inward **J** x **B** force (i.e., magnetic pressure gradient force). Setting the **J** x **B** force equal to the frictional force in eq. (4), and adopting the simple expressions given earlier for the neutral and electron densities, simple expressions [41, 42] were obtained for the magnetic field strength (shown in Figure 2). The plasma density structure is enhanced in a narrow layer right near the cavity boundary, and collisional processes are needed to understand this boundary layer structure (the so-called inner shock) [44].

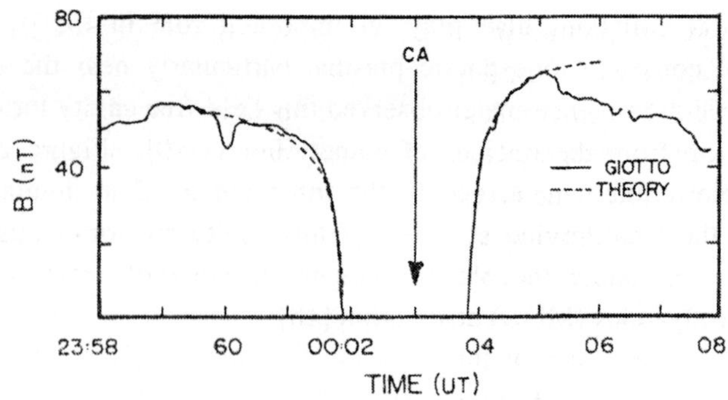

Figure 2. Magnetic field strength versus distance from comet Halley from the Giotto magnetometer experiment [40]. Also shown is the result of a simple model of the plasma dynamics [42]. The diameter of the cavity is close to 9000 km. CA denotes the time of closest approach of the spacecraft to the nucleus.

Ion-neutral collisions are needed to understand the plasma energetics near the cavity boundary. Collisions frictionally heat the ion gas, which subsequently cools via ion-neutral collisions (cf. [45, 34]). In addition, electron-ion and electron-neutral collisions play an important role in the thermal balance of the cometary electron gas [39, 38].

3. Ion-Neutral Collisions and the Water Plume of Enceladus

Enceladus is a small (500 km diameter) icy satellite orbiting Saturn at a distance of 3.95 Saturn radii [46]. Recent observations made by instruments onboard the Cassini spacecraft have revealed that crevices in the icy surface in the southern hemisphere produce a plume of water vapor and ice grains. On 14 July 2005 the spacecraft passed within 168 km of the surface (the E2 flyby) and on 13 March 2008 passed within 50 km (the E3 flyby). The Ion and Neutral Mass Spectrometer (INMS) in its neutral mode observed a plume atmosphere containing mostly water but also some carbon dioxide, carbon monoxide, and nitrogen [47]. Ion species were also detected in the plume, and ion-neutral collisions are important in determining the composition and dynamics of the plume ionosphere [48].

The inner magnetosphere of Saturn contains both neutral H_2O (cf. [49, 50]) and water group (WG) ions (H_2O^+, OH^+, O^+, H_3O^+....) [51, 52]. Cassini measurements made in the inner magnetosphere and near Enceladus support the idea that this satellite is the main source of material for the inner magnetosphere (and the E-ring) [53, 54, 50]. It has been suggested that charge exchange and ion chemistry are important in the gas coma of Enceladus and also in the inner magnetosphere as a whole [55].

During the E3 flyby the Cassini Plasma Spectrometer (CAPS) detected WG ions near Enceladus with a total density of about 80 cm^{-3} and the density did not vary much during the encounter [51]. The plasma observed far outside the plume appears to be moving at the co-rotational speed of 26 km/s relative to Enceladus, but the flow velocity near the plume was reduced to $u \approx 17$ km/s [51]. Just as at comets, ionization of neutral water produces ions that are "picked up" by the flow (in this case the magnetospheric plasma flow). As before, the charged particle motion combines gyration and **E** x **B** drift. Figure 3 is a schematic of a charged particle trajectory illustrating how the external flow encounters the neutral environment of the Enceladus plume.

During the E3 flyby the INMS in its open source ion (osi) mode detected H_3O^+ ions [48] but did not observe other WG ion species. The INMS also apparently detected the hydrated ion species, H_2O^+ - H_2O and H_3O^+ - H_2O. It should be noted that the INMS can only detect ions with velocities with speeds within a couple km/s of the neutral rest frame (i.e., only ions located near the "cusps" of the cycloidal trajectory shown in Fig. 3). The question posed by Cravens *et al.* [48] is what happens to all the other WG ion species that are convected into the plume from the outside? The short answer is that ion-neutral reactions convert these ions into H_3O^+ ions and, just as at comets, reaction (2) is very important.

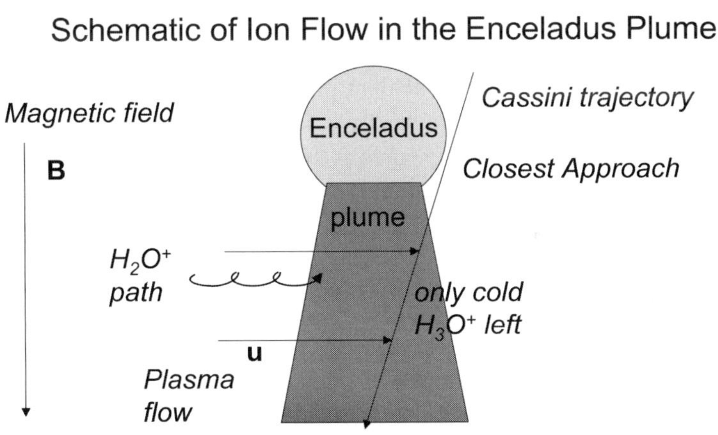

Figure 3. Schematic of the water plume that emanates from the southern hemisphere of Enceladus. External flow velocity vectors and magnetic field vectors are shown, as is an approximate Cassini trajectory for the E3 encounter. A cycloidal pick-up ion trajectory is also shown, roughly to scale.

It is instructive to consider some relevant timescales [48]. The diameter of the plume is about 2 R_{enc} and the transport time is $\tau \approx 2\ R_{enc} / u \approx 30$ s for a bulk flow speed of $u \approx 17$ km/s, where R_{enc} is the radius of Enceladus. This speed is probably even smaller within the plume itself, and the transport time is even higher. The peak gas density in the plume, where Cassini crossed it during the E3 encounter, was measured to be $n_n \approx 10^8$ cm^{-3} (INMS paper in preparation). For a typical ion-neutral reaction, the Langevin collision coefficient is $k_{in} \approx 10^{-9}$ cm^{3}s^{-1} [6], and the associated collision time is $\tau_c \approx 1 / k_{in}\ n_n \approx 10$ s, which is significantly less than the transport time. Hence several collisions can take place as an ion drifts through the plume. The gyroperiod for a water ion in the ambient magnetic field ($B \approx 300$ nT [56]) is about 4 s, which is less than the collision time. Thus, even with collisions, ions should still follow recognizable cycloidal trajectories (Fig. 3). However, near the base of the plume where the neutral density is about ten times higher [57], the collision frequency is much higher resulting in an ion population that is colder and slower than at the spacecraft location.

Another question is whether or not ion production within the plume contributes to the plasma density. The total electron density did not change much during the E2 encounter ([51]; n_e stayed at about 60 to 80 cm^{-3}, but Cassini did not penetrate very deeply into the plume [47]. The photoionization frequency associated with ionizing solar photons is $I_{sun} \approx 5 \times 10^{-9}$ s^{-1} (cf. [58]). The electron impact ionization frequency is about $I_e \approx 1.5 \times 10^{-9}$ s^{-1} due to the hot electron population (0.2 cm^{-3} density and 12 eV thermal energy) seen near Enceladus by CAPS during E2. The density generated by ionization of neutrals as a fluid parcel moves across the plume is: $\Delta n_e \approx (I_{sun} + I_e)\ n_n\ \tau_T \approx 20$ cm^{-3}, which is only about 20% of the incident plasma density. Even if we assumed that the convection speed was lower by a factor of 2 (increasing the passage time), then the dominant plasma source is still the external flow. We can also estimate that electron-ion recombination is not important for this scenario.

Water group ions encounter the plume and undergo ion-neutral collisions (including chemical reactions). Just as at comets, reaction (2) converts the water ions to H_3O^+ [48]. Two other reactions convert O^+ and OH^+ into H_2O^+ or into H_3O^+ [59]:

$$O^+ + H_2O \rightarrow H_2O^+ + O \tag{5}$$

$$\begin{aligned} OH^+ + H_2O &\rightarrow H_2O^+ + OH \\ &\rightarrow H_3O^+ + O \end{aligned} \tag{6}$$

These reactions also take place not just in the plume but throughout the whole inner magnetosphere of Saturn - at much lower rates but over a much larger volume [55].

A possible difficulty exists for the above ion-neutral chemical scheme - the rate coefficients were mostly measured at room temperature (cf. [59]), but the ions entering the plume have energies that range from almost zero up to ≈ 200 eV, depending on where on their cycloidal trajectories they are. Another issue is that the INMS in its ion mode during E3 only measured ions near the neutral rest frame. If the ions, during their collisions with neutrals, are scattered into trajectories that are too different from the unperturbed cycloidal trajectories, then the resulting phase space density would be too low at the v=0 point for the INMS to be able to detect these ions as it did [48]). In other words, the details of the ion-neutral collision process do matter for the INMS measurements.

The book by McDaniel *et al.* [6] provides some insights into the underlying physics of ion-neutral reactions. Chemical reactions become probable for impact parameters less than the "orbiting impact parameter" $b_0 \approx (2\alpha e^2/E)^{1/4}$, where E is the center of mass collision energy and α is the polarizability of the neutral target. For an induced dipole interaction potential and for collision energies less than a few eV the cross section varies inversely as the collision speed such that the resulting "Langevin" rate coefficient ($k \approx \pi (\alpha e^2/\mu)^{1/2}$) is independent of temperature, where μ is the reduced mass. For very low energy ion collisions with water, the H_2O permanent dipole moment enhances the cross sections and the associated rate coefficients.

Laboratory time-of-flight measurements that included deuterated substitutions (e.g., using HDO^+ beams) for water ion collisions with neutral water were made by Lishawa *et al.* [35]. These measurements showed that for collision energies greater than a few eV, the charge exchange channel, rather than the reactive channel (reaction (2)), dominates:

$$H_2O_f^+ + H_2O \rightarrow H_2O_f + H_2O^+, \qquad (7)$$

where the subscript "f" denotes a fast ion or neutral. The lab measurements also indicated that the product ion is left almost at rest. For the Enceladus scenario this means that the ion can start on a fresh cycloidal path. Johnson *et al.* [60] noted that this type of reaction is important in the inner magnetosphere of Saturn for redistributing neutral molecules. The cross section for this process was shown to be $\sigma \approx 10^{-15}$ cm^2 in the energy range 2 eV -- 30 eV [35], and the cross section sharply increases at lower energies, evidently due to the ion interaction with the permanent dipole moment of the water molecule [6, 34].

Returning to the reactive channel, (2), Lishawa *et al.* demonstrated that spectator stripping seemed to be the relevant process (also see discussion by McDaniel *et al.* [6]) such that the product ion is left almost at rest. The proton is transferred to the target water molecule. The cross section for this process exceeds that of the symmetric charge exchange channel only for energies below about 2 eV (see Fig. 4). The implication for Enceladus is that conversion of water group ion species into just H_3O^+ requires that collision energies be less than about 2 eV. Even for a hot pick-up ion distribution this can happen, albeit inefficiently, in the part of the trajectory where the ions are slow, or this can happen very deep in the plume where all ions are slow and cold. Ion-neutral collisions both cool and slow the plasma and thus advance this chemical process.

Ion-neutral collisions also have dynamical consequences for the Enceladus problem. Reaction (6) replaces a fast ion with a slow (or at rest) ion and thus acts to slow down the plasma as a whole. The plasma momentum equation (4) includes this as part of the friction term, as was already discussed for comets.

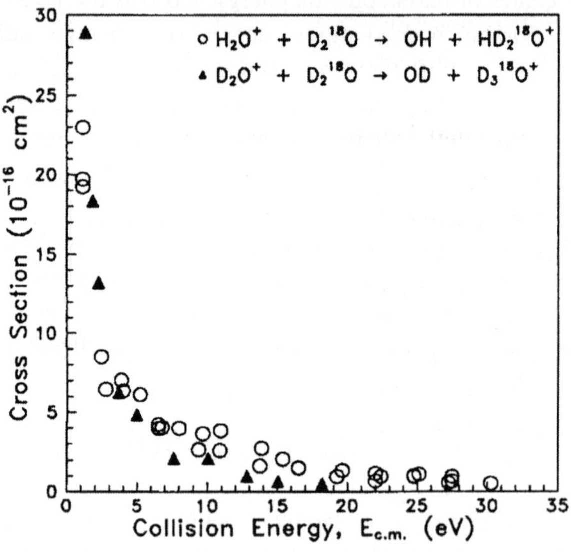

Figure 4. Experimental cross sections for deuterated versions of the reactive channel for water ion collisions with neutral water as functions of energy. From Lishawa *et al.* [35].

For Enceladus, the dense plume ionosphere is linked by magnetic field lines to the surrounding fast flowing magnetospheric plasma (flow speed of $u_0 \approx$ 17 km/s) over a length scale of tens of R_{enc}. Field-aligned currents play an important part in this linkage [61; cf. 10]. The plasma flow in the ionosphere is maintained by **J** x **B** forces and is mainly opposed by ion-neutral collisions. It can be shown that the flow speed in the ionospheric region is given by:

$$u_{pl} \approx u_0 \{\Omega^2 / (\Omega^2 + \nu^2)\} \quad (8)$$

Equation (8) is an approximation to the momentum equation (4) showing that the local flow speed is $u_{pl} \approx u_0$ for the part of the plume with neutral density of $\approx 10^8$ cm^{-3} (e.g., the E3 encounter), because $\Omega/\nu \approx 10$. This neglects the possibility that the plasma incident on the dense plume is forced to flow mostly around the plume. Note that a very extensive region beyond the plume is dynamically attached to Saturn's ionosphere [61]. For the part of the plume very close to the satellite, where the neutral density is about 10^9 cm^{-3} [57], then $\Omega/\nu \approx 1$ and the flow is severely slowed down and highly collisionally-dominated. A coherent ring distribution of ions is unlikely to exist in this very dense plume region.

To recap this section, ion-neutral collisions (and atomic and molecular physics in general) play a key role in controlling the plasma composition and dynamics in the plume ionosphere of Enceladus.

4. Atomic and Molecular Processes in the Ionosphere of Titan

Atomic and molecular processes are also essential for understanding the atmosphere and ionosphere of Saturn's satellite Titan. The Cassini Orbiter has made many encounters over the last four years with this largest satellite of Saturn, traversing the atmosphere at altitudes above approximately 960 km. The atmospheric pressure at the surface is 1.5 bar and the main atmospheric constituents are molecular nitrogen and methane [62, 63]. In common with other atmospheres in the solar system, energy is deposited due to the absorption of solar radiation and due to external particle precipitation, resulting in ionization, heating, and airglow in the upper atmosphere. Saturn's magnetosphere supplies energetic ions and electrons to the atmosphere along the magnetic field lines that intersect Titan's atmosphere. Due to the presence of methane in the atmosphere, a complex chemistry is initiated by this energy deposition and leads to the formation of complex neutral and ionized hydrocarbon and nitrogen-bearing species. This chemistry contributes to the formation of the aerosols populating Titan's haze layer [64]. The ion-neutral chemistry taking place at Titan resembles the chemistry in interstellar clouds [65, 66].

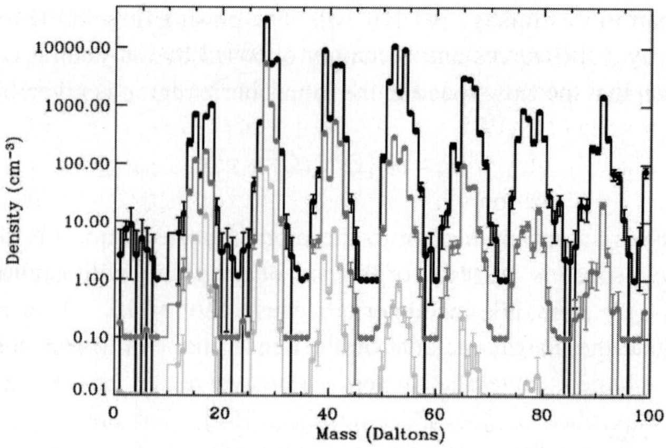

Figure 5. Mass spectrum for ions in the nightside upper atmosphere of Titan, as measured by the INMS in its open source ion mode for 3 altitude ranges. Black: 1027 km – 1200 km, multiplied by a factor of 100. Dark gray: 1200 km – 1400 km, mulitplied by a factor of 10. Light gray: 1400 km – 1600 km, multiplied by a factor of 1. 1σ statistical error bars are shown. From Cravens *et al.* [70].

Densities of both the neutrals and ions were measured by the Cassini INMS. An ion mass spectrum measured in the nightside ionosphere is shown in Figure 5. The spectrum shows a mass spacing of 12 -14 dalton and demonstrates the existence of heavy ion species with multiple carbon and/or nitrogen atoms [67, 68, 69, 70, 71, 72, 73]. The lowest altitude (1050 km) spectrum shows the highest abundances of high-mass species. N_2^+ and CH_4^+ are the ion species generated by the primary ionization, but these species are rapidly converted to higher mass species via ion-neutral chemistry. Recent models include hundreds of reactions, but some of the key reactions include:

$$\begin{array}{l} CH_4^+ + CH_4 \rightarrow CH_5^+ + CH_3 \\ CH_5^+ + C_2H_2 \rightarrow C_2H_5^+ + CH_2 \\ N_2^+ + CH_4 \rightarrow CH_3^+ + N_2 + H \\ CH_3^+ + CH_4 \rightarrow C_2H_5^+ + H_2 \\ C_2H_5^+ + HCN \rightarrow HCNH^+ + C_2H_4 \\ C_2H_5^+ + C_2H_4 \rightarrow C_3H_5^+ + CH_4 \\ HCNH^+ + CH_2NH \rightarrow CH_2NH_2^+ + HCN \end{array} \quad (9)$$

As shown in Figure 5, the species at mass numbers 17, 28, 29, and 30 are particularly abundant (i.e., CH_5^+, $HCNH^+$, $C_2H_5^+$, and $CH_2NH_2^+$, respectively). Aromatic hydrocarbon ions (e.g., $C_6H_7^+$ - protonated benzene at mass number

79) are also present. Waite *et al.* [64] discussed the chemical pathways leading to heavier polycyclic aromatic hydrocarbon (PAH) species.

Negative ion species were also detected in the ionosphere by the Cassini plasma spectrometer (CAPS), some with extremely high mass numbers [74], and these species surely play a role in the formation of heavy PAH species [64]. One issue with negative ion species is their initial formation (cf. [74, 73]), which involves the initial energy deposition processes. PAH species have also been detected in interstellar clouds (e.g., [75]). Alex Dalgarno has pointed out the role of PAHs in interstellar ion chemistry [76, 66].

The energy deposition in Titan's atmosphere by solar radiation, by precipitating energetic electrons (electrons with energies ranging from several eV up to several keV) and by precipitating ions (protons and oxygen ions with energies from a few keV up to MeV) was calculated by Cravens *et al.* [73]. Solar radiation inputs are most important on the dayside, but obviously magnetospheric energetic particle inputs are needed on the nightside. Solar radiation produces an ionosphere with peak densities located near 1000 km [67, 69, 77, 78], but energetic ions deposit energy down to altitudes as low as 500 km. Substantial electron densities were sporadically observed at such low altitudes by the Cassini radio science investigation with the radio occultation technique [79]. Charge exchange and electron loss collisions determine the charge state distribution of the precipitating ions, and for the incident ion energies most relevant to Titan a mixture of O^-, O, O^+, and O^{++} ions is present in the precipitating beam. The O^- ions are particularly interesting because the electron can then attach to other atmospheric species with higher electron affinities (such as HCN ...). Perhaps this can help to explain the presence of negative ions in the ionosphere, as has been observed [74]. Both negative and positive ion species participate in the ionospheric chemistry, but the positive ion chemistry is better understood at this time.

5. Summary

This chapter reviewed a few examples of the application of atomic and molecular processes to our understanding of the upper atmospheres and ionospheres of solar system bodies. Ion-neutral collisional processes at comets, Enceladus, and Titan were emphasized. The advancement of our understanding of the basic atomic and molecular processes and their application to both the solar system and to the interstellar medium has been spearheaded over several decades by Alex Dalgarno, together with his many students and associates [1].

In particular, my own work (some examples of which were presented in this chapter) traces its intellectual origin to those fortunate years when I was one of Alex Dalgarno's Ph.D. students.

Acknowledgments

NASA Planetary Atmospheres grant 06PATM06-0019 and NASA Cassini grant (subcontract through SWRI) 699081KC are acknowledged for their support.

References

1. A. Dalgarno, *Ann. Rev. Astron. Astrophys.* **46**, 1 (2008)
2. H. Zirin, *Astrophysics of the Sun*, Cambridge University Press, Cambridge, U. K. (1988).
3. J. L. Fox and A. Dalgarno, *J. Geophys. Res.* **86**, 629 (1981).
4. A. Dalgarno, *Astrophys. Space Sci.* **285**, 687 (2003).
5. V. Kharchenko, A. Dalgarno, D. Schultz, and P. Stancil, *Geophys. Res. Lett.* **33**, L11105 (2006).
6. E. W. McDaniel, V. Cermak, A. Dalgarno, E. E. Ferguson, and L. Friedman, *Ion-Molecule Reactions*, Wiley-Interscience, New York (1970).
7. V. Kharchenko and A. Dalgarno, *Astrophys. J.* **554**, L99 (2001).
8. V. Kharchenko, A. Dalgarno, and J. L. Fox, *J. Geophys. Res.* **110**, A12306 (2005).
9. V. Kharchenko and A. Dalgarno, *J. Geophys. Res.* **109**, D18311 (2004).
10. T. E. Cravens, *Physics of Solar System Plasmas*, Cambridge Univ. Press, Cambridge, 1997.
11. S. T. Suess, *Rev. Geophys.* **28**, 97 (1990).
12. L. F. Burlaga, N. F. Ness, M. H. Acuna, R. P. Lepping, J. E. P. Connerney, E. C. Stone, and F. B. McDonald, *Science* **309**, 2027 (2005).
13. G. P. Zank, H. L. Pauls, L. L. Williams, and D. Hall, *J. Geophys. Res.* **101**, 21639 (1996).
14. E. Quémerais, R. Lallement, and J.-L. Bertaux, *J. Geophys. Res.* **98**, 15199 (1993).
15. E. Möbius, D. Rucinkski, D. Hovestadt, and B. Kleckler, *Astron. Astrophys.* **304**, 505 (1995).
16. A. J. Hundhausen, H. E. Gilbert, and S. J. Bame, *J. Geophys. Res.* **73**, 5485 (1968).

17. C. M. Lisse *et al.*, *Science* **274**, 205 (1996).
18. C. M. Lisse, D. J. Christian, K. Dennerl, S. J. Wolk, D. Bodewits, R. Hoekstra, M. R. Combi, T. Mäkinen, M. Dryer, C. D. Fry, and H. Weaver, *Astrophys. J.* **635**, 1329 (2005).
19. T. E. Cravens, *Science* **296**, 1042 (2002).
20. V. A. Krasnopolsky, J. B. Greenwood, and P. C. Stancil, *Space Sci. Rev.* **113**, 271 (2005).
21. D. Koutroumpa, R. Lallement, V. Kharchenko, A. Dalgarno, R. Pepino, V. Izmodenov, and E. Quemerais, *Astron. Astrophys.* **460**, 289 (2006).
22. R. Pepino, V. Kharchenko, and A. Dalgarno, .*Astrophys. J.* **617**, 1347 (2004).
23. D. Krankowsky *et al.*, *Nature* **321**, 326 (1986).
24. W. F. Huebner, D. C. Boice, H.U. Schmidt, and R. Wegmann, "Structure of the Coma: Chemistry and Solar Wind Interaction," in *Comets in the Post-Halley Era*, vol. 2, edited by R. L. Newburn Jr. *et al.*, p. 907, Kluwer Acad., Norwell, MA (1991).
25. R. Schwenn, W.-H. Ip, H. Rosenbauer, H. Balsiger, F. Buhler, R. Goldstein, A. Meier, and E. G. Shelley, "Ion Temperature and Flow Profiles in Comet Halley's Close Environment," in *Proc. 20th ESLAB Symposium on the Exploration of Halley's Comet, Eur. Space Agency Spec. Publ.*, **SP-250**, p. 225 (1986).
26. K. Altwegg *et al.*, *Astron. Astrophys.* **279**, 260 (1993).
27. A. A. Galeev, "Plasma Processes in the Outer Coma," in *Comets in the Post-Halley Era*, vol. 2, edited by R. L. Newburn Jr. *et al.*, p. 1145, Kluwer Acad., Norwell, MA (1991).
28. A. A. Galeev, T. E. Cravens, and T. I. Gombosi, *Astrophys. J.* **289**, 807 (1985).
29. G. L. Siscoe, "Solar System Magnetohydrodynamics," in *Solar-Terrestrial Physics: Principles and Theoretical Foundation*, edited by R. L. Carovillano and J. M. Forbes, p. 11, D. Reidel, Dordrecht (1983).
30. T. E. Cravens, "Plasma Processes in the Inner Coma," in *Comets in the Post-Halley Era*, vol. 2, edited by R. L. Newburn Jr. *et al.*, p. 1211, Kluwer Acad., Norwell, MA, (1991).
31. K. R. Flammer, "The Global Interaction of Comets with the Solar Wind," in *Comets in the Post-Halley Era*, vol. 2, edited by R. L. Newburn Jr. *et al.*, p. 1125, Kluwer Acad., Norwell, MA (1991).
32. H. Balsiger *et al.*, *Nature* **321**, 330 (1986).
33. M. Allen *et al.*, *Astron. Astrophys.* **187**, 502 (1987).

34. T. E. Cravens and A. Korosmezey, *Planet. Space Sci.* **34**, 961 (1986).
35. C. R. Lishawa, R. A. Dressler, J. A. Gardner, R. H. Salter, and E. Murad, *J. Chem. Phys.* **93**, 3196 (1990).
36. R. A. Heppner, F. L. Wallis, W. T. Armstrong, and G. H. Dunn, *Phys. Rev. A* **13**, 1000 (1976).
37. P. Eberhardt and D. Krankowsky, *Astron. Astrophys.* **295**, 795 (1995).
38. C. J. Lindgren, T. E. Cravens, and S. Ledvina, *J. Geophys. Res.* **102**, 17395 (1997).
39. L. Gan and T. E. Cravens, *J. Geophys. Res.* **95**, 6285 (1990).
40. F. M. Neubauer, Giotto magnetic field results in the magnetic pile-up region and the cavity boundaries, in *Proc. 20th ESLAB Symposium on the Exploration of Halley's Comet*, edited by B. Battrick, E. J. Rolfe, and R. Reinhard, *Eur. Space Agency Spec. Publ.*, **SP-250**, p. 35 (1986).
41. W.-H. Ip and W. I. Axford, *Nature* **325**, 418 (1987).
42. T. E. Cravens, "The Physics of the Cometary Contact Surface," in *Proc. 20th ESLAB Symposium on the Exploration of Halley's Comet, Eur. Space Agency Spec. Publ.,* **SP-250**, p. 244 (1986).
43. T. E. Cravens, *J. Geophys. Res.* **94**, 15025 (1989).
44. P. Puhl-Quinn and T. E. Cravens, *J. Geophys. Res.* **100**, 21631 (1995).
45. T. E. Cravens, *Geophys. Res. Lett.* **14**, 983 (1987).
46. D. Morrison *et al.*, "Satellites of Saturn: Geological Perspective," in *Saturn*, edited by T. Gehrels and M. S. Matthews, p. 609, Univ. of Arizona Press, Tucson (1984).
47. J. H. Waite, Jr., M. R. Combi, W.-H. Ip, T. E. Cravens, R. L. McNutt Jr., W. Kasprzak, R. Yelle, J. G. Luhmann, H. Niemann, D. Gell, B. Magee, G. Fletcher, J. Lunine, and W.-L. Tseng, *Science* **311**, 1419 (2006).
48. T. E. Cravens, R. L. McNutt Jr., J. H. Waite Jr., I. P. Robertson, J. G. Luhmann, W. Kasprzak, and W.-H. Ip, *Geophys. Res. Lett.* **36**, L08106 (2009).
49. W. H. Ip, *Planet. Space Sci.* **48**, 775 (2000).
50. S. Jurac and J. D. Richardson, *J. Geophys. Res.* **110**, A09220 (2005).
51. R. L. Tokar, R. E. Johnson, T. W. Hill, D. H. Pontius, W. S. Kurth, F. J. Crary, D. T. Young, M. F. Thomsen, D. B. Reisenfeld, A. J. Coates, G. R. Lewis, E. C. Sittler, and D. A. Gurnett, *Science* **311**, 1409 (2006).
52. D. T. Young *et al.*, *Science* **307**, 1262 (2005).
53. M. G. Kivelson, *Science* **311**, 1391 (2006).
54. R. E. Johnson, H. T. Smith, O. J. Tucker, M. Liu, M. H. Burger, E. C. Sittler, and R. L. Tokar, *Astrophys. J.* **644**, L137 (2006).

55. W. L. Tseng and W.-H. Ip, *Geophys. Res. Abstracts* **9**, 1793 (2007).
56. M. K. Dougherty, K. K. Khurana, F. M. Neubauer, C. T. Russell, J. Saur, J. S. Leisner, and M. E. Burton, *Science* **311**, 1406 (2006).
57. C. J. Hansen, L. Esposito, A. I. F. Stewart, J. Colwell, A. Hendrix, W. Pryor, D. Shemansky, and R. West, *Science* **311**, 1422 (2006).
58. T. E. Cravens, J. U. Kozyra, A. F. Nagy, T. I. Gombosi, and M. Kurtz, *J. Geophys. Res.* **92**, 7341 (1987).
59. V. G. Anicich and M. J. McEwan, *Planet. Space Sci.*, **45**, 897 (1997).
60. R. E. Johnson, M. Liu, and E. C. Sittler, Jr., *Geophys. Res. Lett.* **32**, L24201 (2005).
61. D. H. Pontius, Jr. and T. W. Hill, *J. Geophys. Res.* **111**, A09214 (2006).
62. P. M. Hunten, M. G. Tomasko, F. M. Flasar, R. E. Samuelson, D. F. Strobel, and D. J. Stevenson, "Titan," in *Saturn*, eds. J. Gehrels and M. S. Matthews, 671, Univ. of Arizona Press, Tucson (1984).
63. J. H. Waite Jr., T. E. Cravens, W.-H. Ip, W. T. Kasprzak, J. G. Luhmann, R. L. McNutt, H. B. Niemann, R. B. Yelle, I. Meuller-Wodarg, S. A. Ledvina, and S. Scherer, *Science* **307**, 1260 (2005).
64. J. H. Waite, Jr., D. T. Young, T. E. Cravens, A. J. Coates, F. J. Crary, B. Magee and J. Westlake, *Science* **316**, 870 (2007).
65. A. Dalgarno, *PNAS* **103**, 12269 (2006).
66. A. Dalgarno, *Faraday Discuss.* **139**, 9 (2006).
67. C. N. Keller, T. E. Cravens, and L. Gan, *J. Geophys. Res.* **97**, 12117 (1992).
68. C. N. Keller, V. G. Anicich, and T. E. Cravens, *Planet. Space Sci.* **46**, 1157 (1998).
69. J. L. Fox and R. V. Yelle, *Geophys. Res. Lett.* **24**, 2179 (1997).
70. T. E. Cravens, I. P. Robertson, J. H. Waite Jr., R. V. Yelle, W. T. Kasprzak, C. N. Keller, S. A. Ledvina, H. B. Niemann, J. G. Luhmann, R. L. McNutt, W.-H. Ip, V. De La Haye, I. Mueller-Wodarg, J.-E. Wahlund, V. A. Anicich, and V. Vuitton, *Geophys. Res. Lett.* **33**, L07105 (2006).
71. V. Vuitton, R. V. Yelle, and V.G. Anicich, *Astrophys. J.* **647**, L175 (2006).
72. V. Vuitton, R. V. Yelle, and M. J. McEwan, *Icarus* **191**, 722 (2007).
73. T. E. Cravens *et al.*, *Geophys. Res. Lett.* **35** (2008).
74. A. J. Coates, F. J. Crary, G. R. Lewis, D. T. Young, and J. H. Waite, Jr., *Geophys. Res. Lett.* **34**, L22103-L22107 (2007).
75. A. G. G. M. Tielens, *Ann. Rev. Astron. Astro.* **46**, 289 (2008).
76. A. Dalgarno, *J. Tech. Phys.* **41**, 23 (2000).

77. T. E. Cravens, J. Vann, J. Clark, J. Yu, C. N. Keller, and C. Brull, *Adv. Space Res.* **33**, 212, doi:10.1016/J.asr.2003.02.012 (2004).
78. I. P. Robertson, T. E. Cravens, J. H. Waite, Jr., R. V. Yelle, V. Vuitton, A. J. Coates, J. E. Wahlund, K. Agren, K. Mandt, B. Magee, and M. S. Richard,' *Planet. Space Sci.*, revision submitted (2009).
79. A. J. Kliore *et al.*, *J. Geophys. Res.* **113** (2008).

APPROACH TO THERMAL EQUILIBRIUM IN ATOMIC COLLISIONS

PENG ZHANG[1] and VASILI KHARCHENKO[1,2]

[1]*Institute for Theoretical Atomic, Molecular and Optical Physics (ITAMP), Harvard-Smithsonian Center for Astrophysics, Cambridge, Massachusetts 02138, USA*

[2]*Physics Department, University of Connecticut, Storrs, Connecticut 06269, USA*

The energy relaxation of fast atoms moving in a thermal bath gas is explored theoretically. We found that two time scales characterize the equilibration, one a short time, in which the isotropic energy distribution profile relaxes to a Maxwellian shape at some intermediate effective temperature, and the second, a longer time in which the relaxation preserves a Maxwellian distribution and its effective temperature decreases continuously to the bath gas temperature. It is shown that the formation and preservation of a Maxwellian distribution does not depend on the projectile to bath gas atom mass ratio, contrary to predictions of the hard-sphere model. This two-stage behavior is universal. It arises due to the dominance of small angle scattering and small energy transfer in the collisions of neutral particles and reflects a fundamental property of long-range atomic forces. The Boltzmann equation is solved numerically for nitrogen in He and in Ar. The solutions are in close agreement with the experimental measurements of the evolving Doppler profiles of emission from excited initially energetic N atoms traversing bath gases of helium and argon. Our investigation provides the first experimental and theoretical evidence of the formation and preservation of hot Maxwell distributions with a time-dependent effective temperature in actual atomic gases.

1. Introduction

The energy distribution function is critical to the determination of the physical and chemical consequences of the presence of hot atoms in a bath gas, as may occur in applications of plasma physics, atmospheric physics, astrophysics and chemical physics. The approach of the hot atoms to thermal equilibrium depends on the mass ratio of the hot atoms to the bath gas atoms and on the interaction between them [1, 2]. In some cases [1-3], the hot atoms relax to a final equilibrium state by a sequence of Maxwellian distributions characterized by a continuously decreasing effective temperature. In other cases [1-3], the distributions do not follow Maxwellian form. The analyses made use of the hard-sphere approximation or they were based on an interaction potential varying as an inverse power of internuclear distance R.

The translational energy relaxation of a dilute non-equilibrium system of particles as it traverses a uniform bath gas of atoms or molecules is described by a distribution function $f(E, t)$ of kinetic energy E and time t given by the solution of the linear Boltzmann equation. Solving the equation presents a severe challenge and approximations have been made. Two limits have been explored. For one, the Lorentz gas, the mass m_b of the bath gas atoms is much larger than the mass m of the particles and for the other, the Rayleigh gas, the opposite is the case. It has been shown for the hard sphere Rayleigh gas that an initial Maxwellian distribution with an effective temperature T_{eff} larger than the bath gas temperature T remains Maxwellian during the entire thermalization process [1, 2]. The mathematical conditions necessary and sufficient to preserve a Maxwellian shape have been formulated for the asymptotic regime of small energy transfers [1, 2] for which the integral Boltzmann equation may be replaced by the differential Fokker–Planck equation [4, 5]. The energy relaxation may then be described by a single time-dependent function $T_{eff}(t)$ [2]. The effective temperature relaxes exponentially with time to the bath gas temperature.

The approach to equilibrium of an initial anisotropic distribution involves a more complicated interplay of momentum and energy relaxation [6-8]. Shizgal and Blackmore [9] showed that the relaxation of an initially anisotropic hard sphere Lorentz gas occurs by a two-stage process. A fast relaxation to an isotropic distribution is followed by a slow approach to the equilibrium Maxwellian form. These fundamental studies can be questioned because of the assumption that the collisions of the particles with the bath gas can be described by the hard sphere approximation. For collisions of neutral atoms or molecules the hard sphere approximation is a serious misrepresentation of the dependence on velocity and scattering angle and it does not reproduce the strong forward scattering that arises from the attractive long range van der Waals interaction. For a Lorentz gas, because of the small mass ratio, translational energy losses occur by a series of small energy collisions and the Fokker–Planck equation can be used in place of the Boltzmann equation [5, 10, 11]. The Fokker–Planck equation has been solved for scattering by an interaction potential varying as R^{-s} with $s \geq 4$ but the analysis is based on a classical model of the collisions [10]. No precise test of the theory of translational equilibration has ever been made.

The measurement of the thermalization cross section at low energies is a challenging task. Nakayama *et al.* [12] have introduced a precise method for the determination of the time-dependent distribution functions of fast atoms based on measurements of the Doppler shift of laser induced fluorescence and they have reported their results for nitrogen N(^4S) atoms with an initial energy of 0.93 eV, produced by the photolysis of NO_2, traversing bath gases of He, Ar, O_2, and N_2. They presented the detailed measurements of the Doppler profiles of initially energetic N(^4S) atoms as functions of time. For He and Ar as bath gases we can calculate accurate interaction potentials with N(^4S) atoms, and predict the thermalization kinetics. The comparison with precise measurements makes possible a critical assessment of the theory of translational equilibration.

2. Theory and Computational Details

2.1. *Ab Initio Calculations of the Diatomic Potential Energy Curves*

The nuclear dynamics for slow collisions is governed by the adiabatic potential energy curves. The interaction of a close-shell rare gas atom (Rg) with an N(^4S) atom gives rise to one adiabatic state, $^4\Sigma^-$. We evaluated the potentials using the spin unrestricted open shell coupled cluster singles and doubles [13], augmented by triple excitations noniteratively [14], denoted as ROHF-UCCSD(T). A series of augmented Dunning's correlation consistent aug-cc-pVnZ basis sets [15] was used in the calculations (n = Q, 5, 6). The final potential energies at selected internuclear distances were extrapolated to the complete basis (CBS) limit by using the mixed exponential/Gaussian function [16]

$$V(n;R) = V_{CBS}(R) + Be^{-(n-1)} + Ce^{-(n-1)^2}. \tag{1}$$

In the coupled cluster calculations, all electrons were involved in the correlation treatment for N–He, and for N–Ar the 1s orbital of Ar was kept frozen. The MOLPRO2002.6 suite of programs was used throughout for these electronic structure calculations [17].

The calculated ab initio points for the Rg–N interaction potentials were then fitted by analytical expressions of the Degli-Esposti-Werner type [18], composed of the short-range term ($V_{\rm sh}$) and the asymptotic long range-part ($V_{\rm as}$)

$$V(R) = V_{sh}(R) + V_{as}(R), \tag{2}$$

where

$$V_{sh}(R) = \sum_{i=0}^{8} a_i R^i e^{-\alpha(R-\beta)}, \tag{3}$$

and the long-range part was represented by the damped dispersion term,

$$V_{as}(R) = -\frac{1}{2}\left[1+\tanh\left(\gamma+\lambda R\right)\right]\sum_{n=0}^{2}\frac{C_{2n+6}}{R^{2n+6}}. \tag{4}$$

The nonlinear α, β, γ, λ and the linear a_i and C_n parameters were optimized using the Levenberg-Marquardt algorithm with no constraints imposed. The maximum root-mean-square error is 0.2 cm^{-1}, and the RMS error averaged over the internuclear distance does not exceed 0.1 cm^{-1}.

2.2. *Elastic Collision Cross Sections*

The energy transfer process in the present study goes through elastic collisions. Atom-atom elastic scattering theory has been well documented in the literature. We briefly outline here the major steps in the calculations of differential cross sections. The differential cross section for scattering atoms with center of mass (CM) collision energy ε through the polar angle χ into element $d\Omega$ of the CM scattering angle is given by Mott and Massey [19]

$$\frac{d\sigma(\varepsilon,\chi)}{d\Omega} = \left|\frac{1}{k}\sum_{l=0}^{\infty}(2l+1)\sin\eta_l e^{i\eta_l}P_l(\cos\chi)\right|^2, \tag{5}$$

where $k^2 = 2\mu\varepsilon/\hbar^2$ is the wave number, μ is the reduced mass, $\hbar$ is Planck's constant divided by 2π, η_l is the phase shift of the l'th partial wave and P_l denotes the Legendre polynomial. The radial wave function is $R^{-1}u_l(R)$, where $u_l(R)$ satisfies the differential equation

$$\left[\frac{d^2}{dR^2} - \frac{l(l+1)}{R^2} + k^2 - \left(\frac{2\mu}{\hbar^2}\right)V(R)\right]u_l(R) = 0 \tag{6}$$

subject to the to the boundary conditions $u_l(0) = 0$ and

$$u_l(R) \sim kR\left[j_l(kR)\cos\eta_l - n_l(kR)\sin\eta_l\right] \sim \sin\left(kR - \tfrac{l\pi}{2} + \eta_l\right) \tag{7}$$

as $R \to \infty$, where j_l and n_l refer to spherical Bessel and Neumann functions, respectively.

2.3. *Linear Boltzmann Equation*

The energy relaxation of fast atoms due to elastic and inelastic collisions with the bath gas has been considered by Kharchenko *et al.* [20-23]. The rate of

for He. Six vibrational levels were found for NAr with rotational angular momentum $J = 0$, and one level for NHe. At large internuclear distances, the curves lead to the van der Waals interactions, $-5.79/R^6$ for NHe and $-39.42/R^6$ for NAr. For NHe, a value of C_6 of 4.37 had been obtained earlier. We confirm the accuracy of the values we obtained by carrying out direct calculations of C_6, described in the appendix.

Table I. Spectroscopic data (distance in Bohr and energy in cm^{-1}) for NHe and NAr.

	R_e	D_e	D_0	Vibrational levels ($J = 0$) and Inertial Rotation Constant (cm^{-1})						
NHe	6.335	16.3	4.3	v	0					
				$E_{v,J}$	-4.25					
				B_v	0.34					
NAr	6.776	89.3	70.9	v	0	1	2	3	4	5
				$E_{v,J}$	−70.87	−41.12	−20.52	−8.04	−1.99	−0.12
				B_v	−0.12	−0.11	−0.09	−0.07	−0.05	−0.02

3.2. *Differential and Total Cross Sections of Elastic Collisions*

The differential elastic cross sections are shown in Fig. 2. There are large differences in the cross sections for N + He and N + Ar collisions. At the collision energy $\varepsilon = 0.1$ eV, there is much more structure in the differential cross section for the collision with Ar than with He. The deeper potential well, which contains more vibrational levels, is responsible for the structures. Small angle scattering dominates and it causes a sharp peak in the collision kernel as a function of the LF energy. With decrease of the projectile energy the large angle scattering cross section shows a detailed angular dependence, which must be taken into account to describe accurately the thermalization process.

The total elastic cross section and the momentum transfer (diffusion) cross section are shown in Fig. 3 for both systems as functions of the projectile energy. The momentum transfer cross section is defined as

$$\sigma_{transport}(\varepsilon) = \int \frac{d\sigma(\varepsilon,\chi)}{d\Omega}(1-\cos\chi)\, d\Omega\,. \tag{21}$$

The cross sections decrease rapidly with increase of the energy, more so for collisions with Ar. The oscillations in the total cross section reflect glory scattering by a potential with attractive and repulsive regions. Calculations of the l-wave phase shift as $\varepsilon \to 0$, through the Calogerro equation [33]

$$\delta_l'(r) = -k^{-1}V(r)\left[\cos\delta_l(r)\, j_l(kr) - \sin\delta_l(r)\, n_l(kr)\right]^2 \tag{22}$$

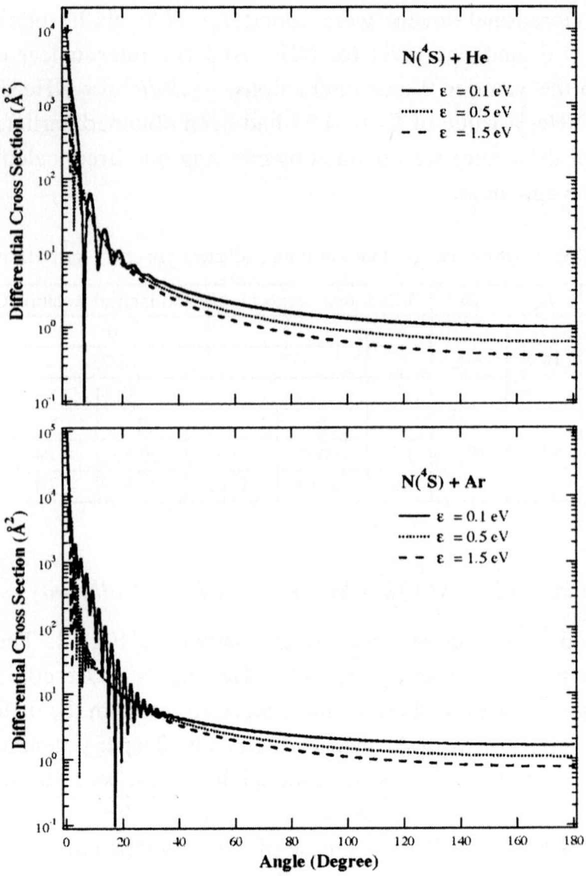

Figure 2. The calculated differential cross sections as functions of scattering angle at projectile energies of 0.1 eV, 0.5 eV and 1.5 eV. (From Ref. [42])

showed $\delta_0(r \to \infty) \approx 6\pi$, confirming that N–Ar has six vibrational levels, as found by solving the radial Schrödinger equation.

The momentum transfer or diffusion cross section may be used for the evaluation of the transport coefficients. The diffusion coefficient can be expressed in terms of the binary collision integral between species A and B [34, 35]

$$\Omega_{A,B}^{(1,1)} = \frac{1}{2\sqrt{\pi}}\left(\frac{\mu}{2k_B T}\right)^{5/2} \int_0^\infty \mathrm{v}^5 \exp\left(-\frac{\mu \mathrm{v}^2}{2k_B T}\right)\sigma(\mathrm{v})d\mathrm{v} \qquad (23)$$

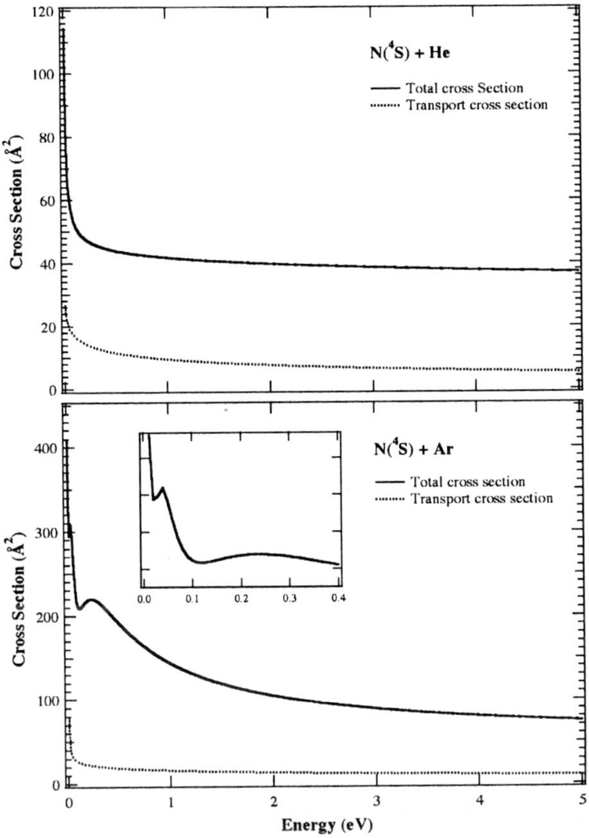

Figure 3. The calculated total cross sections (solid lines) and transport cross sections (dotted lines) as functions of collision energy. (From Ref. [42])

where v is the relative velocity, k_B is the Boltzmann constant and $\sigma(\mathrm{v})$ is given by Eq. (21). Then the binary diffusion coefficient for a total molecular number density $(n_A + n_B)$ is obtained from [35]

$$D_{AB} = \frac{3k_BT}{16\mu\Omega_{A,B}^{(1,1)}} \frac{1}{n_A + n_B} \tag{24}$$

At bath gas temperatures $T > 80$ K, D_{NHe} can be reproduced to within 1% by the expression

$$D_{\mathrm{NHe}} = \frac{1.0\times10^{16}}{n_{\mathrm{He}} + n_{\mathrm{N}}}\left(420.64\times e^{-0.00067T} + 10.463\times T^{0.724} - 422.98\right) \quad (cm^2 s^{-1}) \tag{25}$$

and D_{NAr}

$$D_{\mathrm{NAr}} = \frac{1.0 \times 10^{16}}{n_{\mathrm{Ar}} + n_{\mathrm{N}}} \left(964.72 \times e^{-0.00017T} + 4.380 \times T^{0.725} - 1005.7 \right) \quad (cm^2 s^{-1}) \quad (26)$$

3.3. *Energy Relaxation Kinetics*

In Fig. 4, we present the Boltzmann kernel $B(E|E')$ for the energy relaxation for an initial projectile gas density of 3.27×10^{17} cm^{-3} and a bath gas temperature of 295 K. Sharp maxima occur at $E \sim E'$ because the elastic cross section is dominated by small angle scattering involving small energy losses. We present in Fig. 5 the collision kernel calculated with the hard sphere approximation with the cross sections of 1.8×10^{15} cm^2 for He and 2.3×10^{15} cm^2 for Ar derived by Nakayama *et al.* from experiment [12]. These kernels are dramatically different from results obtained with the hard sphere approximation with rates of small energy transfer $|E - E'| << E'$ larger by orders of magnitude. The lack of the sharp peak in the hard sphere kernel reflects the absence of strong forward scattering in the hard sphere cross section. For collision processes governed by a strong angular and energy dependence considerable error may occur using the HSA in the description of the energy relaxation kinetics [20, 21, 36].

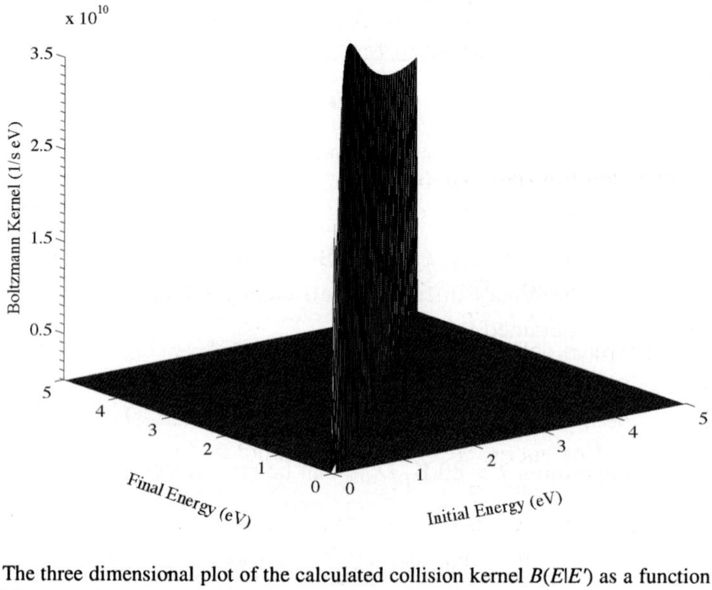

Figure 4. The three dimensional plot of the calculated collision kernel $B(E|E')$ as a function of initial and final laboratory frame energies E and E'. (From Ref. [42])

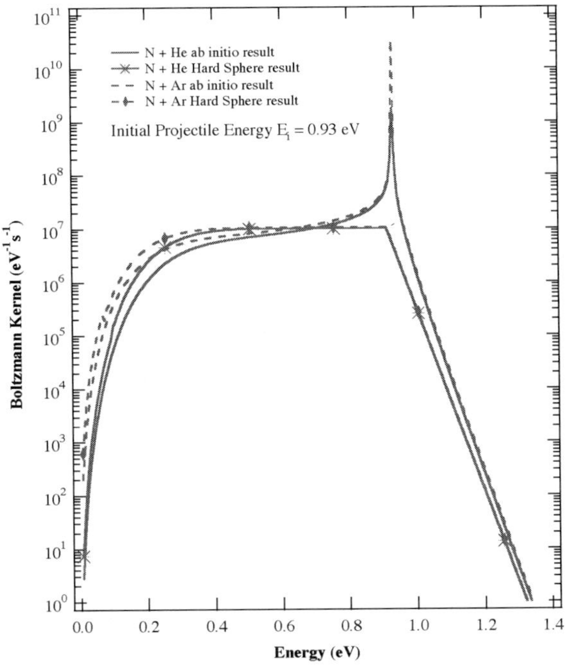

Figure 5. The collision kernel $B(E|E')$ at the initial projectile energy (0.93 eV) as a function of final laboratory frame energy. The plain line and the dashed line (without symbols) were calculated using the differential cross sections evaluated quantum mechanically for the bath gases of He and Ar atoms. The corresponding lines with symbols were calculated using the hard sphere approximation. (From Ref. [42])

The calculated translational energy distribution functions for He and Ar are presented in Fig. 6 as functions of time t. At $t = 0$, both distributions are peaked at the initial energy ε of 0.93 eV, corresponding to the initial CMF energy of the $N(^4S)$ atoms. As time progresses, collisions with bath gases transfer energy and the nitrogen atoms slow down and bring about a bimodal structure which arises because for a typical collision $| E - E' | << E'$ and the energy relaxation of the initial peak is slow. The structure, prominent in the time-derivative of $f(E,t)$, is manifested in the measured profile $\xi(\Delta\nu)$ by its departure from Gaussian. Due to the larger mass of Ar energy transfer is more rapid in Ar gas than in He gas, the double peak appears earlier in time, and the energy distribution when the thermal energy of 0.038 eV is reached is more sharply peaked. At least two time scales are needed to characterise thermalization, one the time to relax to a Maxwellian-like distribution and the second to cool continuously to the bath

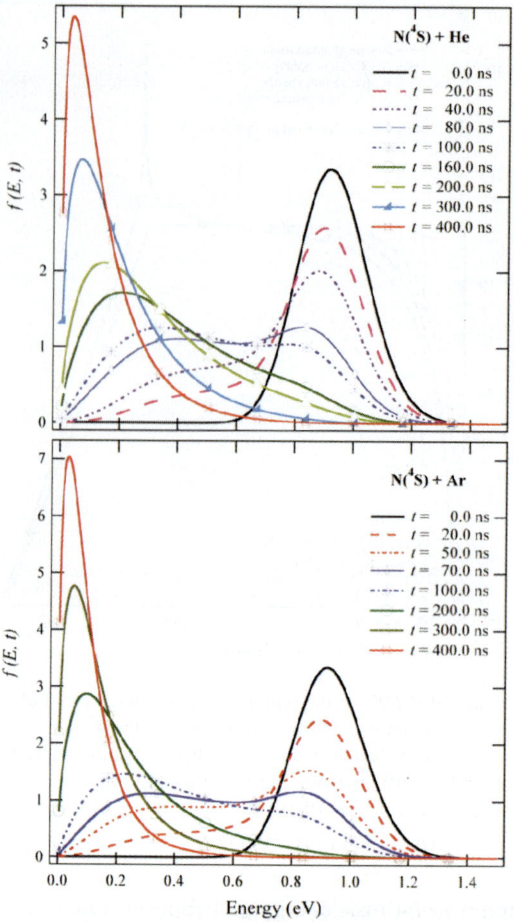

Figure 6. The translational energy distribution function at selected times during the energy relaxation processes for the Ar and He bath gases. (From Ref. [43])

temperature. At higher experimental time resolution, several short relaxation times describing the Boltzmann equation spectrum may be seen, and additional time scales may be required to characterize the first stage. The two-stage evolution applies to projectiles both lighter and heavier than the bath gas species. A detailed comparison of the theoretical and experimental profiles $\xi(\Delta\nu)$ is presented in Fig. 7 for the two bath gases. The agreement demonstrates that a successful measurement of the time-dependent solution of the Boltzmann equation has been achieved. There remain the differences in the wings of the

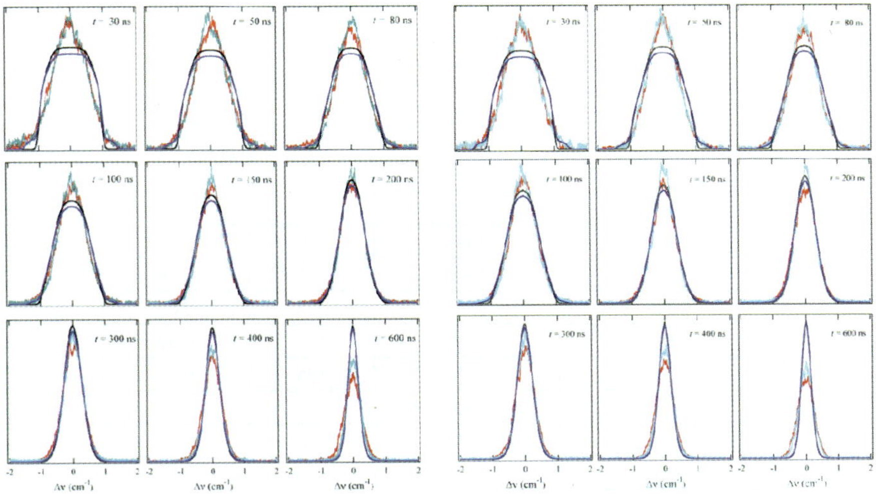

Figure 7. Experimental and theoretical Doppler profiles for He (Right) and Ar (Left) bath gases at selected delay times. The red and teal curves are two sets of experimental data. The black curves are the theoretical Doppler profiles corresponding to an initial energy of 0.93 eV. The blue curves are theoretical Doppler profiles corresponding to a combination of initial energies, 80% with 0.93 eV and 20% with 2.0 eV. The center wavelength is 120.07 nm and $\Delta\nu$ is the Doppler shift. (From Ref. [43])

profiles. They indicate that there is an additional source of N atoms with energies of about 2 eV. The resulting profiles for an admixture of 20% are shown in the figures by the solid blue curves.

Having established that we have a correct solution of the Boltzmann equation we may calculate time scales and mean energies $\langle E(t)\rangle$ for different initial energies E_i. We define the time scales for reaching a Maxwellian shape, τ_g, and for reaching thermal equilibrium at the bath gas temperature, τ, as the times taken to fall within 5% of the intermediate quasi-thermal energy $\frac{3}{2}k_B T_{\text{eff}}(\tau_g)$ and the thermal kinetic energy $\frac{3}{2}k_B T$ of final equilibrium. At τ_g the distribution functions are close to Maxwellian at the intermediate temperature $T_{\text{eff}}(\tau_g)$ which is much higher than the bath gas temperature T. The corresponding times τ_g and τ for a bath gas density of 3.27×10^{16} cm^{-3} and the mean energies $\langle E(\tau_g)\rangle$ expressed as effective temperatures $T_{\text{eff}}(\tau_g)$ are listed in Table II for E_i between 0.5 eV and 3.5 eV. A Maxwellian-like shape of the energy distribution function is attained in a time of 0.1~0.2 μs whereas 3~4 μs elapses before full thermal equilibrium prevails.

Table II. Two-stage relaxation times with different initial projectile energies E_i for N atoms traversing He and Ar bath gases. τ_g is the time to reach the Maxwellian distribution. (From Ref. [43])

He + N			
Initial Energy E_i (eV)	Total Time τ (ns)	τ_g (ns)	$T_{eff}(\tau_g)$ (eV)
0.50	3150	250	0.17
0.93	3250	230	0.29
1.50	3320	210	0.44
2.00	3360	205	0.56
2.50	3380	195	0.69
3.00	3400	190	0.82
3.50	3415	185	0.94
Ar + N			
0.50	2970	210	0.17
0.93	3000	170	0.31
1.50	3040	145	0.49
2.00	3060	135	0.65
2.50	3080	127	0.81
3.00	3100	120	0.99
3.50	3120	115	1.16

Nakayama *et al.* [12] have measured the Doppler profiles of nitrogen atoms with an initial energy ε of 0.93 ± 0.1 eV traversing the bath gases of Ar and He, and they have determined the average kinetic energy of the N atoms as a function of time. They interpreted their data with a Monte Carlo simulation and inferred effective hard sphere cross sections of $2.3\pm0.2 \times 10^{-15}$ cm^{-2} for Ar and $1.8\pm0.2 \times 10^{-15}$ cm^{-2} for He. There is an experimental uncertainty of 0.1 eV in the initial energy. We carried out calculations for ε = 0.83 eV, 0.93 eV and 1.03 eV. The time-dependence of the corresponding average energy is illustrated in Fig. 7 for He and Ar. There are differences between theory and experiment, arising in part from initial conditions. The effective hard sphere thermalization cross sections that approximate the *ab initio* predictions are 1.98×10^{-15} cm^{-2} for Ar and 1.50×10^{-15} cm^{-2} for He, which are slightly smaller than the experimentally derived values. The average energies obtained with the hard sphere approximation are also included in Fig. 8. With an appropriate choice of cross section, the approximation yields a satisfactory description of the energy loss kinetics over a limited range of energy. Fig. 8 also presents a comparison of the experimentally derived time-dependent values of the average energy and the theoretical predictions. The agreement is close with some preference for 0.83 eV as the initial energy. Some of the discrepancy may be due to the interpretation of the experimental data for which it was assumed that the Doppler profiles were at all times Gaussian.

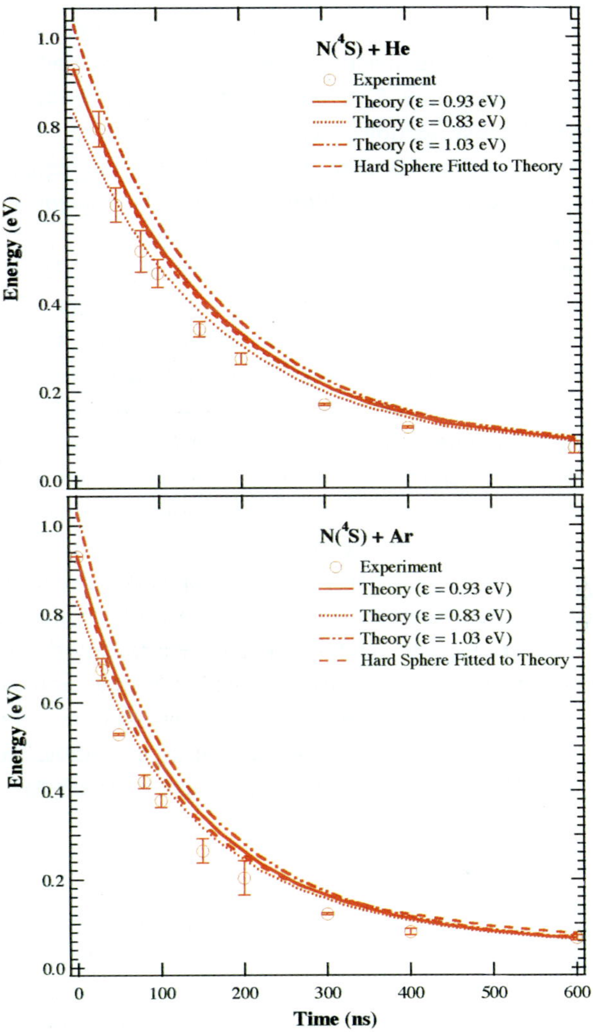

Figure 8. Translational energy degradation versus the time in ns. The open circles are experimental data with error bars. The solid, dotted and dash-dotted lines refer to the theoretical results with different initial kinetic energies $\varepsilon = 0.93$ eV, 0.83 eV and 1.03 eV. The dashed line refers to the hard sphere approximation result fitted the theory result with initial kinetic energy of 0.93 eV. (From Ref. [42])

4. Conclusions

Ab initio calculations of the time-dependent thermalization of initially energetic nitrogen atoms in a buffer gas of He and Ar atoms were performed. Diatomic adiabatic potentials were constructed at the coupled-cluster level of theory with extrapolation to the complete basis limit. The long range interaction coefficients C_6 were obtained from the linear response calculations and using the Páde approximant. The values are 5.79 a.u. for N–He, and 39.42 a.u. for N–Ar. Elastic differential cross sections were evaluated quantum mechanically and adopted to construct the kernel of the Boltzmann equation, and the corresponding time-dependent solutions of the Boltzmann equation were obtained by propagating the LF energy distribution function.

It is shown that the scattering is dominated by small angle collisions driven by the long range van der Waals interaction, and the energy relaxes through a large number of small energy transfer collisions. The energy distribution functions peaked at some initial energy evolve through a bipolar mode into a Maxwellian form at an intermediate effective temperature followed by Maxwellian distributions with effective temperatures which decrease continuously to the temperature of the bath gas. Thus the thermalization process has two time scales. Translational equilibration with the formation and preservation of a Maxwellian-like distribution occurs for both Lorentz and Rayleigh gases and may be characterized by a single parameter, the time-dependent effective temperature. These results are very different from the predictions of the hard-sphere model and they reflect the dominance of small angle scattering and small energy transfer in collisions of neutral particles.

The hard-sphere collisional thermalization cross sections, which reproduce approximately the *ab initio* predictions of the energy relaxation, are respectively 1.50×10^{-15} cm^2 for N with He and 1.98×10^{-15} cm^2 for N with Ar.

Acknowledgments

This research was supported by the Chemical Science, Geoscience and Bioscience Division of the Office of Basic Energy Science, Office of Scicence, U.S. Department of Energy and the NSF under grant ATM-0430506. PZ thanks Professors Dalgarno and Kharchenko for their insight and suggestions during this work. PZ wants to express sincere gratitude to Professor Dalgarno for his great guidance during the collaborations. Professor Dalgarno is a wonderful source of knowledge and exciting ideas. He has a talented ability to look through complicated mathematic expressions and always offers penetrating insight to spot any tiny flaw in the research. He is such as a great mentor that deeply benefits a young scientist from daily communications. His knowledge, kindness and patience are greatly appreciated.

Appendix

We calculated the Cauchy moments [37] using the linear-response theory on the CCSD wave function [38] for He and Ar atoms and on the complete active space self-consistent (CASSCF) wave function [39] for N atom. In the CCSD calculations, all electrons were correlated. The active space in the CASSCF calculation was defined as 7 electrons distributed in 18 orbitals, which include $1s2s2p3s3p4s3d4p$. q-aug-cc-pV6Z and d-aug-cc-pV6Z basis sets were used for CCSD and CASSCF calculations respectively. All these calculations were carried out using the DALTON program [40]. Finally, C_6 was obtained via the $[n, n-1]_\beta$ (upper bound) and $[n, n-1]_\alpha$ (lower bound) Páde approximants [41]. With $n = 7$, $C_6 = 5.79$ a.u. for N–He and 39.42 a.u. for N–Ar in agreement with the values inferred from the calculated potential energy curves.

References

1. H. C. Andersen, I. Opppenheim, K. E. Schuler, and G. H. Weiss, J. Math. Phys. **5**, 522 (1964).
2. K. Andersen and K. E. Shuler, J. Chem. Phys. **40**, 633 (1964).
3. R. L. Liboff, *Kinetic Theory, 3rd Edition* (Springer-Verlag, New York, 2003).
4. E. M. Lifshitz and L. P. Pitaevskii, *Physical Kinetics* (Pergamon, Oxford, 1981).
5. H. Risken, *The Fokker-Planck Equation: methods of solution and applications* (Springer-Verlag, Berlin and New York, 1984).
6. R. Dorfman, Proc. Natl. Acad. Sci. **50**, 804 (1963).
7. M. R. Hoare and C. H. Kaplinsky, J. Chem. Phys. **52**, 3336 (1970).
8. B. D. Shizgal, J. Chem. Phys. **70**, 1948 (1979).
9. B. D. Shizgal and R. Blackmore, Chem. Phys. **77**, 417 (1983).
10. H. Oser, K. E. Shuler, and G. H. Weiss, J. Chem. Phys. **41**, 2661 (1964).
11. R. F. Pawula, Phys. Rev. **162**, 186 (1967).
12. T. Nakayama, K. Takahashi, and Y. Matsumi, Geophys. Res. Lett. **32**, L24803 (2005).
13. P. J. Knowles, C. Hampel, and H.-J. Werner, J. Chem. Phys. **99**, 5219 (1993).

14. J. D. Watts, J. Gauss, and R. J. Bartlett, J. Chem. Phys. **98**, 8718 (1993).
15. T. H. Dunning Jr, J. Chem. Phys. **90**, 1007 (1989).
16. K. A. Peterson, D. E. Woon, and T. H. Dunning Jr, J. Chem. Phys. **100**, 7410 (1994).
17. MOLPRO, version 2002.6, a package of *ab initio* programs, written by H.-J. Werner, P. J. Knowles, R. Lindh, F. R. Manby, M. Schütz, and others.
18. A. Degli-Esposti and H.-J. Werner, J. Chem. Phys. **93**, 3351 (1990).
19. N. F. Mott and H. S. W. Massey, *The Theory of Atomic Collisions* (Oxford Univ. Press, New York, 1965).
20. V. Kharchenko, J. Tharamel, and A. Dalgarno, J. Atoms. Solar-Terr. Phys. **59**, 107 (1997).
21. V. Kharchenko, N. Balakrishnan, and A. Dalgarno, J. Atoms. Solar-Terr. Phys. **60**, 95 (1998).
22. N. Balakrishnan, V. Kharchenko, and A. Dalgarno, J. Phys. Chem. A **103**, 3999 (1999).
23. V. Kharchenko, A. Dalgarno, B. Zygelman, and J. H. Yee, J. Geophys. Res., [Atmos.] **105**, 24899 (2000).
24. I. Opppenheim, K. Schuler, and G. Weiss, *Stochastic Processes in Chemical Physics: The Master Equation* (MIT Press, Cambridge, Mass., 1977).
25. B. D. Shizgal and R. Blackmore, Planet. Space Sci. **34**, 279 (1986).
26. K. Stamnes, O. Lie-Svendsen, and M. H. Rees, Planet. Space Sci. **39**, 1435 (1991).
27. O. Lie-Svendsen, M. H. Rees, and K. Stamnes, Planet. Space Sci. **40**, 1639 (1992).
28. V. I. Schematovich, J. C. Gerard, D. V. Bisikalo, and B. Hubert, J. Geophys. Res. **104**, 4287 (1999).
29. B. D. Shizgal, J. Geophys. Res. **104**, 14833 (1999).
30. V. Kharchenko and A. Dalgarno, J. Geophys. Res. **109**, D18311 (2004).
31. R. N. Zare and D. R. Herschbach, Proc. IEEE **51**, 173 (1963).
32. R. B. Forrey, L. You, V. Kharchenko, and A. Dalgarno, Phys. Rev. A **54**, 2180 (1996).
33. F. Calogerro, *Variable Phase Approach to Potential Scattering* (Academic Press, New York and London, 1967).
34. S. Chapman and T. G. Cowling, *The Mathematical Theory of Non-Uniform Gases, 3rd ed.* (Cambridge University Press, New York, 1970).
35. M. H. Rees, *Physics and Chemsitry of the upper atmosphere* (Cambridge University Press, New York, 1989).

36. N. Balakrishnan, V. Kharchenko, and A. Dalgarno, J. Chem. Phys. **108**, 943 (1998).
37. J. O. Hirschfelder, W. B. Brown, and S. T. Epstein, Adv. Quantum Chem. **1**, 255 (1964).
38. C. Hättig, O. Christiansen, and P. Jørgensen, J. Chem. Phys. **107**, 10592 (1997).
39. P. W. Fowler, P. Jørgensen, and J. Olsen, J. Chem. Phys. **93**, 7256 (1990).
40. DALTON, a molecular electronic structure program, Release 2.0 (see http://www.kiemi.uio.no/software/dalton/dalton.html, 2005).
41. P. W. Langhoff and M. Karplus, J. Chem. Phys. **53**, 233 (1970).
42. P. Zhang, V. Kharchenko and A. Dalgarno, Mol. Phys. **105**, 1487 (2007).
43. P. Zhang, V. Kharchenko, A. Dalgarno, Y. Matsumi, T. Nakayama, and K. Takahashi, Phys. Rev Lett. **100**, 103001 (2008).

HELIOSPHERIC X-RAYS INDUCED BY CHARGE-EXCHANGE COLLISIONS

V. A. KHARCHENKO[1,2]
ITAMP, Harvard-Smithsonian Center for Astrophysics
Cambridge, MA 02138, USA
[1] *E-mail: vkharchenko@cfa.harvard.edu*

Department of Physics, University of Connecticut
Storrs, CT 06269, USA
[2] *E-mail: kharchenko@phys.uconn.edu*

Charge-exchange collisions between heavy solar wind ions and neutral atoms and molecules are considered as the mechanism to explain some astrophysical observations of X-ray and EUV emissions. Theoretical and experimental data on the charge-exchange emission spectra are reviewed. Physical characteristics of heliospheric sources of the charge-exchange X-rays are discussed. Modeling of X-ray and EUV spectra observed from different heliospheric objects is described in detail. Satellite observational data obtained with the Chandra and XMM-Newton X-ray telescopes are compared with our theoretical predictions. Contribution of the charge-exchange mechanism to the observed diffuse soft X-ray background is determined and the computed photon spectra are compared with the results of recent observations. X-ray diagnostics of the solar wind composition and velocity are considered. Examples of a determination of the solar wind ion composition from cometary and planetary X-ray spectra are presented.

Keywords: charge-exchange collisions; X-ray emission; cometary X-rays; heliosphere; soft X-ray background.

1. Introduction

Analysis and interpretation of increasingly complicated data of astrophysical observations require a broad interdisciplinary approach. Professor Alex Dalgarno has recognized this specific feature of astrophysics since the beginning of his scientific career and employed accurate methods from different disciplines to explain observations and evolutionary processes in the Universe, solar system, and planetary atmospheres. During more than four decades of work in astrophysics, Alex has developed advanced methods of solutions of astrophysical problems incorporating the most recent developments in quantum collisional theory, electrodynamics, optics, atomic and molecular physics, physical kinetics, plasma physics, and quantum chemistry. Prof. Dalgarno's results are highly appreciated by researchers investigating astrophysical sources of emission and absorption. Detailed analysis of observed spectra of the electromagnetic radiation induced or absorbed by different cosmic objects may yield unique information regarding physical parameters of astrophysical objects and environments. Observations in modern astronomy provide spectral data in a very broad interval of wavelengths and these spectra are a major source of our knowledge on a state of matter in the Universe. Consistent interpretation of these observational data can be achieved, if accurate information on the excitation and emission mechanisms are available. In astrophysical systems, photons in different energy intervals may be produced by different physical mechanisms and an interpretation of observational data very often includes some uncertainty. Nevertheless, there is the unique mechanism of emission when photons from different wavelength intervals are simultaneously induced in a single process. This mechanism is the production of cascading photons in charge-exchange collisions of highly charged ions with neutral atoms and molecules.

Emission produced by the charge-exchange mechanism has been comprehensively studied for use as a diagnostic of laboratory plasmas[1] . In astrophysical environments, highly charged ions, such as C^{6+}, N^{7+}, O^{7+}, Mg^{9+} and others, are contained in plasma fluxes of the solar and stellar winds. In charge-exchange collisions between highly charged ions and neutral atoms or molecules, an electron is captured into highly excited states of an ion and than relaxes to the ion ground state radiating a cascade of photons. Energies of these photons range from the infrared to the X-ray, but all of them are produced in a single collision. Relative intensities of different emission lines depend on the nature of the colliding particles and the velocity of the collisions. Emission spectra induced in charge-transfer collisions have been used to determine the plasma temperature and ion composition of hot laboratory plasmas and recently have been employed as a diagnostic of astrophysical plasmas. Detailed analysis of cascading emission spectra may also provide information on the composition of the neutral gas.

In this article, I briefly describe astrophysical aspects of the charge-exchange mechanism for producing X-ray emission and show how the theory of charge-exchange collisions has been employed by Prof. Dalgarno and his collaborators to explain intriguing spectral features of X-ray emissions induced by collisions of the solar wind ions and neutral species in cometary and planetary atmospheres and the heliospheric gas.

2. Charge-Exchange Mechanism of X-ray Emission

Astrophysical sources of charge-exchange X-ray emission had not been known until the discovery of cometary X-rays in 1996. An exception was the model of heavy ion auroras proposed to explain intensive fluxes of X-ray photons observed from polar regions of the Jovian atmosphere[2] . According to this hypothesis, the polar X-ray emission is induced in multiple stripping and charge-exchange collisions of energetic sulfur and oxygen ions, precipitating into the Jovian atmosphere. The ion model of Jovian X-rays involves significant acceleration of the heavy ions to energies of a few MeV/amu required for a production of highly charged ions. The solar and stellar wind plasmas naturally include highly charged ions and we may expect production of X-ray emission in any region in which they interact with neutral gas. Although cometary atmospheres provide a perfect condition for a realization of the charge-exchange mechanism of X-ray emission, the discovery of cometary X-rays had happened after decades of X-ray satellite observations. Comets are known as cold and passive objects and it was difficult to expect any intensive X-ray fluxes from them. Nevertheless, in 1996 the ROSAT X-ray telescope observations carried out by Dr. Lisse and his collaborators[3] identified Comet C/Hyakutake 1996 B2 as a strong source of X-ray emission. Analysis of archive data from the ROSAT observations[4] revealed the presence of cometary X-rays in previously detected X-ray emission spectra. These findings established comets as a new class of astrophysical X-ray sources. Several mechanisms for cometary X-ray production were suggested and tested in comparison with observational data. The charge-exchange mechanism proposed in 1997 by Dr. Cravens[5] successfully explained morphology and spectra of observed X-ray emissions and provided a good estimate of the total intensities of cometary X-ray fluxes. The major processes involved in the charge-exchange mechanism are shown in Fig.1. Collisions of highly-charged solar wind ions Z^{q+} with cometary neutral atoms or molecules A populate the highly excited states of the $Z^{(q-1)+}$ ion:

$$Z^{q+} + A \to Z^{*(q-1)+} + A^{+} \to Z^{(q-1)+} + h\nu + A^{+}, \qquad (1)$$

Charge-exchange collisions usually yield ions in several excited states. The set of populated excited levels is schematically shown in Fig.1. The simplest physical case is represented by collisions between bare heavy ions and

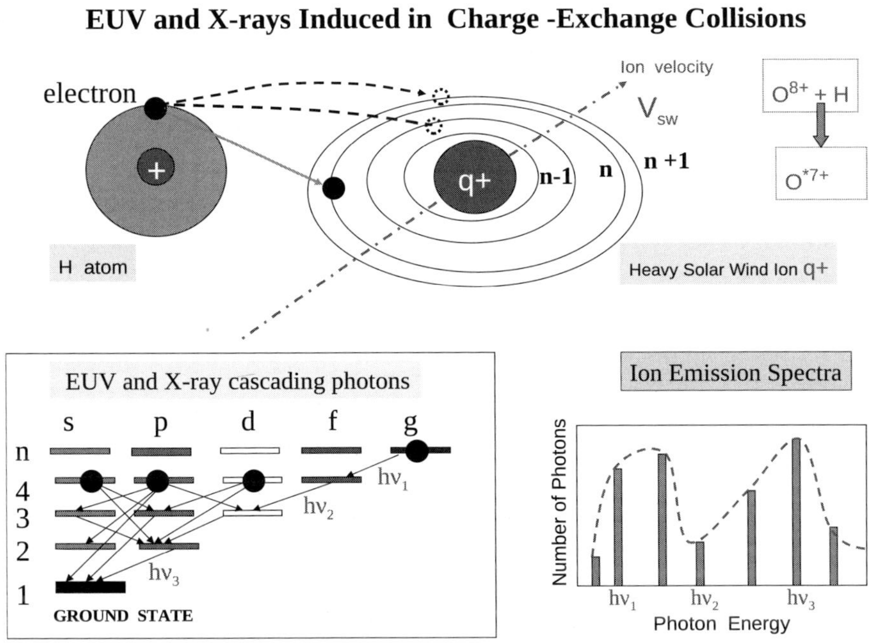

Fig. 1. The scheme of major processes, producing X-ray photons in the charge-exchange collisions between highly charged ions and neutral atoms and molecules. The upper sketch illustrates the electron capture into highly excited states in collisions between highly charged ions and neutral atoms. This example is shown for O^{8+} + H collisions. The schematic representation of populated ion excited states, cascading radiative transitions, and an ensemble of emitted photons (subsequently referred to as a "spectral element") are also shown in the figure.

hydrogen atoms as only one electron is moved. The chain of radiative cascading transitions from the ion excited states produces an ensemble of correlated photons. The relative intensities of all cascading photons, sketched in Fig.1, are fixed for the given initial populations of ion excited states. Initial populations of excited states and thus the ion emission spectra induced in charge-exchange collisions depend on the collisional velocity and nature of the colliding partners. Calculation of the relative photon intensities becomes a complicated problem for heavier ions because of a large number of ion excited states involved in the cascade process. Prof. Dalgarno and his collaborators have developed efficient theoretical methods[6] that predict cascade intensities taking into account all relaxation paths. Fig.2 illustrates some theoretical details of the computation of cascading

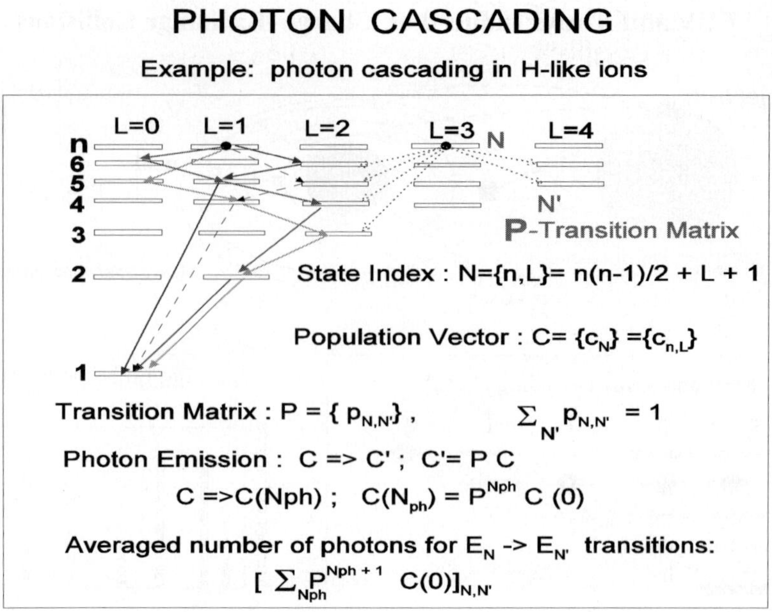

Fig. 2. Production of cascading photon spectra in charge-exchange collisions.

emission spectra. The initial population of exited ion energy levels, which may be involved in the cascade process, is given by the population vector $\mathbf{C}= \{C_1, C_2, C_3, ..., C_N, ..., C_{N_{max}}\}$, where N is the index number of the ion excited state with principle quantum number n and orbital quantum number L, and N_{max} is the highest index number for the excited states considered. The relative intensity of radiative transitions from state N to state N' is given by the branching ratio $p_{N,N'}$. Photon emission alters populations of ion excited states and the corresponding transformation of the population vector $\mathbf{C}$ is described by the transition matrix $\hat{\mathbf{P}} = \{p_{N,N'}\}$. The population vector $\mathbf{C}(k)$ of ion excited states after k cascading steps and the total number of photons ξ_{ph} induced in $E_N \to E'_N$ transitions are given by the formulas

$$\mathbf{C(k)} = \hat{\mathbf{P}}^k \mathbf{C(0)}; \qquad \xi_{ph}(N, N') = p_{N,N'} * \left[\sum_{k=0}^{k=N_{ph}} \hat{\mathbf{P}}^k \mathbf{C(0)} \right]_{N,N'} \tag{2}$$

were N_{ph} is the number of cascading photons induced in the longest relaxation path of radiative transitions.

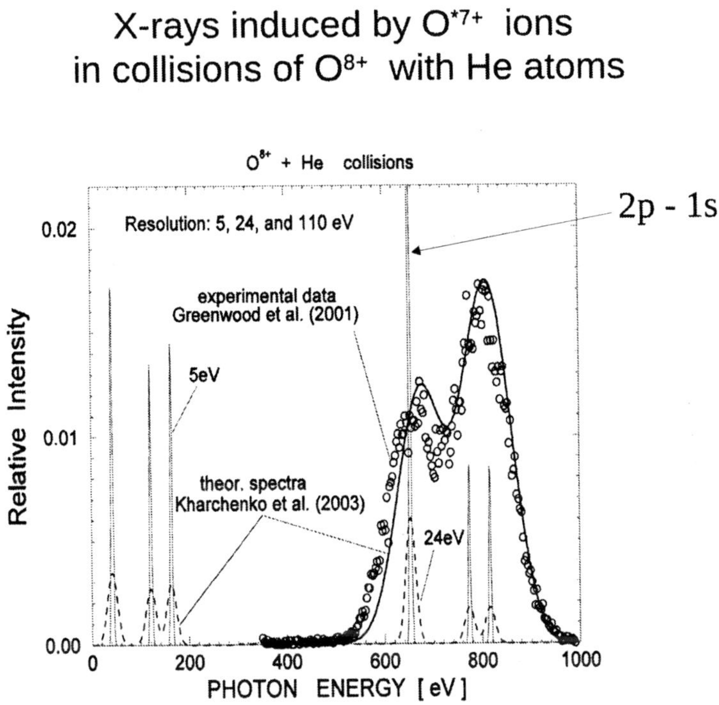

Fig. 3. Comparison between the theoretical X-ray emission spectra and the results of laboratory experiments. Experimental data on the spectra of O^{*7+} X-ray emission are shown by circles[7] and theoretical spectra, computed for different photon energy resolutions, are shown by solid and dashed lines.

The solar wind is composed of different ions and emission spectra induced by the charge-exchange process has to include the weighted ensembles of cascading photons emitted by individual ions. The entire ensemble of cascading photons may be considered as a spectral element (see Fig.1) required for modeling the charge-exchange emission spectra. Theoretical spectra of individual ions have been verified in comparison with recent experimental data. An example of such comparisons between theory and experiments is shown in Fig.3 for the X-ray and EUV emission of excited O^{*7+} oxygen ions produced in collisions of O^{8+} bare ions with hydrogen atoms. The experimental data on the X-ray spectra obtained in the ion-beam experiments conducted by Dr. Chutjian's research group[7] are in very good agreement with the theoretical results. Predicted spectra of photons with energies less than 200eV are also shown in Fig.3.

3. Astrophysical Sources of Charge-exchange X-rays

With respect to the abundance of protons or alpha particles, the abundance of highly charged heavy ions in the solar wind is small, but only these ions are capable of producing X-ray emission. Ion compositions and velocities of the solar wind are variable both in space and time. In general, however, ths solar wind can be described by two components which differ both in composition and in plasma velocity: the slow solar wind with flux velocity around 400km/s; and the fast solar wind with plasma velocity around 700 km/s. The slow solar wind includes a larger fraction of highly charged ions than the fast wind. Charge-exchange X-rays are emitted from regions in which the solar wind plasma and the neutral gas collide. In Fig.4, these interaction regions are schematically shown as sources of the charge-exchange X-ray emission in the Solar System. As shown, the sources of the neutral gas are: planetary atmospheres, cometary atmospheres and the interstellar (heliospheric) gas. All heliospheric objects shown in Fig.4 have been observed as the sources of charge-exchange X-rays[8] . The interaction between the solar wind and interstellar gas takes place throughout the entire heliosphere and this region is seen in observations as a component of the diffuse X-ray background. X-ray photons are induced by heavy solar wind ions and because of that the X-ray emission spectra of all sources described

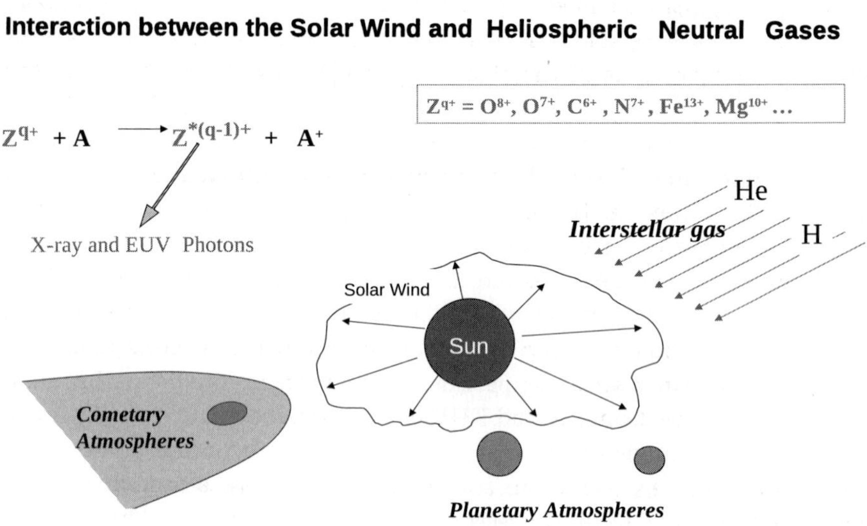

Fig. 4. Sources of the charge-exchange emission in the solar system.

in Fig.4 should be relatively similar. Spatial and temporal fluctuations of the solar wind composition and velocity affect the charge-exchange X-ray spectra induced from different sources. Differences in the neutral gas composition could be an additional reason in variations of the X-ray emission spectra, because populations of ion excited states are sensitive to the ionization potential of neutral atoms or molecules. The interstellar gas consists of H and He atoms, but planetary and cometary gases include both atoms and molecules.

4. X-ray Diagnostics of Astrophysical Plasmas

Sensitivity of the charge-exchange X-ray emission spectra to the solar wind composition can be employed for a determination of its ion composition. In principle, the energies and relative intensities of characteristic X-ray photons induced by different individual ions should provide accurate information on the charge and elemental composition of the solar wind. Doppler shifts of X-ray emission lines and dependence of their relative intensities on ion collisional energies may provide additional opportunities for measurements of the plasma velocity. This type of remote diagnostic of the solar wind could be done with the next generation of high-resolution X-ray detectors, such as X-ray micro-calorimeters. In current and previous observations with X-ray telescopes, a majority of sources of the charge-exchange emission have been seen as extended objects and that has affected the wavelength resolution of satellite spectral detectors. Typical X-ray photon energy resolution in observations of the charge-exchange emission spectra with the Chandra and XMM-Newton X-ray telescope was around 50 eV, but in the ROSAT observations it was about 110 eV. Despite these low resolutions, the average composition of the solar wind may be inferred from observed X-ray emission spectra. In Fig.5 the X-ray spectra observed from different comets are shown together with the theoretical spectra $P(h\nu)$ computed for the charge-exchange mechanism. All spectra are normalized per unit of the solar wind flux (per single solar wind ion), and this type of normalization allows one to compare observational data from different comets. This normalization takes into account a small ($\sim 10^{-4}$) value for the heavy ion abundance. The observational X-ray spectra from different comets, Hale-Bopp, Encke, and McNaught-Hartley are shown in Fig.5 by different point symbols. X-ray spectra observed from Comet Levy are represented by the empirical best-fit curve shown by the dot-dashed line. Cometary X-ray spectra were observed from different comets at different times with different X-ray telescopes, but all observational data are well described by the theoretical

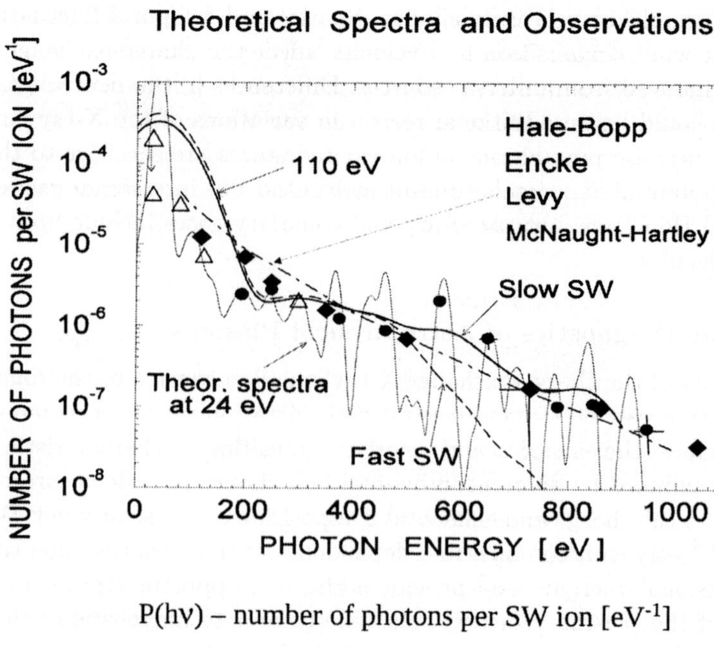

Fig. 5. Comparisons between observed cometary X-ray spectra and theoretical predictions of the change-exchange mechanism. Normalized theoretical spectra of cometary X-ray emission are shown for the slow and fast solar winds.

emission spectra calculated for the slow solar wind (the smooth solid line in Fig.5). There are no any adjustable parameters in the theoretical spectra. The averaged ion composition of the slow solar wind was taken from the results of *in situ* satellite mass-spectrometer measurements and the photon energy resolution of 110 eV was chosen to correspond to that pertaining to the X-ray satellite observations. The same theoretical spectra convoluted with the photon energy resolution of 24 eV are shown in Fig.5 by the thin solid oscillating curve. There is a remarkable agreement between the photon energy positions of observed spectral features and the calculated spectra. The theoretical spectra induced by the fast solar wind are shown by the dashed curve. They are different from the spectra induced by the slow solar at photon energies higher than 500 eV. This difference between the spectra of the slow and fast solar winds reflects a reduced abundance of ions in highest charge states, such as O^{8+}, in the fast solar wind. About 30 comets have been detected as sources of the charge-exchange X-ray emission. Alex

Dalgarno and his collaborators carried out an elaborate analysis of X-ray emission spectra from Comet McNaught-Hartley and inferred the elemental and charge state composition of heavy solar wind ions[9] for this specific case of interaction between the solar wind and the McNaught-Hartley cometary atmosphere. Current observations of cometary X-rays with the Chandra and XMM-Newton telescopes provide spectra sufficient for accurate diagnostics of ion compositions of the solar wind and coronal mass ejections.

5. Planetary X-ray Emission

Observations of X-ray emissions from planetary atmospheres, especially from the Earth and Mars, have significantly improved our understanding of a morphology of X-ray sources described by the charge-exchange mechanism. Observations of the "Dark Moon" X-rays and extended Martian X-ray halo have demonstrated specific features of charge-exchange X-rays induced by the solar wind interacting with planetary coronae[8] . In the case of the "Dark Moon" X-rays, the Moon blocked all sources of X-ray emission except that due to charge-exchange collisions between solar wind ions and neutral atoms of the geocoronal gas. In the case of the Martian halo emission, X-ray spectra include a resolved triplet of $2^3S \rightarrow 1^1S$, $2^3P \rightarrow 1^1S$, and $2^1P \rightarrow 1^1S$ emission lines of the helium like oxygen ions OVII. The relative intensities of the OVII forbidden, intercombination and resonance X-ray emission lines have been accurately described by the charge-exchange model.

Although properties of observed X-ray emissions from the Earth and Mars coronae can be explained by the charge-exchange mechanism, the total X-ray emission from planetary atmospheres has a more complicated nature. The planetary disk emission, originating from regions of relatively dense atmospheric gas, is mostly produced due to scattering and fluorescence of solar X-rays. Differences in the spectra and morphology of the disk and charge-exchange X-ray emissions provide an opportunity to study these mechanisms independently. The disk emissions have been observed from the terrestrial and outer planets. Strong correlations of the disk X-ray emission with the solar X-ray spectra and intensities have been shown in direct observations.

Spectral features and intensities of X-ray emissions from the polar regions of the Jovian atmosphere do not correlate well with the intensity and spectral composition of the solar X-rays. Two hypotheses regarding the mechanism of the Jovian polar X-rays have been discussed: electronic bremsstrahlung emission and precipitation of energetic heavy ions. The

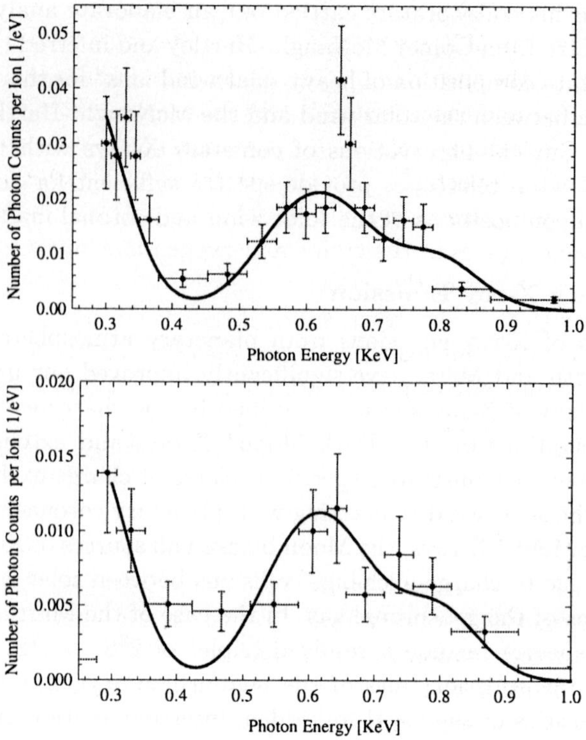

Fig. 6. The spectra of the Jovian X-ray emission from the North and South poles are shown on the top and bottom panels, respectively. Observational data have been obtained with the Chandra X-ray telescope[8] and theoretical spectra have been computed for the charge-exchange emission mechanism of the precipitating O and S ions[9] .

electron aurora mechanism is unable to explain the high intensities of the Jovian X-ray auroras as well as their emission spectra. The second mechanism, the ion precipitation mechanism, is more successful in modeling the X-ray spectra and intensities of the Jovian polar emission. The ion precipitation mechanism includes several essential steps:

(a) acceleration of magnetospheric or solar wind ions up to energies of a few MeV per atomic mass unit (amu);
(b) collisional stripping of energetic heavy ions, precipitating into the planetary atmosphere;
(c) charge-exchange collisions of heavy highly charged ions with subsequent cascade emission of X-ray photons.

Spectra of the Jovian polar X-rays have been analyzed, using the charge-exchange mechanism of X-ray emission induced by energetic sulfur and oxygen ions. In Fig.6 the spectra of Jovian X-ray emission from the North pole (the upper panel) and the South pole (the lower panel)[8] observed with the Chandra X-ray telescope (points with error bars) are shown in comparison with the theoretical spectra[9] . The theoretical spectra (solid lines) include only two adjustable parameters: the initial energy of the precipitating ions and relative abundances of the sulfur and oxygen ions in the precipitating flux. The best fit values of these two parameters allow one to infer from the observed X-ray spectra the initial energy and elemental composition of the precipitating ions[9] . This example illustrates new opportunities for an advanced X-ray diagnostic of precipitating ion fluxes. Inferred parameters of the polar ion fluxes can be used for a determination of the ion-acceleration mechanism.

Acknowledgments

The construction of accurate X-ray and EUV spectra of the charge-exchange emission and analysis of observational data on astrophysical X-ray sources have been long term projects, and I am very grateful to Prof. Alex Dalgarno for the fruitful collaboration during all these years. This work of Dr. Kharchenko has been supported by the NASA grant NNX08AH51G.

References

1. R. K. Janev, H. Winter, *Physics Reports*, **117**, 265 (1985).
2. A. E.Metzger et al., *J. Geophys. Res.*, **88**, 7731 (1983).
3. C. M. Lisse et al., Science, **274**, 205 (1996).
4. K. Dernnel, J. Englhauser, and J. Trumper, *Science*, **277**, 1625 (1997).
5. T. E. Cravens, *Geophys. Res. Lett.*, **24**, 105 (1997).
6. V. Kharchenko, W. Liu, and A. Dalgarno, *J. Geophys. Res.*, **103**, 26687 (1998).
7. J. B.Greenwood et al., *Phys. Rev. A*, **63**, 062707 (2001).
8. A. Bhardwaj A. et al., *Planet. Space Science*, **55**, 1135 (2007).
9. V. Kharchenko, M. Rigazio, A. Dalgarno, and V. Krasnopolsky, *ApJ Lett.*, **585**, L73 (2003).
10. V. Kharchenko et al., *J. Geophys. Res.*, **113**, A08229 (2008).

Session V

Ultra-cold Collisions/ Long-range Interactions

Alex Dalgarno, Elizabeth Rich, Fern Creelan

A TRIBUTE TO ALEX'S LEADERSHIP AND SUPPORT: THE EXAMPLES OF MODEL POTENTIALS AND COLD MOLECULES

FRANÇOISE MASNOU-SEEUWS

Laboratoire Aimé Cotton, CNRS, Université Paris Sud XI, Bât. 505, Campus d'Orsay 91405 Orsay Cedex, France
E-mail: francoise.masnou@Lac.u-psud.fr
www.lac.u-psud.fr

In many research areas, including long range potentials and ultracold collisions, Alex Dalgarno has been a pioneer and a leader. His work is well integrated within a tradition, he has been very supportive of young people or colleagues and has given good advice. The community should be grateful to him for the structuration achievement through the creation of ITAMP and the organization of many workshops.

1. Introduction

Alex has had a very strong influence on my scientific career, although I never published a paper with him. At the very beginning of my career, during the spring of 1968, I heard him for the first time give a beautiful introductory talk at a conference in Belfast. He had just left Queen's University for Harvard, where he was starting a first-class group in Cambridge. Later on, in July 1970, when I met him in Amsterdam for ICPEAC, I realized that while completing my thesis, I was working, alone, on the same subject as his group! Clearly, it was a competition where I was at great disadvantage! Yet I went on desperately, and the expected disaster did not occur : on the contrary, far from being eliminated, I came to know Alex as very supportive, giving much encouragement. This went on throughout my career. Alex was very present at conferences and meetings, discussing thoroughly posters and talks, and giving inspiring comments. In 1978, while I was working on model potentials, I started regular visits to Cambridge, where I found an outstanding research atmosphere, very stimulating. I would present my new results or discuss projects : Alex was always positive; he encouraged me to develop my own research line, and in many occasions to take risks.

I am particularly grateful for his advice when I was hesitating to step into the emerging cold collisions field. In the following paragraphs, I want to develop two aspects of our dialogue : progress on model potentials in the 70's and the blossoming of cold molecules since 1997.

Fig. 1. Alex is "looking far ahead", in Telluride (July 2005).

2. A Long Tradition of Work on Long Range Forces Flourishing into Model Potentials

Since slow atomic collisions are dominated by distant encounters, a precise evaluation of long range potentials is necessary. Early on, Alex and his group developed methods to compute long range forces. It started with mathematical developments for sum rules, in the famous paper by Dalgarno and Lewis.[1] Then the model potential method was introduced : due to the large difference in time scales for the motion of core electrons and valence electron, a separation similar to Born-Oppenheimer approximation can be implemented. For an atom of the first row, the valence electron is moving slowly in the field of a polarizable core, which can be modeled by a potential. The latter can be fit to experimental data for the energy levels, taking advantage of the spectroscopic accuracy which Alex has long recognized as one of the strengths of our field. In the case of a diatomic molecule such as an alkali -rare-gas system, the active electron moves in the field of two polarizable cores. The brilliant work of Chris Bottcher and Alex Dalgarno[2] shows how this leads to three potential terms, since a cross-polarization term has to be introduced. I was happy to work with Pierre Valiron, Ron McCarroll and Martine Philippe in order to develop further this subject, and to be congratulated by Alex when we reached the milestone of 1 cm^{-1} accuracy.[3,4] This became a challenge for our colleagues working in quantum chemistry, who later came to develop pseudo-potential methods taking advantage of spectroscopic data.

3. Going Directly to Key Problems and Structuring an Emerging Research Field

3.1. *Going Directly to Key Problems: The Example of Cold Collisions and Cold Molecules*

The scattering length, which describes the collision between two ultracold atoms at the zero-energy limit, is a key parameter for the formation of Bose Einstein condensates. In collaboration with the experimental group of Randy Hulet, Alex and Robin Côté demonstrated how the scattering length could be extracted from the minima in a photoassociation spectrum, by identifying the position of the nodes of the scattering wavefunction.[6] This is presented in the following talk by Randy Hulet.

As early as 1997, at the ICPEAC conference in Vienna, Alex and Robin Côté were presenting a challenging poster on the possibility of using photoassociation in order to create ultracold molecules:[7] this is developped in the talk by Robin Côté. The first molecules were observed in Orsay later the same year,[8] and we had many discussions at a conference in Boulder that autumn.

3.2. *Structuring a Research Field: The Blooming of Cold Molecules*

Alex and his ITAMP collaborators have been aware of the importance of timely workshops in order to boost an emerging field and to coordinate theoretical and experimental progress. In the case of cold molecules, the European groups have been coordinated within a research training network: Cold Molecules, Formation, Trapping and Dynamics, funded by the European Commission under label HPRN-CT-2002-00290. As the coordinator of this network, I have worked in close collaboration with ITAMP and had many discussions with Alex in order to organize stimulating workshops in this area. Besides photoassociation, Alex and his group have made a strong contribution to the development of collisional cooling of molecules, as described further in the talks of R. Krems and of N. Balakrishnan.

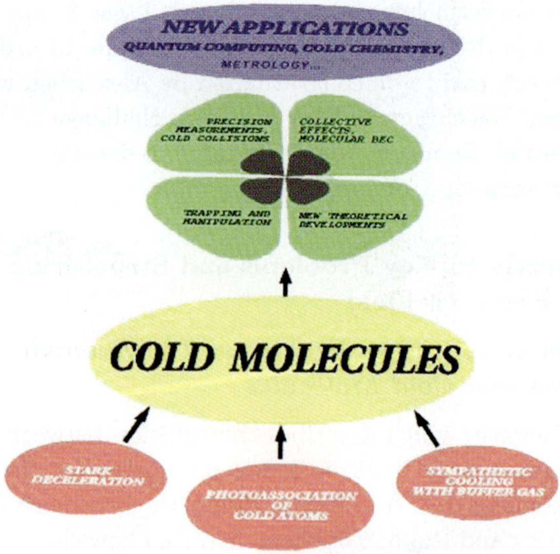

Fig. 2. The projects for the Cold Molecules European network in 2001.

The following table lists some of the main conferences and workshops, that were very stimulating and contributed markedly to the development of the field.

Location (Date)	Title	Organizers
ITAMP (April 1992)	Cold Collisions	Y. Band and P. Julienne
Les Houches (February 1999)	Cold Molecules	A. Dalgarno, F. Masnou
ITAMP (July 1999)	Trapping, spectroscopy and collisions of ultracold molecules	R. Forrey, J. Doyle
Les Houches (March 2002)	Cold Molecules and BEC	O. Dulieu, S. Guibal
ITAMP (January 2004)	Ultracold Polar Molecules	J. Doyle, J. Hutson, R. Krems
Telluride (July 2005)	Theory of Cold Molecules	R. Krems, A. Dalgarno, F. Masnou
Vancouver (July 2007)	Coherent Control of Ultracold Molecules	R. Krems

In particular, the 2004 conference, held while a really ultracold weather had settled in Cambridge, was followed by a special issue of the European Journal of Physics D (vol 31, december 2004) which is highly quoted.

4. Conclusion

I am deeply grateful to Alex for his outstanding hospitality and support, for his leading role in the development of the field of Cold Molecules.

Fig. 3. Involvementment at scientific meetings : before a public lecture in Telluride (left); going for dinner after a DAMOP 2007 session in Calgary (right).

Fig. 4. The field of cold molecules has become main stream : breakfast in UBC, Vancouver July 2007.

References

1. A. Dalgarno and J. T. Lewis, *Proc. Phys. Soc. Lond.* **233**, 70 (1955).
2. C. Bottcher and A. Dalgarno, *Proc. Roy. Soc. Lond.* **340**, 187 (1974).
3. F. Masnou-Seeuws, M. Philippe, P. Valiron,*Phys. Rev. Lett.* **41**, 395 (1978).
4. P. Valiron, R. Gayet, R. Mc Carroll, F. Masnou-Seeuws, M. Philippe *J. Phys. B* **12**,53 (1979).
5. M. Marinescu and A. Dalgarno, *Phys. Rev. A* **52**, 311 (1995)
6. R. Côté, A. Dalgarno, Y. Sun, R. G. Hulet, *Phys. Rev. Lett.* **74**, 3581 (1995).
7. R. Côté and A. Dalgarno, *Chem. Phys. Lett.* **279**, 50 (1997).
8. A. Fioretti, D. Comparat, A. Crubellier, O. Dulieu, F. Masnou-Seeuws and P. Pillet, *Phys. Rev. Lett.* **80**, 4402 (1998).

PHOTOASSOCIATION OF ULTRACOLD ATOMS

R. G. HULET*

Department of Physics and Astronomy, Rice University,
Houston, Texas 77005
E-mail: randy@rice.edu

Photoassociation of ultracold atoms has been used to determine interaction potentials and to extract scattering lengths with great precision, which we discuss here in the context of lithium. We also describe a recent experiment where saturation in the rate of photoassociation of an atomic BEC was observed.

Keywords: photoassociation; diatomic; molecule

1. Introduction

Our collaboration with Alex Dalgarno began in 1994, shortly after we obtained our first photoassocation spectra of lithium atoms confined to a magneto-optical trap (MOT). Our interest in photoassociation was motivated by the need to understand the s-wave interactions of lithium for a Bose-Einstein condensation experiment. It was generally believed that attractive interactions, corresponding to a negative s-wave scattering length, a, would lead to mechanical instability and no condensation. The spectra were complicated, and we were having a hard time making any sense of them. Alex happened to be visiting Rice University at about this time, and I took the opportunity to show the spectra to him. Alex looked at the data for awhile, asked a few questions, and then proceeded to make the following pronouncement: one of the scattering lengths, singlet or triplet, is negative and the other is positive (!). Of course, I was floored by this insight and wanted to know more. Thus began an enjoyable and fruitful collaboration with Alex and Robin Côté, then Alex's student, in which we worked out the singlet and triplet scattering lengths of both lithium isotopes, ^{6}Li and

*I gratefully acknowledge the students and post-doctoral scientists who have contributed to this work, especially E. R. I. Abraham, Y. P. Chen, D. Dries, J. M. Gerton, J. Hitchcock, M. Junker, W. I. McAlexander, and I. D. Prodan.

^{7}Li. I regard this collaboration as one of the most gratifying and productive of my career.

In this chapter, I will discuss how the interactions in ultracold atomic gases depend on the two-body potentials, and how the scattering lengths can be obtained from photoassociation experiments. I will also describe the implications of the results for experiments with the bosons, ^{7}Li, as well as the fermions, ^{6}Li. Finally, I will discuss a new experiment for which we observe saturation in the rate of photoassociation of a BEC of ^{7}Li near a Feshbach resonance. While this is the first observation of saturation in a BEC, interestingly, the saturated rate is much higher than expected.

2. Interactions in Ultracold Atomic Gases

The density is usually small enough in experiments with ultracold gases (typically less than 10^{14} cm^{-3}) that only binary collisions contribute to the interactions of the gas. Furthermore, the de Broglie wavelengths are many orders of magnitude larger than the scale length of the two-body interaction potentials, which are generally about 1 nm. Under these conditions, the effect of interactions can be described by the scattering length a of the dominant partial wave, usually the s-wave, but it could be the p-wave in the case of identical fermions.[1] For low-energy two-body scattering a is the asymptotic displacement of the scattered wave relative to the incoming wave. For a sufficiently small wave vector k, the scattering phase shift $\delta \simeq -ka$.

The implications of the scattering length for quantum gases of ultracold atoms is profound. In the case of bosons, $a > 0$ implies a stable condensate whose thermodynamic properties can be found from a perturbation expansion in na^3, where n is the density,[1] while for a two-component mixture of fermions, a positive means that the atom pairs can bind as bosonic molecules and then condense. The situation is very different for $a < 0$. For bosons, the gas is mechanically unstable and BEC in a gas was thought to be impossible.[2–4] Fermions with negative a can form Cooper pairs at sufficiently low temperature and become superfluid. So in 1994, as we were trying to Bose condense ^{7}Li, we needed to know: is a positive or negative?

3. Photoassociation and the Scattering Length

Photoassociative spectroscopy of ultracold atoms has proven to be one of the most precise and useful methods for determining scattering lengths.[5]

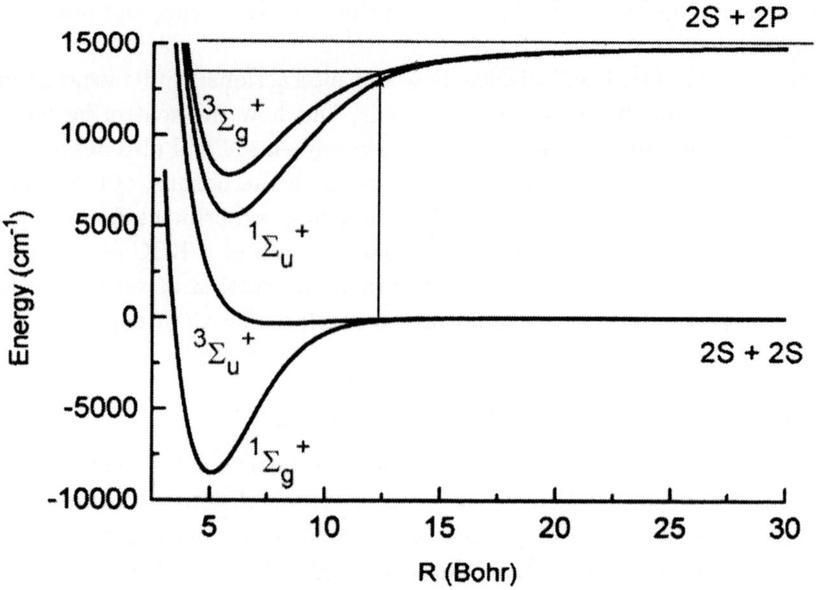

Fig. 1. The molecular potentials of lithium correlating to the 2s+2s ground state and to the 2s+2$p_{1/2}$ excited state. The arrow indicates a free-bound photoassociation resonance.

These methods have been applied to many atomic elements over the past 15 years. In this section, I will describe our work with lithium.

In one-photon photoassociation, a laser beam is passed through a gas of ultracold atoms confined to a trap. As the laser frequency is tuned to a free-bound resonance, diatomic molecules are formed, resulting in a detectable decrease in the number of trapped atoms (see Fig. 1). Figure 2 shows our first spectrum of ^{7}Li, which is the one shown to Alex in 1994. As can be seen, there are a large number of lines over a vast (greater than 1 THz) frequency scale. Alex noted that there were two series of alternating lines, one corresponding to the electronic triplet ($S = 1$) and the other to the singlet ($S = 0$) configuration, and that these two series had very different character. Figure 2 shows that one series is stronger and is visible for nearly 2 THz, while the other is weaker and is not visible beyond ~400 GHz. Alex was able to explain the spectrum qualitatively from the fact that the asymptotic scattering state is proportional to $\sin(kR + \delta) \sim \sin(k(R - a))$ $\sim R - a$, where R is the internuclear distance. If $a > 0$ there will be a node in the scattering state at $R = a$, which results in a node in the

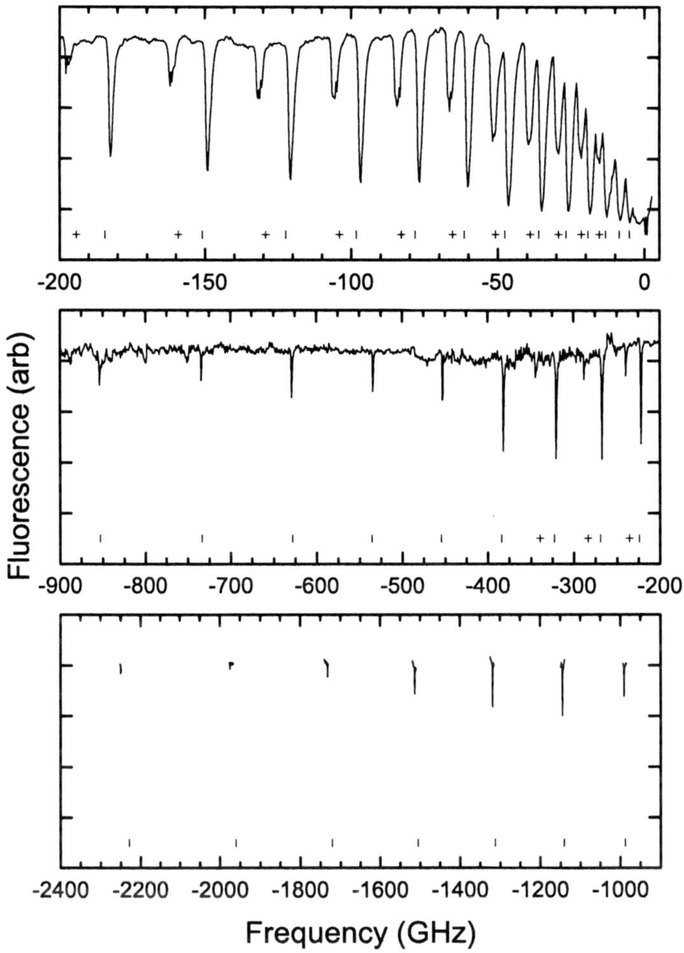

Fig. 2. Photoassociation spectrum of ^{7}Li. The energy scale is relative to the $2s_{1/2}$-$2p_{1/2}$ asymptote. The photoassociation laser is tuned while monitoring the fluorescence signal from a magneto-optical trap. (Reprinted from Ref. 6).

excitation spectrum when the outer turning point (Condon point) of the vibrational state in the excited potential corresponds to a. If $a < 0$, on the other hand, there is no node in the long-range asymptotic scattering state. After more quantitative analysis, we concluded that the ground-state triplet corresponded to $a < 0$, while the singlet was the weaker series with $a > 0$.[7]

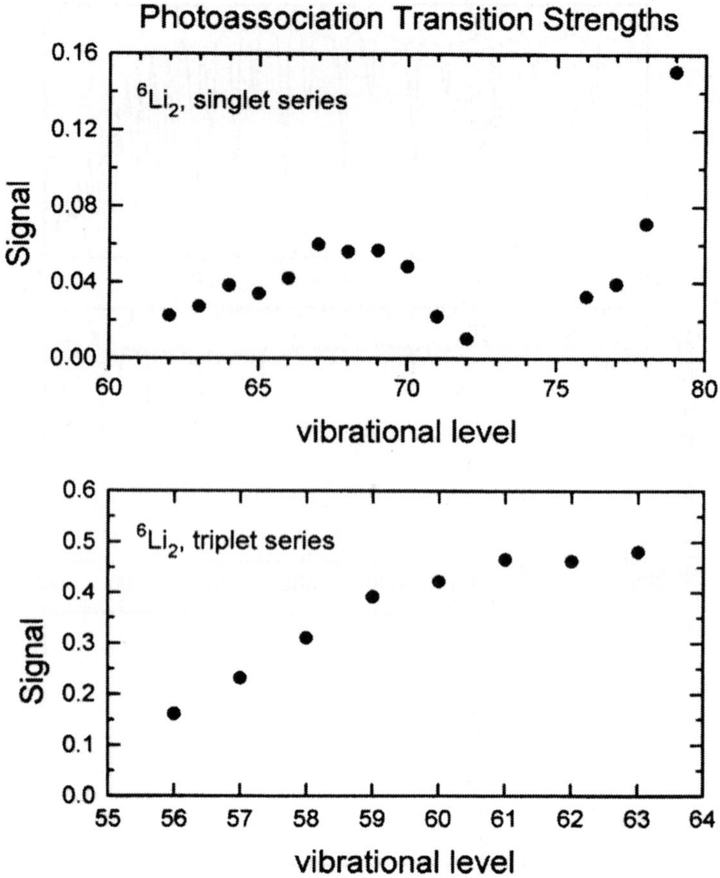

Fig. 3. Photoassociation peak signal strength vs. vibrational quantum number for the singlet (upper) and triplet (lower) levels of ^{6}Li. (Adapted from Ref. 8).

Using the argument of the previous paragraph, Alex predicted that the singlet series would reappear for Condon radius smaller than a, that is, for more deeply bound states. Figure 3 shows the trap loss signal vs. vibrational quantum number v for both the singlet and triplet levels of ^{6}Li. As predicted for the $a > 0$ case (singlets), the signal vanishes for a particular v (72), while for smaller v, the signal reappears. For the triplets, however, $a < 0$, and there is no node in the ground-state wavefunction in the asymptotic region. The identification of the nodal position tightly constrains the ground-state potential, enabling a precise determination of the scattering length.

Table 1. Singlet and triplet scattering lengths in units of a_o, for isotopically pure and mixed gases of lithium.

	^{6}Li	^{7}Li	^{6}Li/^{7}Li
a_T	-2160 ± 250	-27.6 ± 0.5	40.9 ± 0.2
a_S	45.5 ± 2.5	33 ± 2	-20 ± 10

3.1. *Two-photon Photoassociation*

While the ^{6}Li and ^{7}Li interaction potentials are essentially identical (we have never measured a significant difference), their respective scattering lengths are entirely different. Most strikingly, the triplet scattering length for ^{7}Li is -27 a_o, where a_o is the Bohr radius, while for ^{6}Li it is -2200 a_o. The difference, of course, is related to mass. With its lighter mass, the ^{6}Li triplet ground state supports only 10 bound states, while ^{7}Li has 11. The proximity of the last bound state to the dissociation limit largely determines the low-energy scattering properties of the atom. A more precise method for finding scattering lengths, therefore, is to probe the ground-state molecular levels directly. This is particularly true when $a < 0$, since there are no zeroes in the single-photon spectrum, at least in the asymptotic region.

We have used two-photon photoassociation to directly measure the binding energy of the last bound triplet state for both ^{7}Li,[9] and ^{6}Li.[10] In this method, one laser is tuned near the one-photon photoassociation transition, while the second is tuned in resonance back down to the least-bound ground state level. The binding energy is just the frequency difference between the two lasers, which can be measured to extremely high precision using a high-speed photodiode and a counter, or the two lasers may even be phase locked to achieve atomic clock precision. In our final collaborative paper on scattering lengths,[10] we summarized all of our measurements of the scattering lengths of both isotopes of lithium, as well as inter-isotope combinations (see Table 1).

4. Implications of Interactions in Quantum Gases of Lithium

We subsequently performed experiments with both the bosonic isotope, ^{7}Li, and the fermionic one, ^{6}Li. Some of these results are briefly described in this section.

4.1. *Bose-Einstein Condensation*

Despite the prediction that BEC of an attractively interacting gas was not possible, we did indeed observe condensation.[11] The confinement of the trap provides a quantum pressure that counteracts the attractive interactions, at least up to a finite number of atoms.[12] For our trap, this number was ~1200,[13] and for numbers beyond this, the condensate was found to implode.[14,15]

These first experiments were done with the $F = 2, m_F = 2$ (2, 2) state, which is pure triplet, corresponding to $a = -27a_o$. The (2, 2) state is the only stable magnetically-trappable state. This constraint can be relaxed, however, by confining the atoms to an optical trap instead. The (1, 1) state is particularly useful for BEC experiments, because its scattering length can be tuned with a magnetic field via a Feshbach resonance. Evaporative cooling can be made very efficient by making a large and positive, enabling the creation of a large, stable BEC. The Feshbach resonance can also be used to make a small and negative, which the Paris group and our group exploited to make bright, matter-wave solitons.[16,17] We also use this Feshbach resonance in combination with photoassociation to explore photoassociation in the strong coupling regime, as discussed in Sect. 5.

4.2. *Pairing of Fermions*

The identification of the extraordinarily large and negative triplet scattering length in ^{6}Li led us, in collaboration with Henk Stoof, to suggest that ^{6}Li was an ideal candidate to observe Cooper pairing in a gas of atomic fermions.[18] A negative scattering length, corresponding to attractive interactions, is needed to form (unbound) pairs and the large magnitude is important since the transition temperature depends exponentially on $|a|$. We also used the photoassociation data to develop two-body potentials that lead to the prediction of a Feshbach resonance between the $(1/2, 1/2)$ and $(1/2, -1/2)$ atoms in ^{6}Li.[19] This Feshbach resonance was subsequently used by several groups, including ours, to explore the so-called BEC-BCS crossover. In the BCS regime, the attractive interaction between pairs is small so that the pairs are large compared to the average interparticle distance. In the BEC regime, on the other had, the attraction is so strong that the "pairs" are more properly described as tightly bound bosonic molecules.

5. Limits to the Rate of Association

There is currently great interest in creating and trapping ultracold molecules. One pathway to obtain them is to associate ultracold atoms, by either sweeping a magnetic field through a Feshbach resonance[20–24] or by photoassociation. A natural question to ask is what are the fundamental limitations to the rate of association?

Quantum mechanical unitarity always imposes a limit to the scattering cross section, proportional to the square of the deBroglie wavelength. This unitarity limit has been previously observed in non-condensed atomic gases (see for instance, Ref. 25). In the case of a BEC, however, the unitarity limit is expected to be so high that it will be preempted by other mechanisms, and therefore, be irrelevant. Juha Javanainen and collaborators have proposed that at very high rates of association in a BEC, the rate is limited by a process they call "rogue photodissociation", where atoms associated out of the condensate are dissociated into the hot continuum rather than back to the condensate.[26] By energy conservation, the width of this continuum is equal to the effective Rabi frequency of the association process. Javanainen *et al.* have shown that rogue photodissociation gives a limiting rate constant $K_d \sim \hbar n^{-1/3}/m$, where n and m are the atomic density and mass, respectively. More recent theory shows that rogue photodissociation arises from a universal transient response, which is independent of the underlying microscopic association process.[27–29]

Only one previous experiment, performed at NIST with a sodium condensate, has explored high rates of association in a BEC.[30] Although rate constants as high as $\hbar n^{-1/3}/m$ were obtained (just), the rate increased linearly with intensity, and no saturation was observed.

5.1. *Photoassociation Near a Feshbach Resonance*

Since a Feshbach resonance strongly affects the scattering wavefunction, we expect the rate of photoassociation to be correspondingly modified. We have explored this dependence by performing photoassociation on the (1, 1) state of ^{7}Li with fields tuned close to the 737 G Feshbach resonance.[31] Figure 4 shows the photoassociation rate constant, K_p, for a non-condensed gas and for fields near the Feshbach resonance. A maximum in K_p is observed when the field corresponds to the Feshbach resonance, while a minimum is found for a field about 25 G lower. This variation in K_p can be understood by the fact that K_p depends on the magnitude of the scattering state wavefunction at the Condon radius, which varies with magnetic field. As

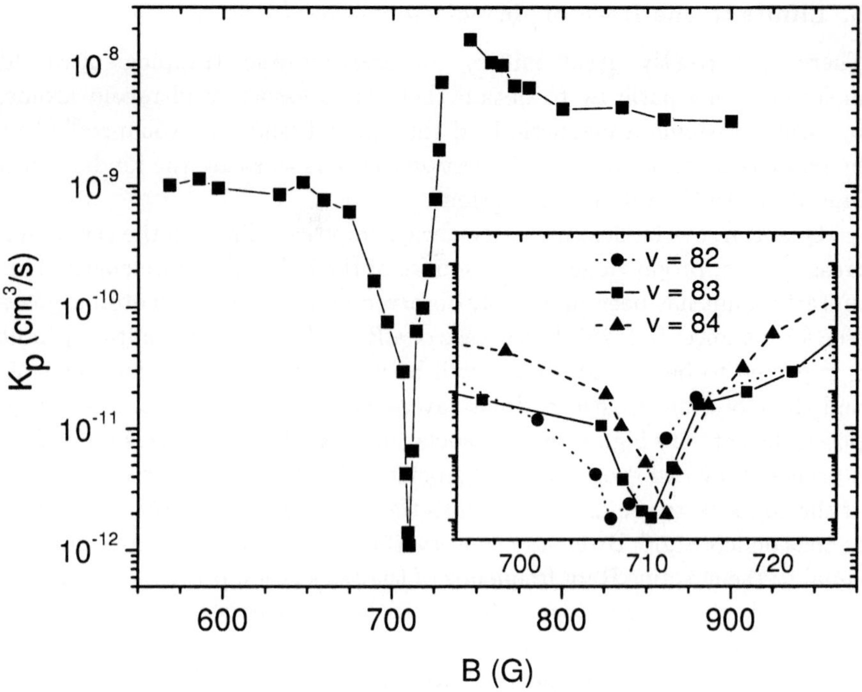

Fig. 4. K_p for the v=83 triplet of ^{7}Li for a thermal gas at ~10 μK. The laser intensity is fixed at 1.65 W/cm^2, while the pulse duration is adjusted between 0.07 and 270 ms to maintain the fractional loss. The inset shows K_p (same scale as main figure) for three different excited state vibrational levels. (Reprinted from Ref. 31).

discussed in Sect. 3, K_p vanishes when a is tuned to the Condon radius, while K_p is maximized when $|a|$ is large. These observations have been discussed theoretically by Pellegrini *et al.*[32]

We exploited this Feshbach enhancement to achieve extremely large rates of photoassociation in a BEC. Figure 5, displaying K_p vs. intensity of the photoassociation laser, clearly exhibits saturation for $K_p \gtrsim 5 \times 10^{-8}$ cm^3/s. The maximum rate constant, K_{max} is nearly 10 times larger than previously observed for any photoassociation experiment. Interestingly, it is also nearly 20 times greater than K_d and is nearly 7 times greater than predicted from the theory of Ref. 29. K_{max} is consistent with unitarity, however, if the de Broglie wavelength is defined by the smallest dimension (the radial width) of our cigar-like condensate.

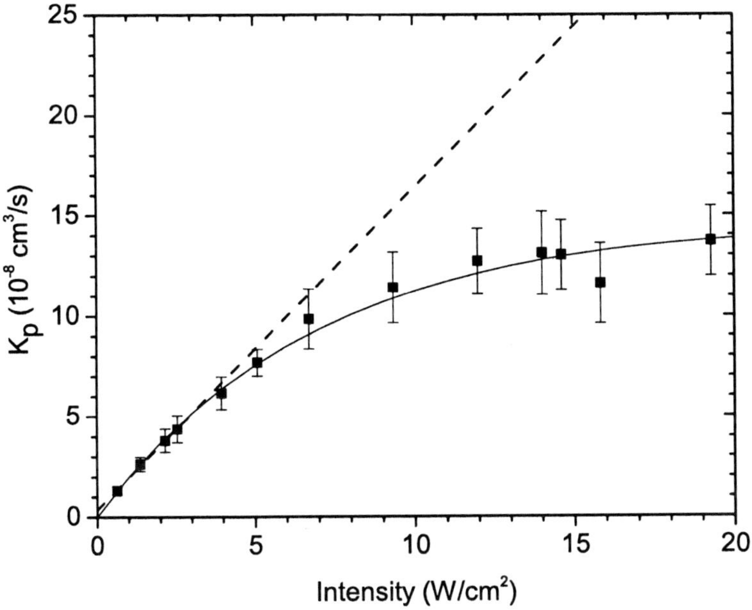

Fig. 5. K_p for a BEC at 732 G (5 G below the Feshbach resonance). The solid line is a fit to a generic saturation function. The dashed line is a linear fit to the low intensity data. The error bars are the statistical uncertainties. (Reprinted from Ref. 31).

The origin of this discrepancy is not presently clear. Mackie *et al.* have directly modeled our experiment with good quantitative results, but they do not determine the origin of the observed rate limit.[33] It is possible that very near the Feshbach resonance, the ground state atoms become weakly bound via three-body recombination collisions. The "photoassociation" laser would then be driving a bound-bound transition rather than free-bound, making the rogue photodissociation and transient response theories inapplicable.

6. Summary

Photoassociation of ultracold atoms is a wonderfully sensitive tool for measuring atomic interaction potentials, and in particular, for determining *s*-wave scattering lengths. The collaboration between Alex Dalgarno, Robin Côté and our group produced interatomic interaction potentials with a quantitative precision that had not previously been achieved.

Photoassociation is also a useful tool for creating ultracold molecules. By tuning a magnetic field near a Feshbach resonance, the rate of photoassociation can be made huge. We have exploited this effect to observe saturation of the rate of association of atoms in a BEC for the first time. The saturation level is much higher than expected from theory. We are looking to Alex to help solve this mystery!

I wish to take this occasion to thank Alex for the opportunity to work with him, and for bringing so much joy in the process. Happy Birthday!

Acknowledgments

This work has been supported by the NSF, ONR, the Keck Foundation, and the Welch Foundation (C-1133).

References

1. K. Huang, *Statistical Mechanics*, 2 ed. (John Wiley & Sons, New York, 1987).
2. N. Bogolubov, *J. of Phys.* **XI**, 23 (1947).
3. L. D. Landau and E. M. Lifshitz, *Statistical Physics*, 1st ed. (Pergamon, London, 1958).
4. H. T. C. Stoof, *Phys. Rev.* A **49**, 3824 (1994).
5. J. Weiner, V. Bagnato, S. Zilio, and P. S. Julienne, *Rev. Mod. Phys.* **71**, 1 (1999).
6. E. R. I. Abraham, N. W. M. Ritchie, W. I. McAlexander, and R. G. Hulet, *J. Chem. Phys.* **103**, 7773 (1995).
7. R. Côté, A. Dalgarno, Y. Sun, and R. G. Hulet, *Phys. Rev. Lett.* **74**, 3581 (1995).
8. E. R. I. Abraham, W. I. McAlexander, J. M. Gerton, and R. G. Hulet, *Phys. Rev.* A **53**, R3713 (1996).
9. E. R. I. Abraham, W. I. McAlexander, C. A. Sackett, and R. G. Hulet, *Phys. Rev. Lett.* **74**, 1315 (1995).
10. E. R. I. Abraham, W. I. McAlexander, J. M. Gerton, R. G. Hulet, R. Côté, and A. Dalgarno, *Phys. Rev.* A **55**, R3299 (1997).
11. C. C. Bradley, C. A. Sackett, J. J. Tollett, and R. G. Hulet, *Phys. Rev. Lett.* **75**, 1687 (1995).
12. P. A. Ruprecht, M. J. Holland, K. Burnett, and M. Edwards, *Phys. Rev.* A **51**, 4704 (1995).
13. C. C. Bradley, C. A. Sackett, and R. G. Hulet, *Phys. Rev. Lett.* **78**, 985 (1997).
14. C. A. Sackett, J. M. Gerton, M. Welling, and R. G. Hulet, *Phys. Rev. Lett.* **82**, 876 (1999).
15. J. M. Gerton, D. Strekalov, I. Prodan, and R. G. Hulet, *Nature* **408**, 692 (2000).
16. L. Khaykovich, F. Schreck, G. Ferrari, T. Bourdel, J. Cubizolles, L. D. Carr, Y. Castin, and C. Salomon, *Science* **296** 1290 (2002).

17. K. E. Strecker, G. B. Partridge, A. G. Truscott, and R. G. Hulet, *Nature* **417**, 150 (2002).
18. H. T. C. Stoof, M. Houbiers, C. A. Sackett, and R. G. Hulet, *Phys. Rev. Lett.* **76**, 10 (1996).
19. M. Houbiers, H. T. C. Stoof, W. I. McAlexander, and R. G. Hulet, *Phys. Rev. A* **57**, R1497 (1998).
20. C. A. Regal, C. Ticknor, J. L. Bohn, and D. S. Jin, *Nature* **424**, 47 (2003).
21. K. E. Strecker, G. B. Partridge, and R. G. Hulet, *Phys. Rev. Lett.* **91**, 080406 (2003).
22. J. Cubizolles, T. Bourdel, S. J. J. M. F. Kokkelmans, G. V. Shlyapnikov, and C. Salomon, *Phys. Rev. Lett.* **91**, 240401 (2003).
23. S. Jochim, M. Bartenstein, A. Altmeyer, G. Hendl, C. Chin, J. Hecker Denschlag, and R. Grimm, *Phys. Rev. Lett.* **91**, 240402 (2003).
24. M. W. Zwierlein, C. A. Stan, C. H. Schunck, S. M. F. Raupach, S. Gupta, Z. Hadzibabic, and W. Ketterle, *Phys. Rev. Lett.* **91**, 250401 (2003).
25. I. D. Prodan, M. Pichler, M. Junker, R. G. Hulet, and J. L. Bohn, *Phys. Rev. Lett.* **91**, 080402 (2003).
26. J. Javanainen and M. Mackie, *Phys. Rev. Lett.* **88**, 090403 (2002).
27. T. Gasenzer, *Phys. Rev.* A **70**, 043618 (2004).
28. P. Naidon and F. Masnou-Seeuws, *Phys. Rev.* A **73**, 043611 (2006).
29. P. Naidon, E. Tiesinga, and P. S. Julienne, *Phys. Rev. Lett.* **100**, 093001 (2008).
30. C. McKenzie *et al.*, *Phys. Rev. Lett.* **88**, 120403 (2002).
31. M. Junker, D. Dries, C. Welford, J. Hitchcock, Y. P. Chen, and R. G. Hulet, *Phys. Rev. Lett.* **101**, 060406 (2008).
32. P. Pellegrini, M. Gacesa, and R. Côté, *Phys. Rev. Lett.* **101**, 053201 (2008).
33. M. Mackie, M. Fenty, D. Savage, and J. Kesselman, *Phys. Rev. Lett.* **101**, 040401 (2008).

FORMING ULTRACOLD MOLECULES

ROBIN CÔTÉ

Physics Department, University of Connecticut,
Storrs, Connecticut 06269-3046, USA
E-mail: rcote@phys.uconn.edu
www.phys.uconn.edu/~rcote

One of the many important contributions of Alex Dalgarno to Atomic, Molecular, and Optical (AMO) Physics relates to his work on ultracold physics. After a brief review of his early work on photoassociation (PA) of ultracold atoms and the formation of ultracold molecules, a short description of new results using Feshbach Optimized Photoassociation (FOPA) will be given. This research points to Alex's influence on the work now performed in my group at UConn.

Keywords: Ultracold atoms; ultracold molecules; photoassociation; Feshbach resonances.

1. Introduction

In what follows, I give a brief account of my early days working with Alex as a graduate student, and discuss how his extraordinary sense of the new frontiers in AMO Physics molded not only my career, but that of many researchers in the world. This will be followed by a short review of some studies done with Alex on photoassociation of ultracold atoms, and on the formation of ultracold molecules in their electronic ground state. Finally, I will describe more recent calculations performed in my group on forming ultracold molecules using a new approach labeled FOPA (Feshbach optimized photoasociation). This work is a direct consequence of Alex's teaching and interests, and shows that his influence on the field keeps growing.

2. My Early Days with Alex

As a young graduate student at M.I.T., I was looking for an advisor to work on something "hot and new". At the time, I was interested in Statistical Mechanics, and especially in phase transitions (the topic of my M.S. thesis at Laval University). A good friend of mine, Daniel Proulx, who was then a student at USC, mentioned that his advisor, Robin Shakeshaft, had just come back from Harvard where Alex had told him that he had an opening for a Ph.D. student. Being curious about that possibility, I decided to contact Alex and set up a meeting.

First, I must admit that I knew nothing of Alex, and that was probably a good thing: I was not at all nervous to meet such a great name in AMO Physics. When I arrived at his office, however, I noticed that he had quite a large room, with a secretary at the entrance and a private back room ... these were hinting at something big! But again, ignorance is bliss, and I started the meeting by asking Alex if he had interesting projects to work on. With a trace of a smile, he began telling me about three potential research topics, each one seemingly more interesting than the other. I then proceeded to ask Alex what he considered reasonable for the duration of a thesis, a few years or a decade ... with a larger smile this time, he answered that 2-3 years should be more than enough, once working on the project full time. My next question got Alex to let out a short laugh; do you have money to support me? Again, not knowing Alex was really a blessing; had I known who he was, I would have never had the nerve to ask something like that!

In any case, the following question was again to the point: when can I start? His answer: tomorrow if you want ... And so I started to work with Alex, just like that! Of the three potential projects, the topic that became my thesis appeared the most interesting: ultracold collisions of identical atoms. One has to remember that this meeting occurred in the Fall of 1991, when only a few people were aware of the potential importance of the topic!

Our initial studies were on collision of ^{7}Li atoms,[1] followed by collisions of Na atoms.[2] This work introduced me to the difficulties of constructing molecular curves that include the proper long-range form while being accurate enough to predict the low-energy scattering properties. It also led to our first paper on photoassociation (PA) of ultracold atoms to form excited molecules.[3] This work was inspired by a visit of Alex to Rice University, where he had talked with Randy Hulet about recent measurements of PA spectra featuring "strange" behavior (at the time). We found that those features were consistent with a negative triplet scattering length (a_T) and a positive singlet scattering length (a_S). We further predicted a simple procedure to determine the value of a positive scattering length from PA spectra (see next section for details).

All this work finally led to a thesis[4] defense on December 2nd 1994, with Alex as my supervisor, Dan Kleppner as my co-supervisor, and Wolfgang Ketterle as the third member of the thesis committee (see Fig.1). Alex was amused at the reason why I wanted to defend my thesis right away – so that I could get my Ph.D. officially before I turned 30! Throughout the defense, Alex made sure that I understood the questions asked by the two other members, in effect "protecting" me from going in the wrong direction when answering questions. This was indeed a very exciting time in AMO; Wolfgang Ketterle had spent the whole night before the defense in the lab, working on his BEC experiment – which finally led to his Nobel Prize in 2001. The defense went well, and in June 1995, Alex accompanied me to the Commencement, a gesture that I still appreciate very much to this day.

Fig. 1. My thesis defense, Dec. 2 1994: starting from the right, Alex Dalgarno, Dan Kleppner, Robin Côté, and Wolfgang Ketterle.

Alex was also instrumental in my "re-entry" into physics after a short interval as a management consultant with McKinsey & Co. in Montréal. I had left physics because there were basically no positions in physics in 1994-1995, and I wanted to explore other options. After a year or so, however, I fully realized that physics was really what I liked to do, and contacted Alex to see if he thought it would be possible to make a "come-back" as a postdoc, and if so, if he knew of positions. After mentioning a few possibilities, he said "It is too bad, you just missed the ITAMP Postdoc Fellow deadline by a few days." He nonetheless thought that I should prepare an application package and send it to ITAMP very quickly. In the meantime, he would check with ITAMP members, especially Rick Heller and Kate Kirby, as well with the search committee and the ITAMP Board members, if they would consider late applicants. To my great relief, they gave the go ahead, and I finally was offered an ITAMP postdoctoral Fellowship in May 1996, and joined ITAMP in October 1996. I really owe to Alex the chance to come back to physics: without his good word on my behalf, I might still be in the business world!

3. Photoassociation: Early Results

As mentioned in the previous section, my early work with Alex involved scattering of ultracold atoms, and photoassociation of ^{7}Li atoms into excited molecular states. Here, I give a brief overview of the theory used, and results obtained. Those form the "spring board" for the recent results presented in the following sections.

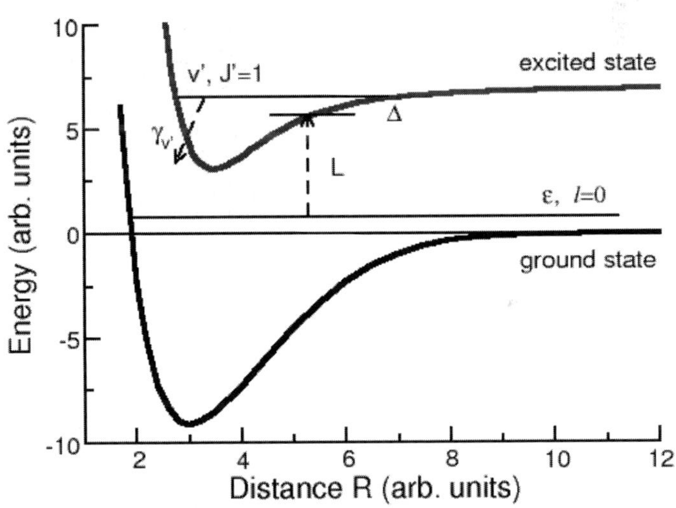

Fig. 2. Schematics of PA: two colliding atoms of energy ε make a transition to a bound level v' (of width $\gamma_{v'}$) induced by a laser L of intensity I and detuning Δ from v'.

In general, a pair of atoms approaching each other along the molecular ground state in the presence of an optical field can make a transition into a molecular bound level v' of an excited electronic state (see Fig. 2). The corresponding photoassociation rate $\mathsf{K}_{v'}$ from a laser $L = \{I, \Delta\}$ (with intensity I and detuning Δ from a bound level (v', J') is given by[5–8]

$$\mathsf{K}_{v'}(T,L) = \left\langle \frac{\pi v_{\rm rel}}{k^2} \sum_{\ell=0}^{\infty} (2\ell+1) |S_{\ell,v'}(\varepsilon, L)|^2 \right\rangle , \tag{1}$$

where $\varepsilon = \hbar^2 k^2/2\mu = \mu v_{\rm rel}^2/2$, μ is the reduced mass, $v_{\rm rel}$ is the relative velocity of the colliding pair, and $S_{\ell,v'}$ represents the scattering matrix element for producing the state v' from the continuum state. Averaging over $v_{\rm rel}$ is implied by $\langle \ldots \rangle$.

At ultracold temperatures, only the s-wave ($\ell = 0$) scattering is relevant (for distinguishable atoms or identical bosons). Assuming a non-degenerate gas, keeping only the $\ell = 0$ contribution, and averaging over a Maxwellian velocity distribution characterized by the temperature T, we obtain[6–8]

$$\mathsf{K}_{v'}(T,L) = \frac{1}{hQ_T} \int_0^{\infty} d\varepsilon \; e^{-\varepsilon/k_B T} |S_{\ell=0,v'}(\varepsilon, L)|^2 , \tag{2}$$

where $Q_T = \left(\frac{2\pi\mu k_B T}{h^2}\right)^{3/2}$. We note that only $(v', J' = 1)$ ro-vibrational levels can be populated for $\Sigma \rightarrow \Sigma$ transitions. The scattering matrix is well approximated by[6,8]

$$|S_{\ell,v'}(\varepsilon, L)|^2 = \frac{\gamma_{v'}\gamma_s(\varepsilon,\ell)}{(\varepsilon - \Delta)^2 + (\gamma/2)^2} , \tag{3}$$

where $\gamma = \gamma_{v'} + \gamma_s$. Here $\gamma_{v'}$ is the width of the bound level v' and γ_s is the stimulated width from the continuum initial state $|\varepsilon, \ell = 0\rangle$ to the target state $|v', J' = 1\rangle$. The stimulated width γ_s can be expressed using the Fermi golden rule as[6–8]

$$\gamma_s = \frac{\pi I}{\epsilon_0 c} |\langle \phi_{v',J'=1} | D(R) | \Psi_{\varepsilon,\ell=0} \rangle|^2 , \tag{4}$$

where ϵ_0 and c are the vacuum permittivity and speed of light, respectively, and $\phi_{v',J'=1}(R)$ and $\Psi_{\varepsilon,\ell=0}(R)$ stand for the wave functions of the excited level v' and the colliding pair of atoms, respectively. Both vary with the internuclear separation R, as does the transition dipole moment $D(R)$. We note that the dipole matrix element appearing in Eq.(4) follows Wigner's threshold law,[9] *i.e.* $|\langle \phi_{v,J=1} | D(R) | \Psi_{\epsilon,l=0} \rangle|^2 = C_{v'}\sqrt{\varepsilon}$, where the coefficient $C_{v'}$ depends on the details of the wave functions.[3,10,11]

If $\gamma_s/\gamma_{v'} \ll 1$, we can approximate $|S_{\ell,v'}|^2$ by $2\pi\gamma_s(\varepsilon,\ell)\delta(\varepsilon-\Delta)$.[6–8,10,11] Substituting this expression in Eq.(2), and using $\gamma_s \propto C_{v'}\sqrt{\varepsilon}$, we find $\mathsf{K} \propto e^{-\Delta/k_BT}\sqrt{\Delta}$, with a maximum value for $\Delta = k_BT/2$. The rate coefficient then takes the simple form[10,11]

$$\mathsf{K}_{v'}(T,I) = \frac{2\pi^2}{h}\frac{I}{\epsilon_0 c}\frac{e^{-1/2}}{Q_T}C_{v'}\sqrt{k_BT/2}\,. \tag{5}$$

Note that we have not accounted for the light polarization in the above expression. The numerical pre-factors are simple to evaluate, and basically scale as $\mu^{-3/2}$; in systems we studied, such as $^7Li_2$, ^7LiH, or NaH,[3,10,11] they are

$$\mathsf{K}_{v'}(T,I) = C_{v'}\frac{I}{T}\times\begin{cases} 1.1\times10^{-24}\ \text{cm}^3/\text{s}\ , \text{ for Li}_2\ , \\ 9.0\times10^{-24}\ \text{cm}^3/\text{s}\ , \text{ for LiH}\ , \\ 8.0\times10^{-24}\ \text{cm}^3/\text{s}\ , \text{ for NaH}\ , \end{cases} \tag{6}$$

with $C_{v'}$ in a.u., I in W/cm^2, and T in Kelvin.

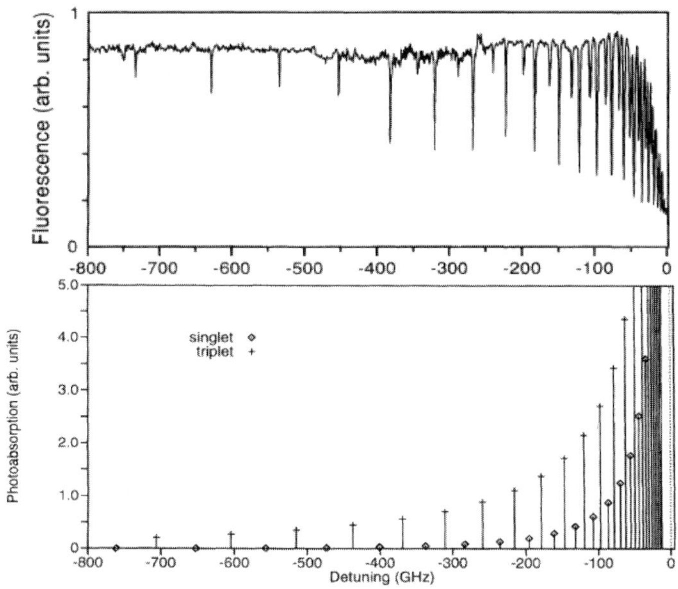

Fig. 3. Top: photoassociation spectrum of ^{7}Li atoms. Bottom: corresponding theoretical rate coefficient $\mathsf{K}_{v'}$ for singlet and triplet transitions. Both plots are from Côté *et al.*[3]

In our early work on ^{7}Li,[3] we calculated $\mathsf{K}_{v'}$ for singlet $X^1\Sigma_g^+ \to A^1\Sigma_u^+$ and triplet $a^3\Sigma_u^+ \to 1^3\Sigma_g^+$ transitions (see Fig.3). We found that both transitions had very different behavior with v': the triplet had strong line strengths even for deep levels v', while the values for the singlet were vanishing rapidly. In that work, we showed[3] that the dipole matrix element has a simple dependence on the scattering length a for high/extended levels v', which can be written as

$$C_{v'} \propto (R_{v'} - a)^2 , \tag{7}$$

where $R_{v'}$ is the classical outer turning point in the potential curve for the level v'. By comparing with the experimental results, we concluded that the triplet scattering length must be negative, while the singlet one must be positive. We predicted that the lines in the singlet spectra should become weak with decreasing v', reaching a minimum as $R_{v'} \sim a$, and then increase again. These predictions were confirmed later by Randy Hulet[12,13] (see Fig. 4), and we used PA spectra to infer the value of the singlet scattering length for both isotopes of lithium.

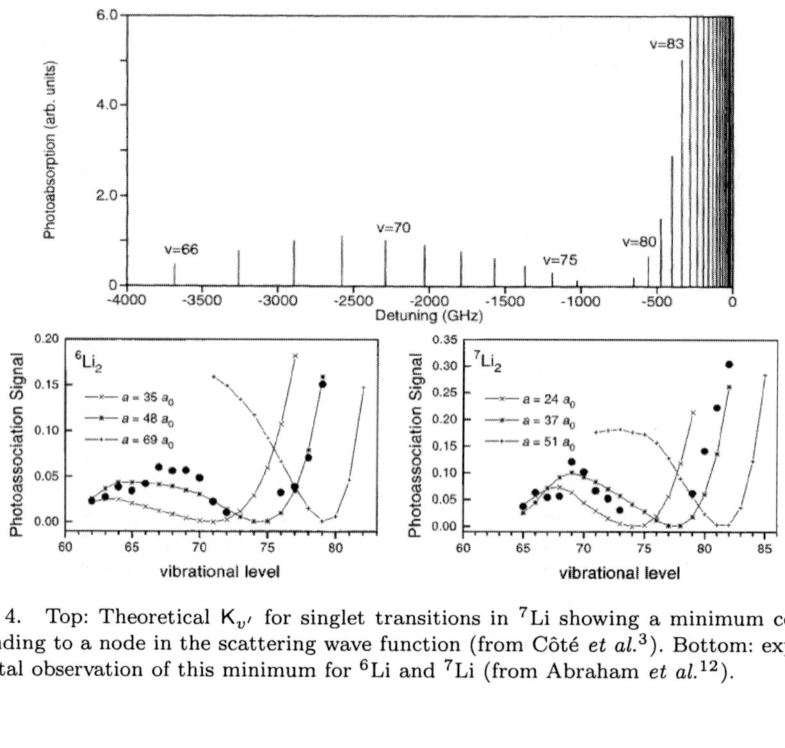

Fig. 4. Top: Theoretical $\mathsf{K}_{v'}$ for singlet transitions in ^{7}Li showing a minimum corresponding to a node in the scattering wave function (from Côté *et al.*[3]). Bottom: experimental observation of this minimum for ^{6}Li and ^{7}Li (from Abraham *et al.*[12]).

The molecules formed by the one-photon process described above are in excited electronic states. To form stable molecules, *i.e.* in singlet or triplet electronic ground states, an additional laser could transfer these excited states to specific ro-vibrational levels of the electronic ground states. The approach to create stable ultracold molecules was first suggested by Alex at the end of 1996, when I returned to Physics as an ITAMP Postdoctoral Fellow, and was published in 1997,[14] with a more complete account published in 1999.[15] Again, Alex was ahead of the field, his suggestion preceding the first report on ultracold Cs_2 molecules[16] by nearly a year.

4. Feshbach Optimized Photoassociation: FOPA

Since the early work on forming ultracold Li_2 molecules,[14,15] we have been exploring other approaches and systems, such as one-photon formation directly from the continuum for LiH and NaH.[10,11] Since those polar molecules have a dipole moment, it is possible to photoassociate a pair of colliding atoms directly into a stable level of the electronic ground state. However, in most cases, we find that the formation rate to deeply bound levels is usually small, since the overlap between the continuum state and a deep target state is also small. This is reflected in a small value of the coefficient $C_{v'}$ in Eq.(5).

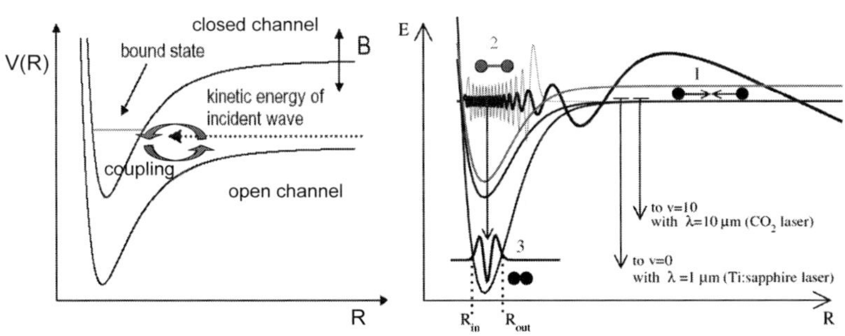

Fig. 5. Left: A Feshbach resonance (tuned by the magnetic B-field) between two coupled channels occurs when the energy of a pair of scattering atoms (in an open channel) is degenerate with the energy of a bound state (in a closed channel). Right: FOPA (from Pellegrini *et al.*[17]): Colliding atoms (1) interact via open and closed channels due to hyperfine interactions. A Feshbach resonance occurs when a bound level (2) (gray wave function) coincides with the continuum state (1) (dark wave function). A photon (wavelength λ) can associate the atoms into a bound level v (3) of the ground state potential.

To enhance the probability density at short range, hence increasing $C_{v'}$ and $\mathsf{K}_{v'}$, we proposed a new PA scheme which uses a magnetically induced Feshbach resonance.[17] A Feshbach resonance occurs when the energy of the colliding atoms coincides with the energy of a bound state in a closed channel[18] (see Fig.5). Feshbach Optimized Photoassociation (FOPA) allows transitions to deeply bound levels (see Fig. 5). We note that PA in the vicinity of a Feshbach resonance has been studied previously.[19,20] However, as opposed to earlier proposals based on the Franck-Condon principle,[21–23] FOPA takes advantage of the whole wave function in a full quantum coupled-channel calculation, making it a more general scheme.

We determine $|\Psi_{\varepsilon,\ell=0}\rangle$ in Eq.(4) by solving the Hamiltonian for two colliding atoms in a magnetic field[17,18] :

$$H = \frac{p^2}{2\mu} + V_C + \sum_{j=1}^{2} H_j^{\text{int}} , \tag{8}$$

where $V_C = V_0(R)P^0 + V_1(R)P^1$ is the Coulomb interaction, decomposed into singlet (V_0) and triplet (V_1) molecular potentials, with the associated projection operator P^0 and P^1. The internal energy of atom j, $H_j^{\text{int}} = \frac{a_{\text{hf}}^{(j)}}{\hbar^2} \vec{s}_j \cdot \vec{i}_j + (\gamma_e \vec{s}_j - \gamma_n \vec{i}_j) \cdot \vec{B}$, consists of the hyperfine and Zeeman contributions, respectively. Here $\vec{s}_j$ and $\vec{i}_j$ are the electronic and nuclear spins of atom j with hyperfine constant $a_{\text{hf}}^{(j)}$, and $\vec{B}$ is the magnetic field. We solve for $|\Psi_{\varepsilon,\ell=0}\rangle$ by using the Mapped Fourier Grid method,[24] and expand it onto the basis constructed from the hyperfine states of both atoms,

$$|\Psi_{\varepsilon,\ell=0}\rangle = \sum_{\sigma=1}^{N} \psi_\sigma(R)\{|f_1, m_1\rangle \otimes |f_2, m_2\rangle\}_\sigma , \tag{9}$$

where $\vec{f}_j = \vec{i}_j + \vec{s}_j$ is the total spin of atom j, and m_j its projection on the magnetic axis. Here, $\psi_\sigma(R)$ stands for the radial wave function associated with channel σ labeled by the quantum numbers f_i, m_i.

As an example, we considered forming LiNa in the ground $X^1\Sigma^+$ state directly from the continuum (Fig. 6) starting with ^{6}Li$(f = \frac{1}{2}, m = -\frac{1}{2})$ and ^{23}Na$(f = 1, m = -1)$.[17] Eight channels with total projection $M = m_1 + m_2 = -\frac{3}{2}$ are coupled by the Hamiltonian (8). We used the potential curves adjusted in our previous work,[25] and found two well-separated Feshbach resonances at 1081 G (narrow) and 1403 G (broad). Fig. 6 displays $\mathsf{K}_{v'}$ as a function of the B-field for different levels ($v', J' = 1$) at $T = 50$ μK and $I = 1$ W/cm^2. $\mathsf{K}_{v'}$ is drastically enhanced near a resonance, by up to five orders of magnitude, even for the lowest ($v' < 10$) levels.

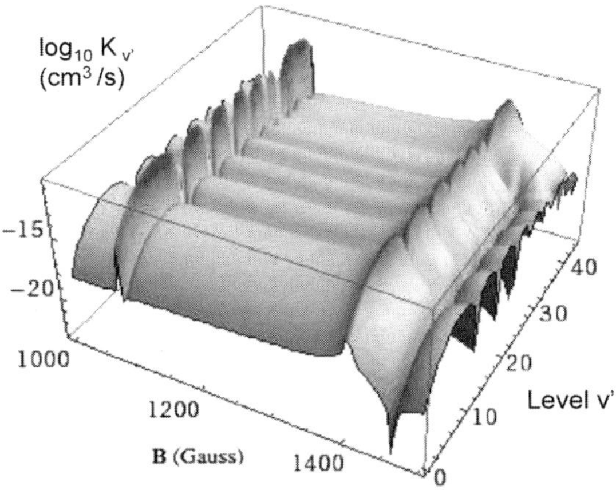

Fig. 6. $K_{v'}$ in cm^3/s vs. the B-field ($T = 50\ \mu\mathrm{K}$, $I = 1\ \mathrm{W/cm}^2$) for various levels ($v', J' = 1$) of the LiNa $\mathrm{X}^1\Sigma^+$ potential, starting from $^6\mathrm{Li}(f = \frac{1}{2}, m = -\frac{1}{2})$ and $^{23}\mathrm{Na}(f = 1, m = -1)$. Two Feshbach resonances at 1081 and 1403 Gauss enhance the PA rates by several orders of magnitude (from Pellegrini *et al.*[17]).

These giant formation rates can be understood by the sharp increase in the amplitudes of the radial wave functions ψ_σ's. In Fig. 7, we show the total probability density $|\Psi_{\varepsilon,\ell=0}(R)|^2$ as a function of B. As the magnetic field B nears either of the resonances at 1081 and 1403 G, $|\Psi_{\epsilon,l=0}(R)|^2$ increases by several orders of magnitude.

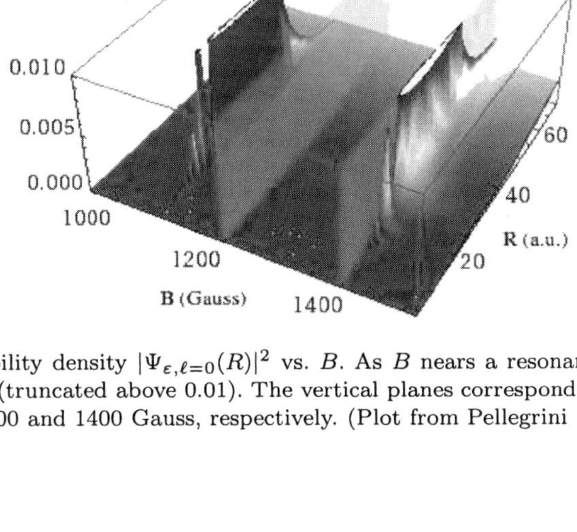

Fig. 7. Probability density $|\Psi_{\varepsilon,\ell=0}(R)|^2$ vs. B. As B nears a resonance, $|\Psi_{\varepsilon,\ell=0}|^2$ increases sharply (truncated above 0.01). The vertical planes correspond to PA off and on resonance at 1200 and 1400 Gauss, respectively. (Plot from Pellegrini *et a.*[17]).

We also obtained analytical results based on a two coupled channel model.[17] We showed that, ignoring saturation effects, the rate coefficient can be expressed in terms of the resonant s-wave phase shift δ as

$$\mathsf{K}_{v'} = \mathsf{K}_{v'}^{\text{off}} \left|1 + C_1 \tan\delta + C_2 \sin\delta\right|^2 , \tag{10}$$

where $\mathsf{K}_{v'}^{\text{off}}$ is the off-resonance rate coefficient ($\delta = 0$) given by Eq.(5), and C_1 and C_2 are ratios of dipole transition matrix elements.[17] We note that to first order in k, the resonant and background s-wave phase shifts are related to the scattering length a by $\tan(\delta + \delta_{\text{bg}}) = -ka$, with $\delta_{\text{bg}} = -ka_{bg}$ and[18]

$$a = a_{bg}\left(1 - \frac{\Delta}{B - B_{\text{res}}}\right) , \tag{11}$$

where a_{bg} is the background scattering length of the pair of atoms (which can vary slowly with B), $B_{\textbf{res}}$ is the position of the resonance, and Δ is its width.[18] In Fig.8, we compare the numerical results with eight coupled channels to the two-channel model: the agreement is good.

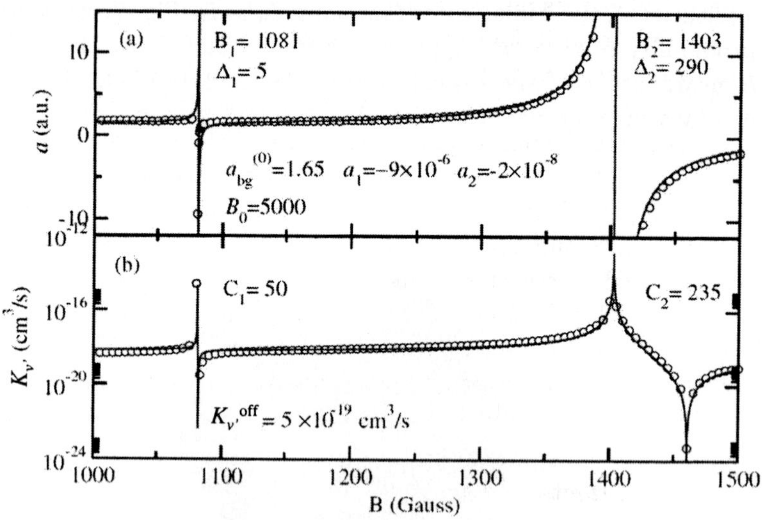

Fig. 8. In (a): scattering length a for full coupled problem (circles) and the fit using $a = a_{\text{bg}}(B)(1 - \frac{\Delta_1}{B-B_1} - \frac{\Delta_2}{B-B_2})$ with $a_{\text{bg}}(B) = a_{\text{bg}}^{(0)} + a_1(B + B_0) + a_2(B + B_0)^2$. in (b): $K_{v'}$ for $v' = 0$ at 50 μK and 1 W/cm^2 (circles) and the formula (10) using $\tan\delta = ka_{\text{bg}}(B)(\frac{\Delta_1}{B-B_1} + \frac{\Delta_2}{B-B_2})$, with C_1 and C_2 B-independent. Numerical parameters are given in each plot (figure from Pellegrini *et al.*[17]).

We also investigated the use of FOPA to access more deeply bound levels in excited electronic states. As an example, we selected PA to the $1^3\Sigma_g^+$ excited molecular state of Li_2 from a gas of ultracold ^{7}Li atoms initially prepared in the $|f=1, m=1\rangle$ hyperfine state: this system has been explored in recent experiments of Hulet and co-workers.[26] We note that the scattering properties for these entrance channels are well understood and have been experimentally investigated.[26–28] We used the ^{7}Li potential curves and transition dipole moment described in Côté *et al.*,[3] adjusted to reproduce the Feshbach resonance at 736 G for the $|f=1, m=1\rangle$ entrance channel[26] (see Fig. 9). With these curves, we calculated $\mathsf{K}_{v'}$ to $v'=83$ of the $1^3\Sigma_g^+$ state as a function of the B-field at $T=10$ μK. We choose these values of v and T to compare with a recent measurement[26] at higher I (see below). Fig. 9 shows $\mathsf{K}_{v'=83}$ for $I = 1$ mW/cm^2, which displays a maximum at resonance ($B \sim 736$ G) and a minimum near 710 G, spanning over eight orders of magnitude. The shape is similar to that obtained for LiNa.

Fig. 9 also illustrates $\mathsf{K}_{v'}$ for many excited levels v' of the $1^3\Sigma_g^+$ state with $I = 1$ mW/cm^2 : they basically behave the same way. The figure also shows another characteristic of FOPA: the sensitivity of $\mathsf{K}_{v'}$ to the exact level v' being probed. In fact, whereas the maximum of $\mathsf{K}_{v'}$ near the resonance at 736 G does not vary much, the exact position of the minimum is very sensitive to v: it follows a "croissant trajectory" as v' decreases, first moving to lower B at large v' and then to higher B at smaller v', even passing to the right of the resonance at lower v'. The reason for this sensitivity resides in the nature of the minimum: it comes from poor overlap of the target level v' and scattering wave functions, reflected in C_1 and C_2 of Eq.(10). Any small difference in the target wave functions will be amplified and will result in a shift of the minimum. It is worth noting that such a shift was observed in recent measurements for $v = 82 - 84$,[26] consistent with our results. This sensitivity to minute changes in the target level wave function could actually be used as a spectroscopic tool for high precision measurements of both the scattering and bound states.

5. Saturation Effects

In the previous section, we neglected the effect of saturation occurring when the stimulated width γ_s becomes large. Observation of saturation has been reported in a mixture of ^{6}Li-^{7}Li,[30] and in an optical lattice loaded with Cs.[31] Theoretical studies and experiments in Bose-Einstein condensates (BECs) have also been conducted, but only PA in a sample of ^{7}Li near the critical

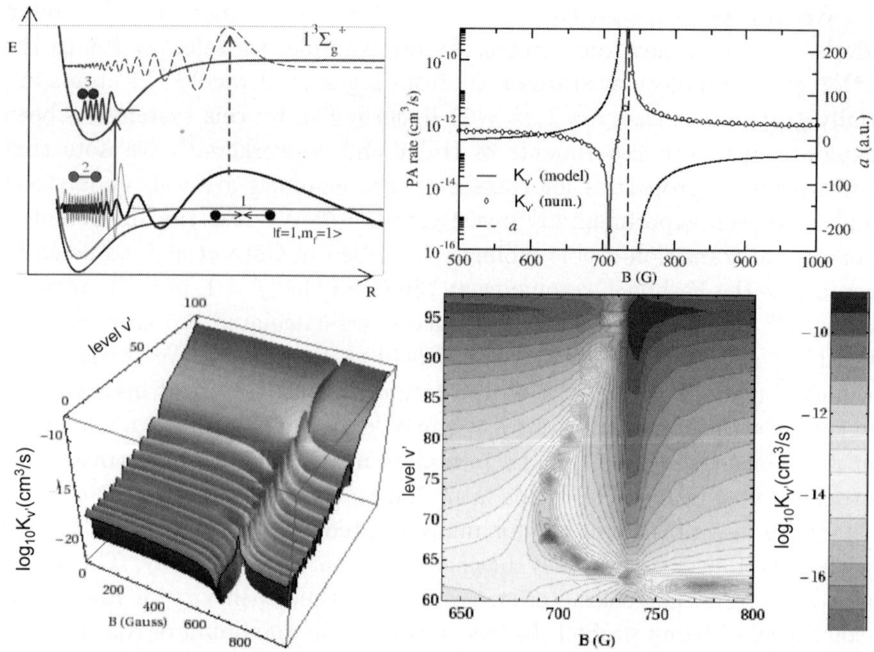

Fig. 9. Top left: FOPA scheme into $1^3\Sigma_g^+$ of $^7Li_2$. Top right: $\mathsf{K}_{v'}$ into $v' = 83$ vs. B for $I = 1$ mW/cm^2 and $T = 10$ μK (circle: full coupled-channel, solid curve: two-channel model). $\mathsf{K}_{v'=83}$ increases by more than four order of magnitude near the resonance. The right axis shows the scattering length of two ^{7}Li atoms in $|f = 1, m = 1\rangle$, which is well fitted by $a = a_{\rm bg}\left(1 - \frac{\Delta}{B - B_{\rm res}}\right)$ with $a_{\rm bg} = -18$ a.u., $\Delta = -200$ G, and $B_{\rm res} = 736$ G. Bottom left: $\mathsf{K}_{v'}$ for all v' vs. B. Bottom right: contour plot showing the maximum always at $B_{\rm res}$ and the minimum following a"croissant shape" trajectory. All plots (except lower left) are from Pellegrini and Côté.[29]

temperature reached saturation,[26,28] whereas PA in a BEC of Na saw no trace of saturation.[32]

We consider the PA rate coefficient $\mathsf{K}_{v'}$ of Eq.(2) with $|S_{\ell,v'}|^2$ evaluated at resonance, *i.e.* $\Delta = 0$ in Eq.(3)

$$\mathsf{K}_{v'}(T, I) = \frac{1}{hQ_T}\int_0^\infty d\varepsilon\; e^{-\varepsilon/k_B T}\frac{\gamma_{v'}\gamma_s}{\varepsilon^2 + (\frac{\gamma_{v'}+\gamma_s}{2})^2}\,. \tag{12}$$

Saturation will occur when γ_s becomes large (of the order of $\gamma_{v'}$), which can take place because one increases the laser intensity I, or if one is near a Feshbach resonance. In fact, we can extend the two coupled-channel treatment previously introduced,[17] and obtain the following expression for the stimulated width γ_s given by Eq.(4)

$$\gamma_s(I,k,B) = \gamma_s^{\text{off}} \left|1 + C_1 \tan\delta + C_2 \sin\delta\right|^2 \ , \tag{13}$$

where $\gamma_s^{\text{off}}(I,k)$ is the stimulated rate off-resonance. For a pair of atoms experiencing s-wave scattering, we have $\gamma_s^{\text{off}}(I,k) \propto Ik$ as $k \to 0$, in agreement with Wigner's threshold law,[9] while $\tan\delta(k) = -ka_{\text{bg}}\frac{\Delta}{B-B_{\text{res}}}$ (see values of a_{bg}, Δ, and B_{res} in Fig. 9 for ^{7}Li). We note that by introducing Eq.(13) into Eq.(12), we obtain the solid curves in Fig. 9, with $\gamma_s^{\text{off}} = 0.8$ kHz, and the fitted values $C_1 = 6$, and $C_2 = 1$. The agreement with the full multi-channel treatment is excellent.

We note that the maximum PA rate coefficient reachable for a given temperature is obtained when the scattering matrix element in Eq.(12) is equal to its maximum value of one: $\gamma_{v'}\gamma_s/[\varepsilon^2 + (\frac{\gamma_{v'}+\gamma_s}{2})^2] = 1$. The integration over the Maxwell-Boltzmann velocity distribution can then be performed so that Eq.(12) gives

$$\mathsf{K}_{v'}^{\text{limit}}(T) = \frac{k_B T}{hQ_T} = \frac{h^2}{(2\pi\mu)^{3/2}} \frac{1}{\sqrt{k_B T}} \ . \tag{14}$$

$\mathsf{K}_{v'}^{\text{limit}}(T)$ for a thermal gas increases as the temperature drops.

To investigate the effect of saturation on the PA rate, we first performed a calculation at $B = 0$ and $T = 500$ nK to compare with experimental results for ^{7}Li atoms in $|f = 2, m = 2\rangle$ (a pure triplet interaction without Feshbach resonance).[28] Fig. 10 shows the results: our calculated (solid curve) and measured (circles) rates are in agreement well past the range of validity of the perturbative prediction (dotted line).[28] Saturation is observed for the highest reported intensities as the rate nears the theoretically predicted unitarity limit of 2.5×10^{-8} cm^3/s. We repeated the calculation for $|f = 1, m = 1\rangle$ (mostly triplet interaction) at $B = 0$ and near the resonance with 740 G: the $B = 0$ result (solid curve) is only slightly different from the pure triplet $|2, 2\rangle$ case, while the PA rate at 740 G (dashed-line) increases rapidly until it reaches the saturation limit at intensities ten times smaller than for $B = 0$, and then slowly decreases. This particular shape can be explained by the integrand in Eq.(12) which takes the form $\gamma_s\gamma_{v'}/(\frac{\gamma_s+\gamma_{v'}}{2})^2$ as $\varepsilon \to 0$; as $\gamma_s \propto I$ increases, so does the integrand, until it reaches its maximum of one at $\gamma_s = \gamma_{v'}$, after which it starts to decrease as $\gamma_s^{-1} \propto I^{-1}$ at large intensities. We note that this behavior has recently been observed experimentally.[26]

One can use this low intensity saturation to probe how it may affect the PA lineshape. In Fig. 11, we display $\mathsf{K}_{v'=83}$ as a function of the intensity I. We observe the appearance of a double minimun structure that

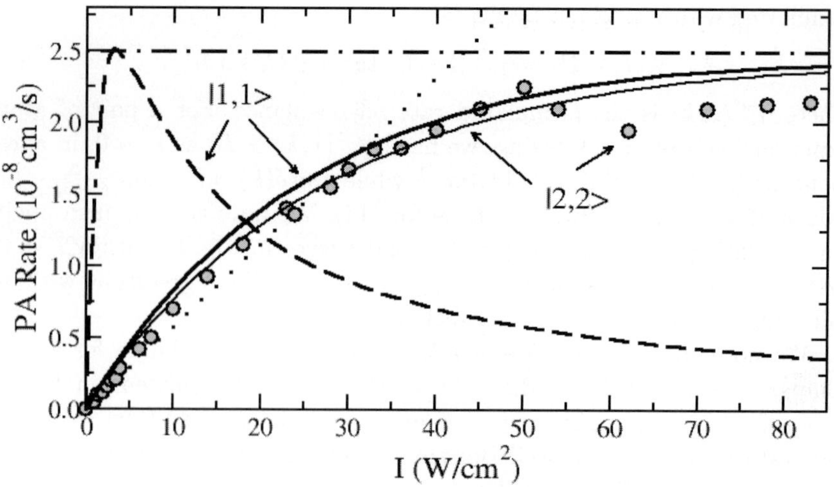

Fig. 10. $\mathsf{K}_{v'=83}$ into $1^3\Sigma_g^+$ at $T = 500$ nK vs. laser intensity, with $B = 0$ (solid curve) and near resonance at $B = 740$ G (dashed curve) for the $|f = 1, m = 1\rangle$ state. We also compare our calculation for $|f = 2, m = 2\rangle$ at $B = 0$ with the experimental data (circles).[28] The dotted line is the perturbative rate of 5.8×10^{-10} (cm^3·s^{-1}/W·cm^{-2}). while the horizontal dot-dashed line illustrates the saturation limit of 2.5×10^{-8} cm^3/s (from Pellegrini and Côté[29]).

becomes more pronounced as I grows. This feature is simple to explain: as I increases, the PA rate reaches its maximum value at $\gamma_s = \gamma_{v'}$, and then decreases at higher I. However, γ_s also varies with B according to Eq.(13), and as B goes through the Feshbach resonance at large enough I, γ_s will reach the value $\gamma_{v'}$ twice, once on the way "up" and once on the way "down" from the resonance. The insets compare the results from the multi-channel treatment and using the simple result (13) for 1 mW/cm^2 and 10 W/cm^2: in both cases, the agreement is good. We note that the double-minima feature visible in Fig. 11 appears quickly, even at rather low intensity.

Finally, we compare our calculation with measurements obtained in a recent experiment.[26] Fig. 12 shows our multi-channel results at $T = 10$ μK and $I = 1.6$ W/cm^2 together with those of Hulet and co-workers,[26] where the atom loss rate K_P is twice our $\mathsf{K}_{v'}$, since two atoms are lost during the PA process. Our results are within or close to the uncertainty of Hulet and co-workers;[26] when taking into account the variation in temperatures between data points ($\sim 9 - 18$ μK[26]), the discrepancy vanishes. Fig. 12 also shows enlargements around the minimum and the maximum. We also

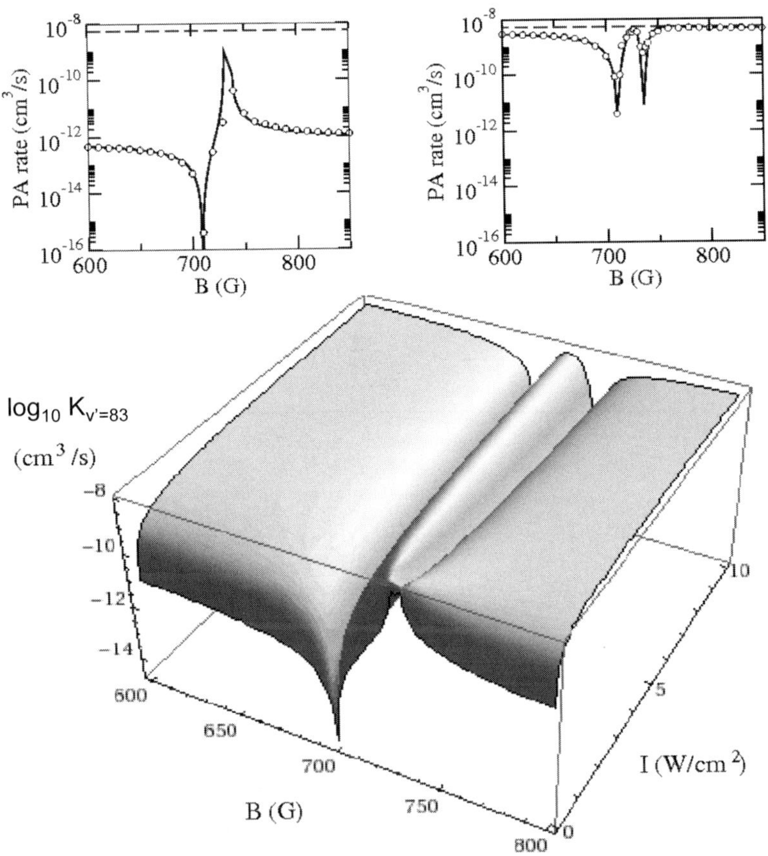

Fig. 11. Variation of $\mathsf{K}_{v'=83}$ with B and I at 10 μK. A double-minima feature appears as I increases near the resonance. The top panels compare the multi-channel and simple treatment (left: 1 mW/cm^2, right: 10 W/cm^2). The same scale illustrates the difference in the background rate: the dashed line is the saturation limit of 5.5×10^{-9} cm^3/s (from Pellegrini and Côté[29]).

note that the experimental points at saturation near the resonance are the least accurate, and that the point at 745 G was particularly challenging to obtain,[26] consistent with the large and negative scattering length a (*i.e.* attractive interaction: see Fig. 9). The extremely good agreement between our results (obtained without any adjustable parameters) and the experimental data shows that our treatment captures all the relevant physics in these binary processes, and thus our predictions made above should be observable.

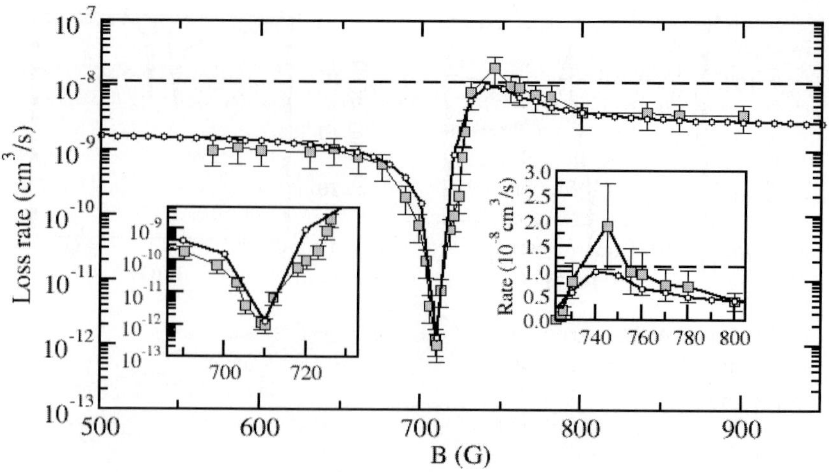

Fig. 12. Comparison of $2K_{v'=83}$ with experimental data[26] at 10 μK and 1.6 W/cm^2. The insets show the minimum and maximum (linear scale); only the point at 745 G is outside the uncertainty of 45%.[26] The dashed horizontal line corresponds to twice the saturation limit of 5.5×10^{-9} cm^3/s (from Pellegrini and Côté[29]).

6. Conclusions

In the previous sections, I have presented recent results on the formation of ultracold molecules using Feshbach resonances, and on the effect of saturation on the formation rate. It was shown that, by photoassociating a pair of atoms in the vicinity of a Feshbach resonance, large molecule formation rates can be obtained. Examples for LiNa and Li_2 were given, for which rates increased by several orders of magnitude. It was also noted that Feshbach optimized photoassociation (FOPA) could provide a new tool for high-precision spectroscopy. Finally, it was shown that saturation effects will play an important role near the resonance, leading to a limiting rate as well as new line shapes as the magnetic field varies.

All this work has been more than influenced by Alex Dalgarno; it is in fact directly related to his mentorship over the years. It all started when I became his student and learned about ultracold physics, and is continuing today, as we still actively discuss future projects.

References

1. R. Côté, A. Dalgarno, and M. J. Jamieson, *Elastic scattering of two ^{7}Li atoms*, Phys. Rev. A **50**, 399 (1994).

2. R. Côté and A. Dalgarno, *Elastic scattering of two Na atoms*, Phys. Rev. A **50**, 4827 (1994).
3. R. Côté, A. Dalgarno, Y. Sun, and R. G. Hulet, *Photoabsorption by Ultracold Atoms and the Scattering Length*, Phys. Rev. Lett. **74**, 3581 (1995).
4. R. J. N. Côté, *Ultra-Cold Collisions of Identical Atoms*, Ph.D. Thesis, Massachusetts Institute of Technology (1995).
5. H. R. Thorsheim, J. Weiner, and P. S. Julienne, *Laser-induced photoassociation of ultracold sodium atoms*, Phys. Rev. Lett. **58**, 2420 (1987).
6. J. L. Bohn and P. S. Julienne, *Semianalytic treatment of two-color photoassociation spectroscopy and control of cold atoms*, Phys. Rev. A **54**, R4637 (1996).
7. R. Côté and A. Dalgarno, *Photoassociation intensities and radiative trap loss in lithium*, Phys. Rev. A **58**, 498 (1998).
8. K. M. Jones, E. Tiesinga, P. D. Lett, and P. S. Julienne, *Ultracold photoassociation spectroscopy: Long-range molecules and atomic scattering*, Rev. Mod. Phys. **78**, 483 (2006).
9. E. P. Wigner, *On the Behavior of Cross Sections Near Thresholds*, Phys. Rev. **73**, 1002 (1948).
10. E. Juarros, P. Pellegrini, K. Kirby, and R. Côté, *One-photon-assisted formation of ultracold polar molecules*, Phys. Rev. A **73**, 041403(R) (2006).
11. E. Juarros, K. Kirby, and R. Côté, *Laser-assisted ultracold lithium-hydride molecule formation: stimulated versus spontaneous emission*, J. Phys. B **39**, S965 (2006).
12. E. R. Abraham, W. I. McAlexander, J. M. Gerton, R. G. Hulet, R. Côté, and A. Dalgarno, *Singlet s-wave scattering lengths of ^6Li and ^7Li*, Phys. Rev. A **53**, R3713 (1996).
13. E. R. I. Abraham, W. I. McAlexander, J. M. Gerton, R. G. Hulet, R. Côté, and A. Dalgarno, *Triplet s-wave resonance in ^6Li collisions and scattering lengths of ^6Li and ^7Li*, Phys. Rev. A **55**, R3299 (1997).
14. R. Côté, and A. Dalgarno, *Mechanism for the production of vibrationally excited ultracold molecules of $^7Li_2$*, Chem. Phys. Lett. **279**, 50 (1997).
15. R. Côté, and A. Dalgarno, *Mechanism for the Production of $^6Li_2$ and $^7Li_2$ Ultracold Molecules*, J. Mol. Spect. **195**, 236 (1999).
16. A. Fioretti, D. Comparat, A. Crubellier, O. Dulieu, F. Masnou-Seeuws, and P. Pillet, *Formation of Cold Cs_2 Molecules through Photoassociation*, Phys. Rev. Lett. **80**, 4402 (1998).
17. P. Pellegrini, M. Gacesa, and R. Côté, *Giant Formation Rates of Ultracold Molecules via Feshbach-Optimized Photoassociation*, Phys. Rev. Lett. **101**, 053201 (2008).
18. T. Köhler, K. Góral, and P. S. Julienne, *Production of cold molecules via magnetically tunable Feshbach resonances*, Rev. Mod. Phys. **78**, 1311 (2006).
19. F. A. van Abeelen and B. J. Verhaar, *Time-Dependent Feshbach Resonance Scattering and Anomalous Decay of a Na Bose-Einstein Condensate*, Phys. Rev. Lett. **83**, 1550 (1999).
20. S. J. J. M. F. Kokkelmans, H. M. J. Vissers, and B. J. Verhaar, *Formation of a Bose condensate of stable molecules via a Feshbach resonance*, Phys. Rev. A **63**, 031601 (2001).

21. B. Laburthe-Tolra, N. Hoang, B. TJampens, N. Vanhaecke, C. Drag, A. Crubellier, D. Comparat, and P. Pillet, *Controlling the formation of cold molecules via a Feshbach resonance*, Europhys.Lett. **64**, 171 (2003).
22. W. C. Stwalley, *Efficient Conversion of Ultracold Feshbach-Resonant Polar Molecules into Ultracold Ground State ($X^1\Sigma_g^+ v = 0, J = 0$) Molecules*, Eur. Phys. J D **31**, 221 (2004).
23. F. A. van Abeelen, D. J. Heinzen, and B. J. Verhaar, *Photoassociation as a probe of Feshbach resonances in cold-atom scattering*, Phys. Rev. A **57**, R4102 (1998).
24. V. Kokoouline, O. Dulieu, R. Kosloff, and F. Masnou-Seeuws, *Mapped Fourier methods for long-range molecules: Application to perturbations in the $Rb_2(0_u^+)$ photoassociation*, J. Chem. Phys. **110**, 9865 (1999).
25. M. Gacesa, P. Pellegrini, and R. Côté, *Feshbach resonances in ultracold $^{6,7}Li$+^{23}Na atomic mixtures*, Phys. Rev. A **78**, 010701(R) (2008).
26. M. Junker, D. Dries, C. Welford, J. Hitchcock, Y. P. Chen, and R. G. Hulet, *Photoassociation of a Bose-Einstein Condensate near a Feshbach Resonance*, Phys. Rev. Lett. **101**, 060406 (2008).
27. K. E. Strecker, G. B. Partridge, A. G. Truscott, and R. G. Hulet, *Formation and Propagation of Matter Wave Soliton Trains*, Nature **417**, 150 (2002).
28. I. D. Prodan, M. Pichler, M. Junker, R. G. Hulet, and J. L. Bohn, *Intensity Dependence of Photoassociation in a Quantum Degenerate Atomic Gas*, Phys. Rev. Lett. **91**, 080402 (2003).
29. P. Pellegrini and R. Côté, *Probing the unitarity limit at low laser intensities*, New J. Phys. **11** 055047 (2009).
30. U. Schlöder, C. Silber, T. Deuschle, and C. Zimmermannn, *Saturation in heteronuclear photoassociation of $^6Li^7Li$*, Phys. Rev. A **66**, 061403(R) (2002).
31. S.D. Kraft, M. Mudrich, M. U. Staudt, J. Lange, O. Dulieu, R. Wester, and M. Weidemüller *Saturation of Cs_2 photoassociation in an optical dipole trap*, Phys. Rev. A **71**, 013417 (2005).
32. C. McKenzie, J. H. Denschlag, H. Häffner, A. Browaeys, L. E. E. de Araujo, F. K. Fatemi, K. M. Jones, J. E. Simsarian, D. Cho, A. Simoni, E. Tiesinga, P. S. Julienne, K. Helmerson, P. D. Lett, S. L. Rolston, and W. D. Phillips, *Photoassociation of Sodium in a Bose-Einstein Condensate*, Phys. Rev. Lett. **88**, 120403 (2002).

QUANTUM THEORY OF ATOMIC AND MOLECULAR SCATTERING BASED ON THE FULLY UNCOUPLED SPACE-FIXED REPRESENTATION

R. V. KREMS
Department of Chemistry, University of British Columbia
Vancouver, BC V6T 1Z1, Canada
E-mail: rkrems@chem.ubc.ca
http://www.chem.ubc.ca/faculty/krems

We describe a quantum theory of atomic and molecular collisions in the presence of external electromagnetic fields based on the fully uncoupled space-fixed basis representation of the scattering wave functions. The fully uncoupled basis leads to simple expressions for the matrix elements of the Hamiltonian providing all operators of fine and hyperfine interactions as well as the operators describing interatomic and intermolecular interactions are represented as direct products of spherical tensors defined in the laboratory-fixed coordinate system. We present a general expression for the electrostatic interaction operator of two atoms in arbitrary electronic states in terms of uncoupled products of space-fixed spherical tensors and describe recent studies of molecular collisions in external fields.

Keywords: Scattering Theory, Open-Shell Atoms, Electric Fields, Magnetic Fields, Ultracold Atoms, Ultracold Molecules

1. Introduction

Dynamics of atomic and molecular systems are generally described by solutions to the Schrödinger equation. For problems without time-dependent perturbations, the Schrödinger equation can be numerically solved using a close coupling formalism.[1] The theory is based on the basis set expansion of the eigenfunction of the total Hamiltonian in terms of the wave functions describing different fragments of the dynamical system. For example, the collision problem of two atoms is most conveniently described by representing the wave function of the diatomic system as a close coupling expansion over the wave functions of the separated atoms. The substitution of this basis set expansion in the Schrödinger equation leads to a system of coupled differential equations, which can be solved for the expansion coefficients subject to scattering boundary conditions.[1] The number of differential equations N is equal to the number of basis functions. The computing time to solve a system of coupled differential equations scales as N^3. The number of basis functions N is determined by the internal degrees of freedom of the dynamical system such as states of non-zero rotational, nuclear spin, electronic spin and electronic orbital angular momenta. In order to reduce the complexity of the problem, Arthurs and Dalgarno[1] proposed to couple all angular momenta of the system to form the total angular momentum $\boldsymbol{J}$ and expand the eigenfunction of the total Hamiltonian in terms of the eigenfunctions of $\boldsymbol{J}^2$ and J_Z. The total angular momentum basis is called the Arthurs-Dalgarno representation.

In the absence of external fields, the total angular momentum of dynamical systems is conserved and the Arthurs-Dalgarno representation factorizes the system of coupled differential equations into a set of smaller problems that can be solved independently. The Arthurs-Dalgarno representation has therefore been widely used for quantum scattering calculations of atomic and molecular collision properties as well as the dissociation dynamics of diatomic and triatomic molecules. The total angular momentum representation, however, leads to cumbersome expressions for the matrix elements of the Hamiltonian, especially for atoms and molecules with multiple degrees of freedom. For example, the matrix elements of the electrostatic interaction potential describing two atoms in states of non-zero electronic spin and orbital angular momenta are proportional to an 18-j symbol[2] if expressed in the total angular momentum basis.

The research field of atomic and molecular physics has recently been revolutionized by experimental work leading to the creation of ultracold atoms and molecules.[3] The cooling experiments rely on thermal isolation of atomic and molecular ensembles in external field traps and energy exchange in atomic and molecular collisions. The presence of external trapping fields modifies the dynamical properties of ultracold atoms and molecules. In order to predict the possibility of cooling specific atomic or molecular species to ultracold temperatures, it has therefore been necessary to develop a quantum scattering theory that would account for the effects of external electromagnetic fields on atomic and molecular collisions in the temperature regime where perturbations due to external fields are more significant than the energy of the translational motion. Such a theory could be formulated in the Arthurs-Dalgarno representation. However, external fields induce couplings between different total angular momenta and the total angular momentum representation does not reduce the complexity of the scattering problem in the presence of electromagnetic fields. In this chapter, we will show that the scattering theory of atomic and molecular collisions in external fields is most conveniently formulated in the fully uncoupled space-fixed basis set representation, leading to simple expressions for the matrix elements of the total Hamiltonian. The following discussion focuses on interactions of atoms in arbitrary electronic states and diatomic molecules in Σ electronic states. The theory described in Section 2 was first presented by Krems and Dalgarno.[4,5] It is based to a large extent on the work of Volpi and Bohn.[6]

2. Close Coupling Scattering Theory

The total Hamiltonian of two atoms or molecules (A and B) with non-zero electron spin can be generally written as

$$\hat{H} = -\frac{1}{2\mu R}\frac{\partial^2}{\partial R^2}R + \frac{\boldsymbol{l}^2}{2\mu R^2} + \hat{U} \tag{1}$$

where

$$\hat{U} = \hat{H}_{\mathrm{A}} + \hat{H}_{\mathrm{B}} + \hat{V}_{\mathrm{AB}}, \tag{2}$$

R is the Jacobi coordinate joining the centers of mass of the colliding particles, $\boldsymbol{l}$ is the angular momentum describing the rotation of the vector $\boldsymbol{R}$, μ is the reduced mass of the colliding particles, H_{A} and H_{B} denote the Hamiltonians of the isolated particles A and B, and $\hat{V}_{\mathrm{AB}}$ is the potential for the electronic interaction between A and B. The interaction potential $\hat{V}_{\mathrm{AB}}$ depends on the electronic spin and orbital angular momenta of the colliding particles. It is independent of the nuclear spin states.

The Hamiltonians of the separated particles can be generally represented as

$$\begin{aligned} \hat{H}_{\mathrm{A}} &= \hat{H}_{\mathrm{A}}^{S_{\mathrm{A}}} + \hat{H}_{\mathrm{A}}^{I_{\mathrm{A}}} + \hat{H}_{\mathrm{A}}^{L_{\mathrm{A}}} + \hat{V}_{\mathrm{A}}^{I_{\mathrm{A}},S_{\mathrm{A}}} + \hat{V}_{\mathrm{A}}^{S_{\mathrm{A}},L_{\mathrm{A}}} + \hat{V}_{\mathrm{A}}^{I_{\mathrm{A}},L_{\mathrm{A}}} + \hat{V}_{\mathrm{A}}^{I_{\mathrm{A}},L_{\mathrm{A}},S_{\mathrm{A}}} \\ H_{\mathrm{B}} &= \hat{H}_{\mathrm{B}}^{S_{\mathrm{B}}} + \hat{H}_{\mathrm{B}}^{I_{\mathrm{B}}} + \hat{H}_{\mathrm{B}}^{L_{\mathrm{B}}} + \hat{V}_{\mathrm{B}}^{I_{\mathrm{B}},S_{\mathrm{B}}} + \hat{V}_{\mathrm{B}}^{S_{\mathrm{B}},L_{\mathrm{B}}} + \hat{V}_{\mathrm{B}}^{I_{\mathrm{B}},L_{\mathrm{B}}} + \hat{V}_{\mathrm{B}}^{I_{\mathrm{B}},L_{\mathrm{B}},S_{\mathrm{B}}}, \end{aligned} \tag{3}$$

where S_{A} and S_{B} denote the electron spin angular momenta, I_{A} and I_{B} are the nuclear spin angular momenta, and L_{A} and L_{B} denote the electron orbital angular momenta for open-shell atoms or the nuclear rotational angular momenta for Σ-state molecules. The first three terms in Eq. (3) depend only on S, I or L, respectively, and the last four terms describe the interactions that give rise to fine and hyperfine structures of A and B. The terms $\hat{H}_{\rm A}^{S_{\rm A}}$, $\hat{H}_{\rm A}^{I_{\rm A}}$, $\hat{H}_{\rm B}^{S_{\rm B}}$, and $\hat{H}_{\rm B}^{I_{\rm B}}$ include the interaction with external magnetic fields and the terms $\hat{H}_{\rm A}^{L_{\rm A}}$ and $\hat{H}_{\rm B}^{L_{\rm B}}$ include the interaction of molecules with external electric fields.

Instead of using the total angular momentum representation for the eigenfunction of the Hamiltonian (1), Krems and Dalgarno[4,5] proposed to expand the total wave function in *direct* products of eigenfunctions of $\boldsymbol{l}^2$ and $\boldsymbol{l}_z$, and the eigenfunctions of each of $\hat{H}_{\rm A}^{S_{\rm A}}$, $\hat{H}_{\rm A}^{I_{\rm A}}$, $\hat{H}_{\rm A}^{L_{\rm A}}$, $\hat{H}_{\rm B}^{S_{\rm B}}$, $\hat{H}_{\rm B}^{I_{\rm B}}$, and $\hat{H}_{\rm B}^{L_{\rm B}}$ defined in the space-fixed quantization frame with the z-axis determined by the direction of an external field. The close coupling expansion can for example be written as

$$\Psi = R^{-1} \sum_i F_i(R) \phi_i, \tag{4}$$

where

$$\phi_i = |I_{\rm A} M_{I_{\rm A}}\rangle |I_{\rm B} M_{I_{\rm B}}\rangle |L_{\rm A} M_{L_{\rm A}}\rangle |L_{\rm B} M_{L_{\rm B}}\rangle |S_{\rm A} M_{S_{\rm A}}\rangle |S_{\rm B} M_{S_{\rm B}}\rangle |l m_l\rangle, \tag{5}$$

and m_l, $M_{S_{\rm A}}$, $M_{S_{\rm B}}$, $M_{I_{\rm A}}$, $M_{I_{\rm B}}$, $M_{L_{\rm A}}$, $M_{L_{\rm B}}$, are the projections of $\boldsymbol{l}$, $\boldsymbol{S}_{\rm A}$, $\boldsymbol{S}_{\rm B}$, $\boldsymbol{I}_{\rm A}$, $\boldsymbol{I}_{\rm B}$, $\boldsymbol{L}_{\rm A}$, and $\boldsymbol{L}_{\rm B}$ on the space-fixed quantization axis. We assume for simplicity that the diatomic molecules contain only one atom with non-zero nuclear spin.[7] The generalization of the theory to describe molecules with two atoms carrying non-zero nuclear spin is straightforward.

When $R = \infty$, U approaches an asymptotic Hamiltonian $H_{\rm as}$. The matrix $\mathbf{H}_{\rm as}$ is generally not diagonal in the basis (5) so Krems and Dalgarno[4] introduced a transformation $\mathbf{C}$ that diagonalizes the matrix $\mathbf{H}_{\rm as}$. It is assumed that the interaction of $\boldsymbol{l}$ with other angular momenta and the external fields can be neglected. The asymptotic matrix $\mathbf{H}_{\rm as}$ is always diagonal in l and m_l. The matrix of the transformation $\mathbf{C}$ is constructed numerically but its structure follows from the form of the operators in Eq. (3). The solution of the close coupled equations at a fixed total energy E

$$\left[\frac{d^2}{dR^2} - \frac{l(l+1)}{R^2} + 2\mu E\right] F_{\alpha l m_l}(R) = 2\mu \sum_{\alpha' l' m_l'} [\mathbf{C}^{\rm T}\mathbf{U}\mathbf{C}]_{\alpha l m_l;\alpha' l' m_l'} F_{\alpha' l' m_l'}(R), \tag{6}$$

subject to the boundary conditions

$$\begin{aligned}
&F^{\alpha l m_l}_{\alpha' l' m'_l}(R \to 0) \to 0 \\
&F^{\alpha l m_l}_{\alpha' l' m'_l}(R \to \infty) \sim \delta_{\alpha\alpha'}\delta_{ll'}\delta_{m_l m'_l} \exp\left[-\mathrm{i}(k_\alpha R - \pi l/2)\right] \\
&\qquad - \left(\frac{k_\alpha}{k_{\alpha'}}\right)^{1/2} S_{\alpha' l' m'_l;\alpha l m_l} \exp\left[\mathrm{i}(k_{\alpha'} R - \pi l'/2)\right]
\end{aligned} \tag{7}$$

gives the S-matrix or the probability amplitudes for transitions between the eigenstates of the $\mathbf{C}^{\mathrm{T}}\mathbf{H}_{\mathrm{as}}\mathbf{C}$ matrix labeled by the indexes α, l, m_l. The notation k_α is used for the wave-number corresponding to channel α. If the external fields are all directed along the same axis, the total angular momentum projection M is conserved and the matrix $\mathbf{U}$ does not contain couplings between the states with different values of M. The computations can therefore be carried out independently for different values of M.

The cross sections for elastic and inelastic scattering are computed from the S-matrix according to

$$\sigma_{\alpha\to\alpha'} = \frac{\pi}{k_\alpha^2} \sum_M \sum_l \sum_{m_l} \sum_{l'} \sum_{m'_l} |\delta_{ll'}\delta_{m_l m'_l}\delta_{\alpha\alpha'} - S^M_{\alpha l m_l;\alpha' l' m'_l}|^2. \tag{8}$$

3. The Wigner-Eckart Theorem

Given the above formulation of the close coupling theory, the scattering problem reduces to the evaluation of the matrix elements of the operators in Eq. (2) in the basis (5) and the numerical integration of Eqs. (6). The matrix elements can be most easily evaluated if the interaction operators are all represented as uncoupled products of spherical tensors defined in the space-fixed coordinate system. The integrals over the basis functions (5) then factorize into products of simple matrix elements, which can be evaluated using the Wigner-Eckart theorem. For example, the spin-orbit interaction for an atom A can be generally represented as

$$\begin{aligned}
\hat{V}_{\mathrm{A}}^{S_{\mathrm{A}},L_{\mathrm{A}}} &= a\boldsymbol{L}_{\mathrm{A}} \cdot \boldsymbol{S}_{\mathrm{A}} \\
&= -a\sqrt{3}\sum_q \langle 1q1-q|00\rangle \hat{T}^1_q(L_{\mathrm{A}})\hat{T}^1_{-q}(S_{\mathrm{A}}),
\end{aligned} \tag{9}$$

where $\langle 1q1-q|00\rangle$ is a Clebsch-Gordan coefficient and $\hat{T}^1_q(L_{\mathrm{A}})$ and $\hat{T}^1_{-q}(S_{\mathrm{A}})$ denote the spherical tensors[8] of rank 1 acting on the space of the $|L_{\mathrm{A}} M_{L_{\mathrm{A}}}\rangle$ and $|S_{\mathrm{A}} M_{S_{\mathrm{A}}}\rangle$ functions, respectively. The matrix elements of the operator

(9) in the basis (5) have the following form:

$$\langle\phi_i|\hat{V}_{\rm A}^{S_{\rm A},L_{\rm A}}|\phi_{i'}\rangle = -a\delta_{ll'}\delta_{m_l m_l'}\delta_{M_{I_{\rm A}}M'_{I_{\rm A}}}\delta_{M_{I_{\rm B}}M'_{I_{\rm B}}}\delta_{M_{S_{\rm B}}M'_{S_{\rm B}}}\delta_{M_{L_{\rm B}}M'_{L_{\rm B}}}\times$$
$$\sqrt{3}\sum_q\langle 1q1-q|00\rangle\langle L_{\rm A}M_{L_{\rm A}}|\hat{T}_q^1(L_{\rm A})|L_{\rm A}M'_{L_{\rm A}}\rangle\langle S_{\rm A}M_{S_{\rm A}}|\hat{T}_{-q}^1(S_{\rm A})|S_{\rm A}M'_{S_{\rm A}}\rangle \tag{10}$$

The last two integrals in Eq. (10) can be evaluated using the Wigner-Eckart theorem, which generally reads as follows:[8]

$$\langle\alpha jm|\hat{T}_q^k|\alpha' j'm'\rangle = (-1)^{j-m}\begin{pmatrix} j & k & j' \\ -m & q & m' \end{pmatrix}\langle\alpha' j||\hat{T}^k||\alpha j'\rangle, \tag{11}$$

where the symbol in brackets is a $3j$-symbol and the last term denotes the reduced matrix element of the spherical tensor $\hat{T}_q^k$. The reduced matrix elements of all operators describing intra- and inter-atomic or molecular interactions can be expressed in terms of measured spectroscopic constants or evaluated directly using the rules of angular momentum algebra.[8]

4. Uncoupled Spherical Tensor Representation of $\hat{V}_{\rm AB}$

Most of the leading terms giving rise to the interaction of atoms and molecules with external fields as well as fine and hyperfine structure of atoms and molecules in Σ electronic states are scalar products of two rank-1 tensors or one rank-2 and two rank-1 spherical tensors. These operators can therefore be easily decomposed into direct products of space-fixed spherical tensors as in the above example of the spin-orbit interaction. In order to make use of the simplicity of the space-fixed uncoupled basis representation (5) and the Wigner-Eckart theorem for the complete formulation of the scattering problem, it is also necessary to express the electronic interaction potentials $\hat{V}_{\rm AB}$ in terms of uncoupled products of spherical tensors defined in the space-fixed coordinate system. This is a more complicated task. The electronic potentials describing interatomic and intermolecular interactions are naturally defined and computed in the molecule-fixed coordinate frame. For example, the electronic interaction potentials of two open-shell atoms are characterized by the projection of the total electronic orbital angular momentum on the interatomic axis. Krems, Groenenboom and Dalgarno presented a general theory[2] providing expressions for the electronic interaction potential between two atoms in arbitrary electronic states in the space-fixed coordinate frame. This theory is described in the following section. Section 4.2 discusses the space-fixed spherical tensor expansion of molecule - molecule interaction potentials.

4.1. *Interaction of Two Open-Shell Atoms*

The goal of the derivation in this section is to obtain a general form of the operator $\hat{V}_{\mathrm{AB}}$ describing the electronic interaction between two atoms in arbitrary electronic states expressed in terms of uncoupled spherical tensors in the space-fixed coordinate frame. The electrostatic interaction conserves the total electron spin (S) so the operator $\hat{V}_{\mathrm{AB}}$ can be decomposed into a sum of terms corresponding to the total electron spin of the diatomic molecule AB. Based on general principles, the different spin terms must be of the form

$$\hat{V}^S_{\mathrm{AB}} = \sum_{t_1,t_2,t_3} \sum_{u_1,u_2,u_3} \hat{V}_{t_1,t_2,t_3}(R)\hat{T}^{t_1}_{u_1}(\hat{R})\hat{T}^{t_2}_{u_2}(A)\hat{T}^{t_3}_{u_3}(B), \tag{12}$$

where the spherical tensors $\hat{T}^{t_2}_{u_2}(A)$ and $\hat{T}^{t_3}_{u_3}(B)$ are defined in the space of the wave functions of atoms A and B, respectively, and the tensor $\hat{T}^{t_1}_{u_1}(\hat{R})$ is defined in the space spanned by the $|lm_l\rangle$ functions. The work of Krems, Groenenboom and Dalgarno[2] provided the exact form of the tensors in Eq. (12) and the relation between the expansion coefficients in Eq. (12) and the adiabatic electronic potentials of the molecule AB.

Consider two interacting atoms A and B in the states with electronic orbital ($\boldsymbol{L}_{\mathrm{A}}$ and $\boldsymbol{L}_{\mathrm{B}}$) and spin ($\boldsymbol{S}_{\mathrm{A}}$ and $\boldsymbol{S}_{\mathrm{B}}$) angular momenta. The molecular states $|L\Lambda\rangle|S\Sigma\rangle$ are defined in the limit of large interatomic distance R by the Clebsch-Gordan theorem

$$|L\Lambda(L_{\mathrm{A}}L_{\mathrm{B}})\rangle = \sum_{\Lambda_{\mathrm{A}}}\sum_{\Lambda_{\mathrm{B}}} |L_{\mathrm{A}}\Lambda_{\mathrm{A}}\rangle|L_{\mathrm{B}}\Lambda_{\mathrm{B}}\rangle\langle L_{\mathrm{A}}\Lambda_{\mathrm{A}}L_{\mathrm{B}}\Lambda_{\mathrm{B}}|L\Lambda\rangle \tag{13}$$

and the analogous equation for $|S\Sigma\rangle$. The products $|L\Lambda\rangle|S\Sigma\rangle$ are related to molecular Born-Oppenheimer functions $|c\Lambda S\Sigma\rangle$ as follows

$$|c\Lambda S\Sigma\rangle = \sum_{L} |L\Lambda(L_{\mathrm{A}}L_{\mathrm{B}})\rangle|S\Sigma(S_{\mathrm{A}}S_{\mathrm{B}})\rangle U^{\Lambda}_{L,c}, \tag{14}$$

where Λ and Σ specify the projections of $\boldsymbol{S}$ and $\boldsymbol{L}$ on the interatomic axis and the quantum number c is used to distinguish between molecular states corresponding to the same values of Λ and S. The transformation (14) is valid at large interatomic separation where the atomic wave functions do not overlap and the exchange interaction is absent. The transformation coefficients $U^{\Lambda}_{L,c}$ are, therefore, independent of spin S. Krems, Groenenboom and Dalgarno[2] proposed to use the transformation (14) at arbitrary values of R which leads to an effective representation of the electronic potential appropriate for the description of slow atom - atom collisions and the dissociation of diatomic molecules.

The matrix elements of the electronic Hamiltonian

$$\langle c\Lambda S\Sigma|\hat{V}_{\rm AB}|c\Lambda S\Sigma\rangle = \hat{V}_{c\Lambda S}(R) \tag{15}$$

are the non-relativistic Born-Oppenheimer potentials of the molecule AB. The operator $\hat{V}$ in Eq. (15) can be expanded as

$$\hat{V}_{\rm AB} = \sum_{c\Lambda S\Sigma} |c\Lambda S\Sigma\rangle \hat{V}_{c\Lambda S}(R)\langle c\Lambda S\Sigma|, \tag{16}$$

which can be re-written in the effective potential form

$$\hat{V}_{\rm AB} = \sum_{S}\sum_{\Sigma} |S\Sigma\rangle\langle S\Sigma|\hat{V}^{S}_{\rm AB}, \tag{17}$$

where

$$\hat{V}^{S}_{\rm AB} = \sum_{L}\sum_{L'}\sum_{\Lambda} |L\Lambda(L_{\rm A}L_{\rm B})\rangle\langle L'\Lambda(L_{\rm A}L_{\rm B})|V^{S}_{L\Lambda;L'\Lambda}(R) \tag{18}$$

and

$$V^{S}_{L\Lambda;L'\Lambda}(R) = \sum_{c} U^{\Lambda,*}_{L,c}\hat{V}_{c\Lambda S}(R)U^{\Lambda}_{L',c}. \tag{19}$$

The transformation $U^{\Lambda}_{L,c}$ is an identity matrix when one of the interacting atoms is in an S-state or when the quantum number L can be assigned to each state $\hat{V}_{c\Lambda S}$. Otherwise, it can be determined by numerical diagonalization of the leading term in the interaction potential multipole expansion at long range.

The summation over L and L' in Eq. (18) runs from $|L_{\rm A}-L_{\rm B}|$ to $L_{\rm A}+L_{\rm B}$ and S takes the values from $|S_{\rm A}-S_{\rm B}|$ to $S_{\rm A}+S_{\rm B}$. $S_{\rm A}$, $S_{\rm B}$, $L_{\rm A}$ and $L_{\rm B}$ are kept fixed. Using the definition of the spherical tensor operators

$$\hat{T}^{k}_{Q}(L,L') = \sum_{\Lambda}\sum_{\Lambda'} |L\Lambda(L_{\rm A}L_{\rm B})\rangle\langle L'\Lambda'(L_{\rm A}L_{\rm B})|(-1)^{L-\Lambda}\begin{pmatrix} L & k & L' \\ -\Lambda & Q & \Lambda' \end{pmatrix}(k)^{1/2}, \tag{20}$$

and the orthogonality of $3j$-symbols, Eq. (18) can be re-written as a series over $\hat{T}^{k}_{Q}(L,L')$

$$\hat{V}^{S} = \sum_{L}\sum_{L'}\sum_{k}\sum_{Q}(k)^{1/2}\hat{T}^{k}_{Q}(L,L')\sum_{\Lambda}(-1)^{L-\Lambda}\begin{pmatrix} L & k & L' \\ -\Lambda & Q & \Lambda \end{pmatrix} V^{S}_{L\Lambda;L'\Lambda}. \tag{21}$$

The symbols in parentheses are $3j$-symbols and $(k)^{1/2}$ is a shorthand notation for $(2k+1)^{1/2}$.

Because the functions $|L\Lambda(L_A L_B)\rangle$ are referred to the molecular axis, Eq. (20) defines the tensor $\hat{T}^k_Q(L, L')$ in the body-fixed frame. It can be expressed in terms of the tensor $\hat{T}^k_q(L, L')$ defined in the space-fixed frame as follows

$$\hat{T}^k_Q(L, L') = \sum_q \hat{T}^k_q(L, L') D^k_{qQ}(\hat{R}). \tag{22}$$

The $\hat{V}_{AB}$ operator does not couple states with different values of Λ so that $Q = 0$ and the Wigner rotation matrix elements $D^k_{q,Q=0}$ are proportional to spherical harmonics $Y^*_{kq}(\hat{R})$ that depend on the polar angles of $\hat{R}$.

The tensor $\hat{T}^k_q(L, L')$ is related to the tensorial product of tensors $\hat{T}^{k_1}_{q_1}(L_A)$ and $\hat{T}^{k_2}_{q_2}(L_B)$ describing the rotation of electrons in atoms A and B, respectively:

$$\hat{T}^k_q(L, L') = \sum_{k_1} \sum_{k_2} \begin{Bmatrix} L_A & L_A & k_1 \\ L_B & L_B & k_2 \\ L & L' & k \end{Bmatrix} [(k_1)(k_2)(L)(L')]^{1/2} \left[\hat{T}^{k_1}(L_A) \otimes \hat{T}^{k_2}(L_B)\right]^{(k)}_q, \tag{23}$$

where the symbol in curly brackets is a $9j$-symbol and

$$\left[\hat{T}^{k_1}(L_A) \otimes \hat{T}^{k_2}(L_B)\right]^{(k)}_q = \sum_{q_1} \sum_{q_2} \langle k_1 q_1 k_2 q_2 | kq \rangle \hat{T}^{k_1}_{q_1}(L_A) \hat{T}^{k_2}_{q_2}(L_B). \tag{24}$$

The tensors $\hat{T}^{k_1}_{q_1}(L_A)$ and $\hat{T}^{k_2}_{q_2}(L_B)$ are defined as

$$\hat{T}^{k_1}_{q_1}(L_A) = \sum_{M_A} \sum_{M'_A} |L_A M_A\rangle\langle L_A M'_A| (-1)^{L_A - M_A} \begin{pmatrix} L_A & k_1 & L_A \\ -M_A & q_1 & M'_A \end{pmatrix} (k_1)^{1/2}, \tag{25}$$

and

$$\hat{T}^{k_2}_{q_2}(L_B) = \sum_{M_B} \sum_{M'_B} |L_B M_B\rangle\langle L_B M'_B| (-1)^{L_B - M_B} \begin{pmatrix} L_B & k_2 & L_B \\ -M_B & q_2 & M'_B \end{pmatrix} (k_2)^{1/2}, \tag{26}$$

where M_A and M'_A and M_B and M'_B are the projections of $\boldsymbol{L}_A$ and $\boldsymbol{L}_B$ on the space-fixed quantization axis.

Using Eqs. (21), (22) and (23), the effective potential $\hat{V}^S_{\rm AB}$ can be represented as a tensorial expansion in the space-fixed coordinate frame

$$\hat{V}^S_{\rm AB} = (4\pi)^{1/2} \sum_{k_1} \sum_{k_2} \sum_{k} V^S_{k_1,k_2,k}(R) \times \sum_{q_1} \sum_{q_2} \sum_{q} (-1)^{k_1-k_2} \begin{pmatrix} k_1 & k_2 & k \\ q_1 & q_2 & q \end{pmatrix} \hat{T}^{k_1}_{q_1}(L_{\rm A}) \hat{T}^{k_2}_{q_2}(L_{\rm B}) Y_{kq}(\hat{R}) \tag{27}$$

with the expansion coefficients

$$V^S_{k_1,k_2,k}(R) = \sum_{L} \sum_{L'} \sum_{\Lambda} V^S_{L\Lambda;L'\Lambda}(R)(-1)^{L-\Lambda} \begin{pmatrix} L & k & L' \\ -\Lambda & 0 & \Lambda \end{pmatrix} \times [(k_1)(k_2)(k)(L)(L')]^{1/2} \begin{Bmatrix} L_{\rm A} & L_{\rm A} & k_1 \\ L_{\rm B} & L_{\rm B} & k_2 \\ L & L' & k \end{Bmatrix}. \tag{28}$$

Eq. (27) is the space-fixed representation of the effective electronic potential for interaction between two atoms in arbitrary states and Eq. (28) provides the relation between the expansion coefficients $V^S_{k_1,k_2,k}$ and the matrix elements $V^S_{L\Lambda,L'\Lambda}$ which can be obtained from quantum chemistry calculations using Eq. (19).

Since the operator of the total interaction potential is diagonal also in the $|SM_S\rangle$ representation, Eq. (17) can be re-written as

$$\hat{V}_{\rm AB} = \sum_{S} \sum_{M_S} |SM_S\rangle\langle SM_S| \hat{V}^S_{\rm AB}, \tag{29}$$

thus providing the space-fixed effective representation of the total interaction potential.

As the tensors $\hat{T}^{k_1}_{q_1}(L_{\rm A})$, $\hat{T}^{k_2}_{q_2}(L_{\rm B})$ and $Y_{kq}(\hat{R})$ act only on the functions $|L_{\rm A}M_{L_{\rm A}}\rangle$, $|L_{\rm B}M_{L_{\rm B}}\rangle$ and $|lm_l\rangle$, respectively, and the operator $|SM_S\rangle\langle SM_S|$ acts only on the product functions $|S_{\rm A}M_{S_{\rm A}}\rangle|S_{\rm B}M_{S_{\rm B}}\rangle$, expansion (27) and relation (29) provide the most suitable representation of the interatomic interaction potential for the evaluation of the matrix elements of $\hat{V}_{\rm AB}$ in the uncoupled space-fixed basis (5).

4.2. *Molecule – Molecule Interaction*

The interaction potential between two molecules in Σ electronic states can also be represented by Eq. (29). The $\hat{V}^S_{\rm AB}$ terms are functions of the length and orientations of the vectors $\boldsymbol{r}_{\rm A}$ and $\boldsymbol{r}_{\rm B}$ representing the interatomic distances in the molecules A and B, and the vector $\boldsymbol{R}$. They can be expanded in spherical harmonics as follows[9]

$$V^{\rm S}_{\rm AB}(\mathbf{R},\mathbf{r}_{\rm A},\mathbf{r}_{\rm B}) = (4\pi)^{3/2} \sum_{\lambda_{\rm A}\lambda_{\rm B}\lambda} V^{S}_{\lambda_{\rm A}\lambda_{\rm B}\lambda}(R, r_{\rm A}, r_{\rm B})$$
$$\times \sum_{m_{\lambda_{\rm A}} m_{\lambda_{\rm B}} m_{\lambda}} \begin{pmatrix} \lambda_{\rm A} & \lambda_{\rm B} & \lambda \\ m_{\lambda_{\rm A}} & m_{\lambda_{\rm B}} & m_{\lambda} \end{pmatrix} Y_{\lambda_{\rm A} m_{\lambda_{\rm A}}}(\hat{r}_{\rm A}) Y_{\lambda_{\rm B} m_{\lambda_{\rm B}}}(\hat{r}_{\rm B}) Y_{\lambda m_{\lambda}}(\hat{R}). \quad (30)$$

The quantum numbers $m_{\lambda_{\rm A}}$, $m_{\lambda_{\rm B}}$ and m_{λ} can be viewed as space-fixed projections of $\lambda_{\rm A}$, $\lambda_{\rm B}$ and λ. The expression (30) can therefore be used directly for the evaluation of matrix elements of the interaction potential in the basis (5).

Expansion (30) has a similar form to expansion (27) of the interaction potential between two open-shell atoms. However, there is an important difference between Eq. (27) and Eq. (30). A close inspection of Eq. (27) reveals that the expansion of the atom - atom interaction potential may include terms that cannot occur in the spherical harmonics expansion (30). The non-equivalence of the expansions (27) and (30) arises from the different behavior of the tensors $\hat{T}^{k_1}_{q_1}(L_{\rm A})$ and $\hat{T}^{k_2}_{q_2}(L_{\rm B})$ and the spherical harmonics under inversion of the coordinate system. As follows from the definitions (25) and (26), the tensors $\hat{T}^{k_1}_{q_1}(L_{\rm A})$ and $\hat{T}^{k_2}_{q_2}(L_{\rm B})$ are invariant under the inversion, whereas the spherical harmonics $Y_{k_1 q_1}$ and $Y_{k_2 q_2}$ acquire the factors $(-1)^{k_1}$ and $(-1)^{k_2}$. The invariance of the tensors $\hat{T}^{k_1}_{q_1}(L_{\rm A})$ and $\hat{T}^{k_2}_{q_2}(L_{\rm B})$ under inversion is critical for the complete representation of the potential (27). This important observation was first made by Gerrit Groenenboom.

5. Recent Work Based on Uncoupled Representations

The original formulation of the close coupling scattering theory in the fully uncoupled space-fixed representation[4] was developed to analyze the collision dynamics of molecules in magnetic fields. This study focused on collision-induced spin depolarization and neglected the hyperfine interactions. The simplicity of the uncoupled representation (5) allowed for several immediate extensions of the theory. In particular, Timur Tscherbul and the present author used the uncoupled representation to develop the formalism for quantum calculations of cross sections for elastic and inelastic molecular collisions in the presence of combined electric and magnetic fields.[10] This work led to the observation of a new mechanism for controlling molecular collisions with combined electric and magnetic fields. The mechanism is based on inducing interactions between molecular states of different symmetry, which can be used to study controlled chemical reactions of cold molecules in a magnetic trap.[11] Abrahamsson and coworkers[12] used the theory to analyze the effects of relative field orientations on collisions of molecules in combined electric and magnetic fields. Tscherbul and coworkers[7] also extended the work to study collisions of molecules with large hyperfine interactions. The work of Tscherbul and coworkers demonstrated that elastic and inelastic collisions of molecules can be effectively controlled by external electric and magnetic fields not only at ultracold temperatures, but also at temperatures near 1 Kelvin. It is much easier to create molecular ensembles at these "warmer" temperatures.

In Ref. [5], Krems and Dalgarno used the uncoupled space-fixed representation to derive analytical expressions for the energy transport cross sections based on the expressions for the angle-resolved differential scattering probabilities. Tscherbul[13] extended this work to examine differential scattering of cold polar molecules in combined electric and magnetic fields. He found that electric fields may modify scattering shape resonances by inducing couplings between different partial waves of the collision system. He also showed that measurements of angle-resolved cross sections for differential scattering at low temperatures may be used to identify and characterize scattering shape resonances.

Gonzalez-Martinez and Hutson recently extended their publicly available computer program MOLSCAT to allow calculations of atom - molecule scattering in the uncoupled space-fixed representation.[14] They used the program to identify magnetically tunable Feshbach resonances in collisions of $NH(^3\Sigma)$ molecules with He atoms. In particular, they discovered that Feshbach resonances in collisions of molecules in metastable excited states may be significantly suppressed due to efficient inelastic processes. The resonance peaks in the elastic cross sections for collisions of molecules in excited states were found to be much smaller than in collisions of atoms or molecules in the absolute ground state.

Using the uncoupled space-fixed representation as a starting point, Tscherbul developed a theory and computer programs for rigorous quantum calculations of probabilities of chemical reactions in the presence of external electromagnetic fields.[15] The theory is based on the Fock-Delves hyperspherical coordinates and adiabatic partitioning of the total Hamiltonian of the reactive system. The adiabatic channel wave functions are expanded in basis sets of hyperangular functions corresponding to different reaction arrangements and the interactions with external fields are included in each chemical arrangement separately. The theory was applied to examine the effects of electric fields on the chemical reactions of LiF molecules with H atoms and HF molecules with Li atoms at low temperatures. It was shown that electric fields may enhance the probability of chemical reactions and modify reactive scattering resonances.

Groenenboom, Chu and Krems[16] have recently used the spherical tensor expansion (27) in order to analyze the long range interaction potentials of open-shell atoms in states of high electronic angular momenta. They derived expressions for the first order electrostatic and second order dispersion and induction anisotropic interaction coefficients in terms of the multipole moments and dynamic polarizabilities of the atoms and showed that a complete description of the second order interaction requires odd rank or "out-of-phase" polarizabilities, which is consistent with the presence of odd-rank terms in the general expansion (27). They also showed that the tensorial expansion (27) may be used to derive relations between the van der Waals coefficients of the interatomic interaction potentials of different symmetries. Based on this theory, Chu and Groenenboom[17] developed a numerical approach for accurate numerical calculations of the van der Waals interaction coefficients for complex atoms in states of non-zero electronic orbital angular momenta.

Maxwell and coworkers[18] have recently extended the formalism described in this chapter to explore the collision dynamics of Bi atoms with He atoms in the presence of strong magnetic fields. The lowest-energy electronic term of Bi is classified as an S state with a spin multiplicity of 4. According to Eq. (27), the interaction between S-state atoms is isotropic, and inelastic collisions between an S-state atom and helium at low temperatures should be generally forbidden. The LS-coupling scheme is, however, insufficient for heavy atoms with strong spin-orbit interactions. This combined experimental and theoretical study demonstrated that the Zeeman relaxation of Bi in the ground electronic state is very efficient due to the admixture of electronic excited states. This is important because this indicates that dense ensembles of heavy relativistic atoms in magnetic traps will generally be unstable.

In conclusion, the close coupling formulation of scattering theory in the fully uncoupled space-fixed basis representation provides a versatile method for numerical studies of a large variety of problems. In fairness, it must be noted that a partially coupled representation in terms of basis functions corresponding to approximately conserved quantum numbers may be more convenient for certain problems as, while being more complex, it may provide more physical insight into the mechanisms of collisional energy transfer.[14] A partially coupled representation may also provide a better basis for bound-state calculations leading to well-defined zeroth-order solutions.

Much of the research described in this Chapter was catalyzed by ideas of Alex Dalgarno. His influence on the development of the research field of cold atoms and molecules is so immense that solving any significant problem in this research field feels like walking in his footprints. His inspirational guidance and contributions have defined new research directions in molecular physics for years to come. This contribution is a testament to Alex's vast legacy that symbolizes research ingenuity and productivity.

References

1. A. M. Arthurs and A. Dalgarno, "The theory of scattering by a rigid rotor", Proceedings of the Royal Society, London, Ser. A 256, 540 (1960).
2. R. V. Krems, G. C. Groenenboom, and A. Dalgarno, "Electronic interaction anisotropy between atoms in arbitrary angular momentum states", Journal of Physical Chemistry 108, 8941 (2004).
3. J. Doyle, B. Friedrich, R. V. Krems, and F. Masnou-Seeuws, "Quo vadis, cold molecules?", European Physical Journal D 31, 149 (2004).
4. R.V. Krems and A. Dalgarno, "Quantum mechanical theory of atom - molecule and molecular collisions in a magnetic field: Spin depolarization", Journal of Chemical Physics 120, 2296 (2004).
5. R. V. Krems and A. Dalgarno, "Collisions of atoms and molecules in external magnetic fields", Fundamental World of Quantum Chemistry, chapter 14, Vol. 3, pp. 273-294, eds: E. Kryachko and E. Brandas (Kluwer 2004).
6. A. Volpi and J. L. Bohn, "Magnetic-field effects in ultracold molecular collisions", Physical Review A **65** 052712 (2002).
7. T. V. Tscherbul, J. Klos, L. Rajchel, and R. V. Krems, "Fine and hyperfine interactions in cold YbF-He collisions in electromagnetic fields", Physical Review A **75**, 033416 (2007).
8. R. N. Zare, *Angular momentum* (John Wiley and Sons, Inc., New York, 1988).
9. S. Green, "Rotational excitation in $H_2 - H_2$ collisions: Close-coupling calculations", Journal of Chemical Physics **62**, 2271 (1975).
10. T. V. Tscherbul and R. V. Krems, "Controlling spin relaxation of cold molecules with electric fields", Physical Review Letters 97, 083201 (2006).
11. R. V. Krems, "Cold controlled chemistry", Physical Chemistry Chemical Physics **10**, 4079 (2008).
12. E. Abrahamsson, T. V. Tscherbul, and R. V. Krems, "Inelastic collisions of cold polar molecules in non-parallel electric and magnetic fields", Journal of Chemical Physics **127**, 044302 (2007).
13. T. V. Tscherbul, "Differential scattering of cold molecules in superimposed electric and magnetic fields", Journal of Chemical Physics **128**, 244305 (2008).

14. M. L. Gonzalez-Martinez and J. M. Hutson, "Ultracold atom-molecule collisions and bound states in magnetic fields: Tuning zero-energy Feshbach resonances in He-NH($^3\Sigma^-$)", Physical Review A **75**, 022702 (2007).
15. T. V. Tscherbul and R. V. Krems, "Quantum theory of chemical reactions in the presence of electromagnetic fields", Journal of Chemical Physics **129**, 034112 (2008).
16. G. C. Groenenboom, X. Chu, and R. V. Krems, "Electronic anisotropy between open shell atoms in first and second order perturbation theory", Journal of Chemical Physics **126**, 204306 (2007).
17. Xi Chu and Gerrit Groenenboom, private communication.
18. S. E. Maxwell, M. T. Hummon, Y. Wang, A. A. Buchachenko, R. V. Krems, and J. M. Doyle, "Spin-orbit interaction and large inelastic rates in bismuth-helium collisions", Physical Review A **78**, 042706 (2008).

COLLISIONS AND REACTIONS IN ULTRACOLD GASES

N. BALAKRISHNAN

Department of Chemistry, University of Nevada Las Vegas
Las Vegas, NV 89154, USA
E-mail: naduvala@unlv.nevada.edu

Recent years have witnessed exciting advances in the development of novel methods to cool and trap molecules and collisions and interactions involving ultracold atoms and molecules have become topics of considerable experimental and theoretical interest. In this article, we provide an overview of recent theoretical investigations of inelastic and reactive collisions in atomic and molecular samples at cold and ultracold temperatures.

Keywords: ultracold molecules, vibrational relaxation, chemical reaction

1. Introduction

Fundamental studies of atomic and molecular collisions have dramatically enhanced our understanding of molecular processes and chemical reactions occurring in a variety of environments. Atmospheric and astrochemistry models include rate coefficients and cross section data for a large number of chemical reactions and the accuracy of the model predictions is limited by the accuracy of the underlying molecular and chemical data. Theoretical studies based on classical, semiclassical, and quantum mechanical models have played a central role in evaluating the relevant cross sections and rate coefficients. For atmospheric chemistry models, most data pertain to room temperature or processes occurring at elevated temperatures though temperatures as low as 150 K are attained in the mesosphere. In astrophysical environments, molecular and chemical data at temperatures as low as 10 K are required. At these low temperatures, quantum effects become important and results of molecular collisions become very sensitive to the details of the interaction potential, especially long-range intermolecular forces.

Over the last 10 years, there have been a renewed interest in understanding molecular collisions and energy transfer processes occurring at temperatures lower than 1.0 K.[1] This was motivated by new experimental breakthroughs achieved in cooling and trapping of atoms and subsequently diatomic molecules at mK and μK temperatures. Several methods have been successfully applied to cool and trap molecules. Photoassociation[2–7] and Feshbach resonance[8–11] methods enable the creation of ultracold ground state molecules in high lying vibrational levels. These methods have been successfully applied to several homonuclear and heteronuclear alkali-metal dimers and important progress is being made in creating ultracold molecules in the absolute rovibrational ground state. Other techniques such as the buffer-gas cooling method[12,13] and the Stark-deceleration method[14–16] permit cooling and trapping of molecules with permanent magnetic or electric moments and are applicable to a wider range of molecules. Novel methods to study ultracold chemical reactions involving ion - molecule systems in a linear Paul trap have also been proposed.[17] The experimental breakthroughs in creating dense samples of cold and ultracold molecules motivated the possibility of investigating chemical reactions and molecular processes at temperatures close to absolute zero where quantum effects play an important role.

The formation of molecules in highly excited vibrational levels as in photoassociation spectroscopy and Feshbach resonance methods motivated the investigation of vibrational and rotational energy transfer processes in atom-diatom collisions at cold and ultracold temperatures.[18–22] These studies revealed that vibrational and rotational relaxation may occur with large rate coefficients at ultracold temperatures. Subsequent studies of atom-diatom chemical reactions showed that barrier reactions, dominated by tunneling at low temperatures, may also occur with measurably large rate coefficients at ultracold temperatures.[18–20,23] The studies of atom-exchange reactions in alkali-metal atom-dimer systems[24–30] also indicated that they may occur with large rate coefficients at cold and ultracold temperature. Recent studies of barrierless non-alkali metal atom-diatom chemical reactions show reactivity comparable to alkali metal systems at ultracold temperatures.[31]

In this article, we give an overview of recent theoretical studies of ultracold reactive and nonreactive atom-diatom and diatom-diatom collisional calculations with specific examples from our own group. For more detailed discussions, we refer to several review articles that were published in the last few years.[1,18–22]

First we will briefly review some of the properties of low energy scattering and the concept of scattering length, and discuss how it can be extended to describe inelastic scattering in the limit of zero-temperature. In particular, we will discuss the relation between imaginary part of the scattering length and the zero-temperature quenching or reaction rate coefficient and how it may be used to predict lifetimes of weakly bound complexes.

2. Properties of Low Energy Scattering

The scattering length is an important parameter in the quantum mechanical description of ultracold collisions and its sign and magnitude have important consequences on the stability of Bose-Einstein condensates and trapped atoms and molecules. Here we discuss the scattering length in the context of inelastic scattering in atom-molecule systems. We denote the initial vibrational quantum number of the molecule by v and the rotational quantum number by j. The kinetic energy in the incident channel $E_{vj} = \hbar^2 k_{vj}^2/(2\mu)$ where k_{vj} is the wave vector and μ is the reduced mass of the atom-molecule system. The scattering length is defined as

$$a_{vj} = -\lim_{k_{vj}\to 0} \frac{\delta(k_{vj})}{k_{vj}}, \tag{1}$$

where δ_{vj} is the phase shift associated with the incident channel wave function. For single-channel scattering where only elastic scattering occurs the scattering length is real but it has an imaginary part when inelastic or reactive channels are present.[32] The complex scattering length may be defined as

$$a_{vj} = \alpha_{vj} - i\beta_{vj} \tag{2}$$

with the real and imaginary parts of the scattering length given in terms of the elastic component of the S−matrix, S_{vj}^{el}, according to

$$\alpha_{vj} = -\lim_{k_{vj}\to 0} \frac{\mathrm{Im}(S_{vj}^{\mathrm{el}})}{2k_{vj}} \tag{3}$$

and

$$\beta_{vj} = \lim_{k_{vj}\to 0} \frac{1 - \mathrm{Re}(S_{vj}^{\mathrm{el}})}{2k_{vj}}. \tag{4}$$

A large and positive value for the real part of the scattering length is an indicator of the presence of a near-zero energy bound/quasibound state while a large and negative value of α indicates the presence of a virtual state. A virtual state will become a bound state as the strength of interaction potential is increased, during which the scattering length goes through infinity.

The elastic cross section attains a finite value in the zero energy limit and its magnitude is given by

$$\sigma_{vj\to vj}(E_{vj} \to 0) = 4\pi(\alpha_{vj}^2 + \beta_{vj}^2). \tag{5}$$

Furthermore, in the limit of zero temperature, the inelastic (reactive and nonreactive) cross sections vary inversely with the velocity in accordance with Wigner's law.[32,33] In this limit, the imaginary part of the scattering length is related to the total inelastic cross section, σ_{vj}^{in}:[32]

$$\beta_{vj} = \lim_{k_{vj}\to 0} \frac{k_{vj}\sigma_{vj}^{\text{in}}}{4\pi}. \tag{6}$$

The total inelastic cross section is given by

$$\sigma_{vj}^{\text{in}} = \frac{\pi}{k_{vj}^2(2j+1)} \sum_{J=0}^{J_{\max}} (2J+1)P_{vj}^J(E_{vj}), \tag{7}$$

where $P_{vj}^J(E_{vj})$ is the reaction probability for a given value of the total angular momentum quantum number, J. From eq.(6) it is clear that rate coefficients for inelastic processes become finite in the limit of zero temperature:[32]

$$R_{vj}(T \to 0) = 4\pi\beta_{vj}\hbar/\mu. \tag{8}$$

It has been shown that the scattering length formalism can be applied to predict pre-dissociation lifetimes of quasibound states that lie close to the channel threshold:[32,34]

$$\tau_{vj} = \frac{\mu|a_{vj}|^4}{\alpha_{vj}R_{vj}(T \to 0)}. \tag{9}$$

Quéméner et al.[22] have recently presented an exhaustive review of reactive and nonreactive collisional calculations at cold and ultracold temperatures and they have tabulated values of zero-temperature quenching and reaction rate coefficients for several atom-molecule and molecule-molecule systems.

3. Nonreactive Collisions

Until recently, theoretical studies of ultracold nonreactive collisions have mostly focused on rotational and vibrational energy transfer in atom-diatom systems.[19,22,34,35] Recently, these studies have been extended to diatom-diatom systems using both the rigid rotor approximation,[36,37] and full-dimensional quantum calculations.[38] Before we discuss the molecule-molecule calculations, first we will review some key features of atom-diatom collisions at cold and ultracold temperatures.

3.1. *Atom-Diatom Systems*

Vibrational relaxation in atom-diatom systems at cold and ultracold temperatures has been the topic of a number of recent investigations.[19,22,34,35,39–42] The studies revealed that the temperature dependence of rate coefficients for vibrational and rotational relaxation at low temperatures is dramatically different from that at high temperatures. The corresponding cross sections attain a minimum value at collision energies comparable to the depth of the van der Waals interaction potential. The cross sections increase with subsequent decrease in kinetic energy and eventually attain the inverse velocity dependence in the s-wave limit. For systems that form van der Waals complexes, the relaxation cross section exhibits sharp resonances corresponding to the decay of the metastable van der Waals complexes. This is illustrated in Fig. 1 for the vibrational relaxation of the CO molecule in the $v = 1, j = 0$ level in collisions with ^{4}He.[35] The sharp resonances in the energy dependence of the cross section are attributed to the decay of quasibound states of the He$\cdots$CO van der Waals complexes. Calculations showed that vibrational and rotational relaxation processes at low energies are dramatically influenced by the presence of quasibound van der Waals complexes if they lie very close to the channel threshold.[35] This is illustrated in Fig. 2 in which we show rate coefficients for the vibrational relaxation of CO($v = 1, j = 0$) and CO($v = 1, j = 1$) in collisions with ^{4}He as functions of the temperature. While the relaxation rate coefficients are very similar for temperatures above 30 K, significant differences appear for temperatures less than 10 K. In the Wigner threshold regime where the rate coefficients attain finite values, the vibrational quenching rate coefficient for the $v = 1, j = 1$ level is about two orders of magnitude larger than that of the $v = 1, j = 0$ level. This difference is attributed to the presence of a quasibound state of the He-CO van der Waals complex that lie just below the threshold of the $v = 1, j = 1$ level. The decay of the van der Waals complex leads to a Feshbach resonance in the energy dependence of the elastic cross section in the $v = 1, j = 0$ channel below the opening of the $v = 1, j = 1$ level. The Feshbach resonance is shown in Fig. 3. Calculations of vibrational relaxation of $H_2(v)$ in collisions with He atoms by Balakrish-

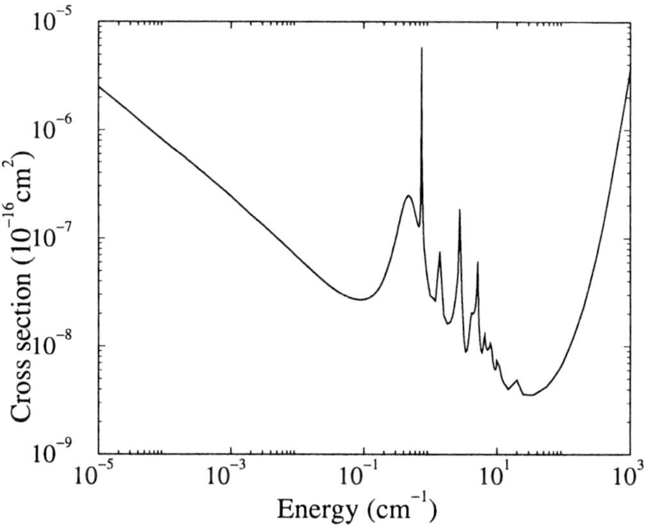

Fig. 1. Cross section for the relaxation of the $v = 1, j = 0$ level of CO in collisions with ^{4}He as a function of the kinetic energy. Reproduced with permission from Balakrishnan et al.[35]

nan, Forrey and Dalgarno[34] revealed that the zero-temperature vibrational relaxation rate coefficients are strongly dependent on the initial vibrational level of the H_2 molecule. The rate coefficients were found to increase by about three orders of magnitude when the vibrational quantum number of the H_2 molecule is raised from 1 to 10. The enhancement was attributed in part to the increase in the number of de-excitation channels as well as the increase in the effective area of interaction as the vibrational quantum number is increased.

Experiments performed nearly two decades ago[43,44] indicated that vibrational relaxation of rotationally excited molecules leads to highly efficient and selective transitions in which vibrational and rotational quantum numbers of the molecule change according to $\Delta j = -4\Delta v$ or $\Delta j = -2\Delta v$. These transitions were termed "quasiresonant transitions" (QR) and were found to be largely insensitive to the details of the interaction potential. Interestingly, the quasiresonant transitions were found to be even more dominant in ultracold collisions. Calculations by Forrey et al.[40,45] showed that vibrational relaxation in He+$H_2(v, j)$ collisions proceed via quasiresonant transitions for initial vibrational levels $v > 12$. Forrey et al. [46–49] have subsequently observed QR transitions in a number of atom-diatom systems.

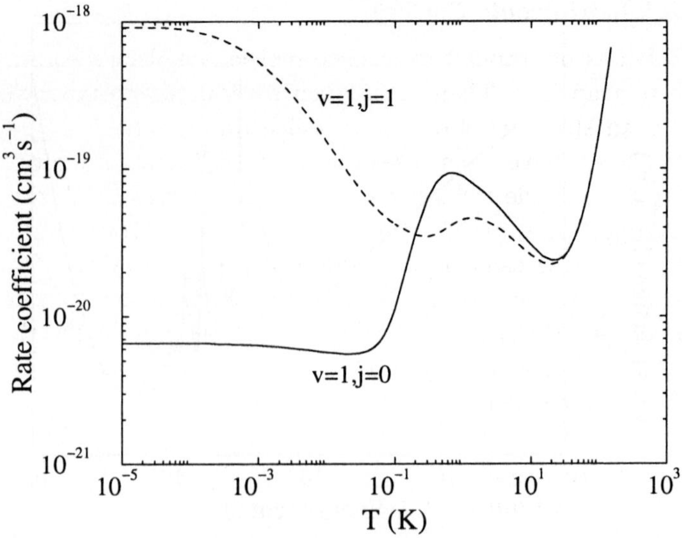

Fig. 2. Rate coefficients for the quenching of CO($v = 1, j = 0, 1$) by collisions with ^{4}He as functions of the temperature. Reproduced with permission from Balakrishnan et al.[35]

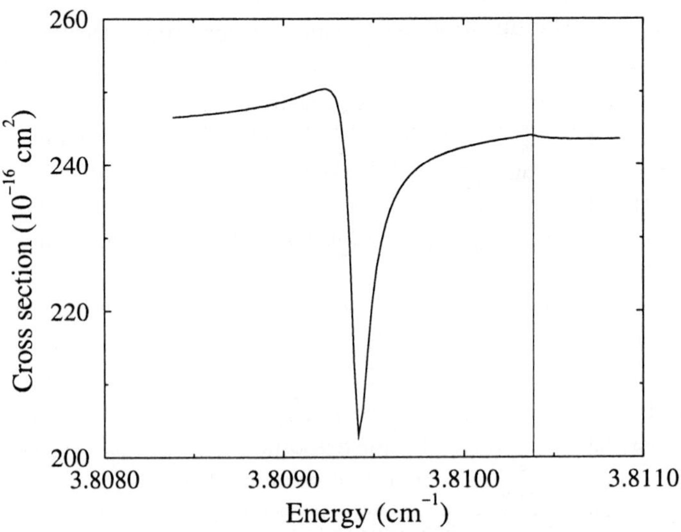

Fig. 3. Feshbach resonance in the elastic scattering cross section between ^{4}He and CO($v = 1, j = 0$). The resonance occurs just below the opening of the $v = 1, j = 1$ level shown by the vertical line. The energy is relative to the $v = 1, j = 0$ level of the CO molecule. Reproduced with permission from Balakrishnan et al.[35]

3.2. *Molecule-Molecule Systems*

At high densities of trapped molecules, molecule-molecule collisions also contribute to trap loss. There have been several recent experiments involving trapped alkali-metal dimers in which rate coefficients of molecule-molecule collisions have been measured.[50–52] Unlike atom-molecule collisions, molecule-molecule collisions are more challenging for theoretical investigations due to the proliferation in the number of rovibrational levels to be included in the calculations. This is especially the case for heavier molecules with high density of states of rovibrational levels. Due to the light mass and relatively fewer number of rovibrational levels, the H_2-H_2 system has served as a prototype for recent quantum mechanical investigations of molecule-molecule collisions at ultracold temperatures. Forrey et al.[36] and Lee et al.[37] adopted the rigid rotor approximation to study molecule-molecule collisions. In a recent work, Quéméner et al.[38] reported full-dimensional quantum calculations of rotational and vibrational transitions in H_2-H_2 collisions without using any decoupling approximations. The calculations showed that molecule - molecule collisions may involve highly efficient near-resonant energy transfer if they are accompanied by simultaneous conservation of the total rotational angular momentum and total internal energy of the molecules. This mechanism was also found to be independent of the initial vibrational excitation of the molecules. Fig. 4

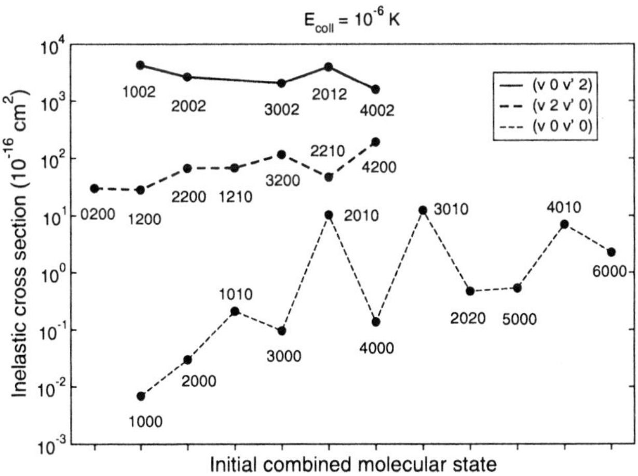

Fig. 4. Inelastic cross sections as functions of the initial combined molecular state (CMS) for a collision energy of 10^{-6} K. Solid line ($v0v'2$); dashed bold line ($v2v'0$); dashed line ($v0v'0$). The cross section values are labeled by the corresponding initial CMSs and the vibrational energy increases along the horizontal axis.

shows the vibrational relaxation cross sections for different combinations of rovibrational levels of the two H_2 molecules at a collision energy of 10^{-6} K.

4. Reactive Collisions

The possibility of observing and controlling chemical reactivity in ultracold gases is an important driving force for current research in ultracold molecules. At temperatures lower than 1 K, perturbations introduced by external electric or magnetic fields are typically larger than the collision energies involved and there is considerable interest in understanding the role of external fields in controlling chemical reactivity in ultracold gases.[18,20] Here we will discuss some salient features of atom-diatom chemical reactions at cold and ultracold temperatures that were performed in the absence of electric fields. These calculations include both barrier reactions which proceed via tunneling at low temperatures and also barrierless reactions.

4.1. *Barrier Reactions*

Atom-diatom exchange reactions involving energy barriers in entrance or exit channels have been the topic of a large number of experimental and theoretical investigations. In recent years there have been considerable interest in studying the influence of van der Waals interaction on chemical reactivity. Studies of F+H_2, F+HD, Cl+H_2, Cl+HD and related systems[23,53–61] have shown that the van der Waals interaction plays an important role in the dynamics of these reactions and that the decay of quasibound van der Waals complexes formed in the entrance or exit channels of the reaction lead to sharp resonances in the energy dependence of the reaction probabilities. Calculations by Balakrishnan and Dalgarno[23] showed that the F+H_2 reaction is characterized by a relatively large rate coefficient at ultracold temperatures. Using the potential energy surface developed by Stark and Werner[62] they reported a zero-temperature limiting value of 1.25×10^{-12} cm^3 s^{-1} for the rate coefficient of the F+$H_2(v = 0, j = 0)$ reaction. As shown in Fig. 5 the reaction yields HF molecules primarily in the $v = 2$ vibrational level, in agreement with results at thermal energies. Calculations on the F+HD reaction showed that[56] in the zero-temperature limit the HF+D channel is more probable than the DF+H channel due to the enhanced tunneling of the H atom compared to the D atom at low temperatures. Subsequent studies by Bodo et al.[58] showed that, in the Wigner regime, the cross sections of the F+$D_2(v = 0, j = 0)$ reaction is nearly two orders of magnitude smaller than that of the F+$H_2(v = 0, j = 0)$ reaction.

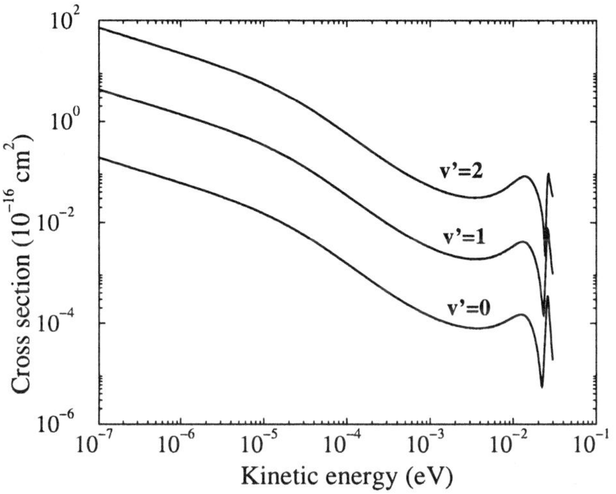

Fig. 5. $J = 0$ cross sections for F + $H_2(v = 0, j = 0) \rightarrow HF(v')$ + H reaction for $v' = 0 - 2$ as functions of the incident kinetic energy. Reproduced with permission from Balakrishnan and Dalgarno.[23]

This is illustrated in Fig. 6 where we compare $J = 0$ cumulative reaction probabilities and cross sections for the two systems. Since such large difference cannot be attributed to tunneling alone, Bodo et al. argued that the enhanced reactivity of the F+H_2 system is due to the presence of a virtual state close to the channel threshold of the F+$H_2(v = 0, j = 0)$ system. The virtual state is absent for the F+D_2 reaction and its presence for the F+H_2 reaction was further confirmed by the existence of a Ramsauer minimum in the elastic scattering cross section of the F+H_2 system at low energies.[58]

There have been several other studies of tunneling dominated atom-diatom chemical reactions in the last several years. Balakrishnan[57] showed that in the reaction between ground state Cl atom and vibrationally excited HD($v = 1, j = 0$) molecule, chemical reaction leading to the formation of the HCl+D channel dominates over that leading to the formation of the DCl+H channel as well as pure nonreactive vibrational quenching leading to the formation of HD($v = 0$). He also showed that the low energy values of the reaction probability are characterized by a number of sharp resonances due to the decay of quasibound Cl$\cdots$HD(v, j) van der Waals complexes. Subsequent studies of the Li+HF$\rightarrow$LiF+H and the reverse process by Weck and Balakrishnan [59,60] showed that the van der Waals interaction potential plays a significant role in the reactivity at low temperatures. Interestingly,

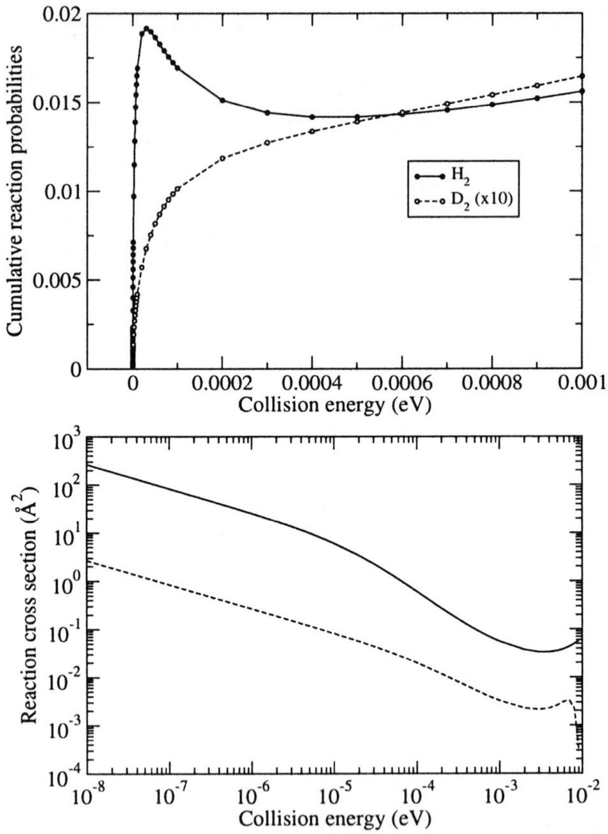

Fig. 6. Comparison of the $J = 0$ cumulative reaction probabilities (upper panel) and cross sections (lower sub-panel) of F + H_2 and F + D_2 reactions as functions of the incident kinetic energy. Reproduced with permission from Bodo et al.[58]

at low temperatures, the Li+HF/LiF+H reactions occur by tunneling of the relatively heavy F atom as the LiH+F channel is energetically not open at low temperatures. Recently, Tscherbul and Krems[63] have investigated the dynamics of the Li+HF reaction in the presence of an external electric field and they showed that the reactivity can by modified by several orders of magnitude by varying the strength of the electric field.

4.2. *Barrierless Reactions*

Until recently, studies of barrierless reactions at cold and ultracold temperatures have mostly focused on homonuclear alkali-metal atom-dimer systems.

Recently, several experimental studies reported measurements of reactivity in molecule-molecule and heteronuclear atom-dimer systems involving alkali-metal atoms. Due to the relatively heavy mass and the large density of states associated with these systems, accurate theoretical calculations are extremely challenging. Quantum calculations of vibrational relaxation and chemical reactivity in $Li+Li_2(v)$,[24–27] $Na+Na_2(v)$[28,29] and $K+K_2(v)$[30] systems have been reported by several investigators in the last several years. The calculations showed that vibrational relaxation and chemical reactivity occur with rate coefficients on the order of $10^{-11} - 10^{-10}$ cm^3/s for these systems.

Due to the development of techniques such as the Stark decelerator method and the buffer gas cooling method a broader range of chemically interesting molecules have been able to cool and trap at sub Kelvin temperatures. The OH molecule has attracted important attention in this regard due to its easy detection, non-zero dipole moment, and interest in atmospheric and astrochemistry. The reaction between OH radicals and oxygen atoms has been considered to be a key step in oxygen chemistry in interstellar medium and significant uncertainty exists in its rate coefficient at temperatures lower than 200 K. Though the lowest temperatures of interest in interstellar medium is about a few Kelvin, studies of chemical reactivity in the cold and ultracold regime can provide important insights into dynamics of chemical reactions at astrophysically relevant temperatures. Indeed, recent calculations of the O+OH reaction by Quéméner et al.[31,64] have shown that, to reliably predict rate coefficients of the O+OH reaction at temperatures of about 10 K of interest in interstellar medium, reaction probabilities at energies as low as 10^{-7} eV (0.001 K) are need. Quéméner et al. also explored the sensitivity of the rate coefficient to the energy grid of the reaction probabilities. They found that a fine mesh of collision energies is needed to accurately perform the integration of the reaction probability over the Boltzmann distribution of kinetic energies in evaluating the rate coefficients. This is because, the collision between OH and O results in the formation of many deeply bound HO_2 complexes, and their subsequent decay leads to a large number of sharp resonances in the energy dependence of the reaction probability. Figure 7 compares probability of the $O+OH(v = 0, j = 0)$ reaction evaluated using two representations of the electronically adiabatic PES for the HO_2 system. The upper panel shows the results of Quéméner et al.[64] using the diatomics-in-molecule (DIM) potential energy surface by Kendrick and Pack (referred to as DIMKP) and the lower panel compares the results on the XXZLG

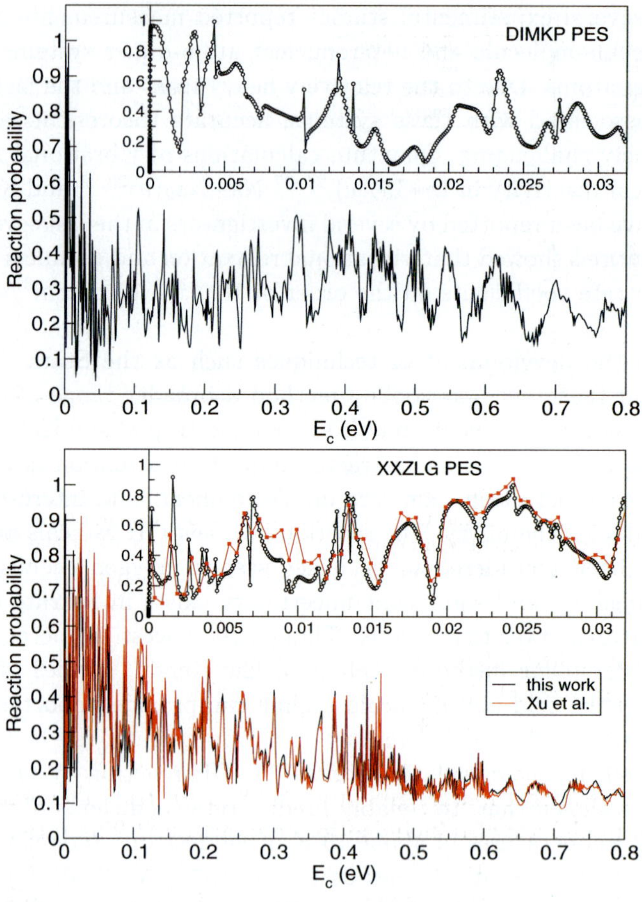

Fig. 7. $J = 0$ reaction probabilities of O + OH($v = 0, j = 0$) → H + O_2 reaction on the DIMKP PES (upper panel) and the XXZLG PES (lower panel). As discussed in[66] results of Xu et al.[65] on the XXZLG PES have been shifted back by 0.00409 eV to make a one-to-one comparison with our results. Reproduced with permission from Quéméner et al.[64]

PES obtained by Xu et al.[65] and Quéméner et al.[64] It is seen that the results are sensitive to the choice of the potential energy surface. While the results on the XXZLG PES show excellent agreement between the calculations of Xu et al. and Quéméner et al. additional resonances are present (see inset) in the results of Quéméner et al. which employed a denser energy grid. Quéméner et al.[64] have performed a systematic analysis of the sensitivity of the rate coefficients to the energy resolution of the reaction

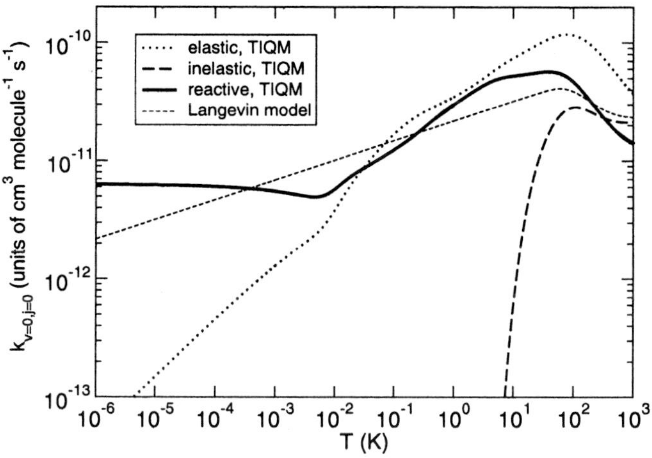

Fig. 8. Rate coefficients for elastic (dotted curve), inelastic (long-dashed curve) and reactive (solid curve) collisions in O + OH($v = 0, j = 0$) system evaluated using the DIMKP PES. The prediction of a classical Langevin model (short-dashed curve) is also shown. Reproduced with permission from Quéméner et al.[31]

probability and they found that the low temperature values of the rate coefficients are strongly influenced by the resolution of the energy grid. In Fig. 8 we compare rate coefficients computed using a J-shifting approximation for elastic, inelastic and reactive scattering in O+OH($v = 0, j = 0$) collisions for temperatures ranging from 10^{-6} to 10^3 K.[31] The predictions of a Langevin model are also included for comparison.[31] The rate coefficients attain the Wigner regime for temperatures below $T \approx E_c \approx 10^{-4}$ K with a limiting value of 6.2×10^{-12} cm^3 molecule^{-1} s^{-1}. In experiments using the stark decelerator methods, OH molecules were cooled to $T = 10 - 100$ mK. In this temperature range, the predicted values of the elastic rate coefficients are comparable to the reactive ones. This suggests that sympathetic cooling of OH by collisions with O atoms may not be feasible in this temperature range. The relatively large rate coefficient for the reaction in the zero-temperature limit indicates that barrierless exothermic reactions occur at significant rates at ultracold temperatures, in agreement with similar results for alkali-metal atom - diatom reactions.[26,28–30]

5. Summary and Perspective

In this article, we provided an overview of nonreactive and reactive collisions in atom-molecule and molecule-molecule systems at cold and ultracold temperatures. The dramatic progress being made in cooling and trapping of a wide varieties of atomic and molecular systems has spurred both experimental and theoretical investigations of energy transfer and chemical reactivity in a number of molecular systems. Cold and ultracold molecules serve as excellent candidates for controlled studies of molecular encounters and have provided new impetus to the field of coherent control and controlled chemistry. These studies are also finding applications in areas such as atmospheric chemistry and astrochemistry by providing improved molecular potentials and spectroscopic and molecular data from high precision experiments performed on trapped molecules. Molecular calculations at cold and ultracold temperatures can also be used to test the sensitivity of the rate coefficients to details of the interaction potentials and develop more reliable potential energy surfaces for the collisional system.

Acknowledgments

This work was supported by NSF grants # PHY-0855470, ATM-0635715, AST-0607524, and Nevada NASA EPSCoR. I thank Alex Dalgarno, Robert Forrey, Roman Krems and Goulven Quéméner on productive collaborations on many aspects of the work reported here.

References

1. J. Doyle, B. Friedrich, R. V. Krems, and F. Masnou–Seeuws, *Eur. Phys. J. D* **31**, 149 (2004).
2. J. T. Bahns, W. Stwalley, and P. L. Gould, *Adv. At. Mol. Opt. Phys.* **42**, 171 (2000).
3. F. Masnou–Seeuws and P. Pillet, *Adv. At. Mol. Opt. Phys.* **47**, 53 (2001).
4. A. J. Kerman, J. M. Sage, S. Sainis, T. Bergeman, and D. DeMille, *Phys. Rev. Lett.* **92**, 153001 (2004).
5. J. M. Sage, S. Sainis, T. Bergeman, and D. DeMille, *Phys. Rev. Lett.* **94**, 203001 (2005).
6. D. Wang , E. E. Eyler, P. L. Gould, and W. C. Stwalley, *J. Phys. B: At. Mol. Opt. Phys.* **39**, S849 (2006).
7. H. K. Pechkis, D. Wang, Y. Huang, E. E. Eyler, P. L. Gould, and W. C. Stwalley, and C. P. Koch, *Phys. Rev. A* **76**, 022504 (2007).
8. E. A. Donley, N. R. Claussen, S. T. Thompson, and C. E. Wieman, *Nature* **417**, 529 (2002).
9. M. Greiner, C. A. Regal, D. S. Jin, *Nature* **426**, 537 (2003).
10. Regal, C.A., Ticknor, C., Bohn, J. L., Jin, D. S., *Nature* **424**, 47 (2003).
11. M. Junker, D. Dries, C. Welford, J. Hitchcock, Y. P. Chen, and R. G. Hulet, *Phys. Rev. Lett.* **101**, 060406 (2008)
12. J. M. Doyle, B. Friedrich, J. Kim., and D. Patterson, *Phys. Rev. A* **52**, R2515 (1995)

13. J. D. Weinstein, R. deCarvalho, T. Guillet, B. Friedrich, and J. M. Doyle, *Nature* **395**, 148 (1998).
14. G. Meijer, *ChemPhysChem.* **3**, 495 (2002).
15. H. L. Bethlem and G. Meijer, *Int. Rev. Phys. Chem.* **22**, 73 (2003).
16. S. Y. T. Van De Meerakker, H. L. Bethlem, and G. Meijer G., *Nature Physics*, **4**, 595 (2008).
17. S. Willitsch, M. T. Bell, A. D. Gingell, S. R. Procter, and T. P. Softley, *Phys. Rev. Lett.* **100**, 043203 (2008).
18. R. V. Krems, *Int. Rev. Phys. Chem.* **24**, 99 (2005).
19. P. F. Weck and N. Balakrishnan, *Int. Rev. Phys. Chem.* **25**, 283 (2006).
20. R. V. Krems, *Phys. Chem. Chem. Phys.* **10**, 4079 (2008).
21. J. M. Hutson and P. Soldán, *Int. Rev. Phys. Chem.* **25**, 497 (2006).
22. G. Quéméner, N.Balakrishnan, and A. Dalgarno, *Inelastic collisions and chemical reactions of molecules at ultracold temperatures*, in *Cold Molecules: Theory, Experiment and Applications*, eds. R. V. Krems, J. M. Doyle, G. Meijer and B. Friedrick (CRC press, 2009, in press).
23. N. Balakrishnan and A. Dalgarno, *Chem. Phys. Lett.* **341**, 652 (2001).
24. M. T. Cvitaš, P. Soldán, J. M. Hutson, P. Honvault, and J.-M. Launay, *Phys. Rev. Lett.* **94**, 033201 (2005).
25. M. T. Cvitaš, P. Soldán, J. M. Hutson, P. Honvault, and J.-M. Launay, *Phys. Rev. Lett.* **94**, 200402 (2005).
26. G. Quéméner, J.-M. Launay, and P. Honvault, *Phys. Rev. A* **75**, 050701(R) (2007).
27. M. T. Cvitaš, P. Soldán, J. M. Hutson, P. Honvault, and J.-M. Launay, *J. Chem. Phys.* **127**, 074302 (2007).
28. P. Soldán, M. T. Cvitaš, J. M. Hutson, P. Honvault, and J.-M. Launay, *Phys. Rev. Lett.* **89**, 153201 (2002).
29. G. Quéméner, P. Honvault, and J.-M. Launay, *Eur. Phys. J. D* **30**, 201 (2004).
30. G. Quéméner, P. Honvault, J.-M. Launay, P. Soldán, D. E. Potter, and J. M. Hutson, *Phys. Rev. A* **71**, 032722 (2005).
31. G. Quéméner, N. Balakrishnan, and B. K. Kendrick, *Phys. Rev. A* (in press).
32. N. Balakrishnan, V. Kharchenko, R. C. Forrey, and A. Dalgarno, *Chem. Phys. Lett.* **280**, 5 (1997).
33. E. P. Wigner, *Phys. Rev.* **73**, 1002 (1948).
34. N. Balakrishnan, R. C. Forrey, and A. Dalgarno, *Phys. Rev. Lett.* **80**, 3224 (1998).
35. N. Balakrishnan, R. C. Forrey, and A. Dalgarno, *J. Chem. Phys.* **113**, 621 (2000).
36. R. C. Forrey, *Phys. Rev. A* **63**, 051403(R) (2001).
37. T.-G. Lee, N. Balakrishnan, R. C. Forrey, P. C. Stancil, D. R. Schultz, and G. J. Ferland, *J. Chem. Phys.* **125**, 114302 (2006).
38. G. Quéméner, N. Balakrishnan, and R. V. Krems, *Phys. Rev. A* **77**, 030704(R) (2008).
39. R. C. Forrey, N. Balakrishnan, V. Kharchenko, and A. Dalgarno, *Phys. Rev. A* **58**, R2645 (1998).

40. R. C. Forrey, N. Balakrishnan, A. Dalgarno, M. R. Haggerty, and E. J. Heller, *Phys. Rev. Lett.* **82**, 2657 (1999).
41. R. C. Forrey, V. Kharchenko, N. Balakrishnan, and A. Dalgarno, *Phys. Rev. A* **59**, 2146 (1999).
42. C. Zhu, N. Balakrishnan, and A. Dalgarno, *J. Chem. Phys.* **115**, 1335 (2001).
43. B. Stewart, P. D. Magill, T. P. Scott, J. Derouard, and D. E. Pritchard, *Phys. Rev. Lett.* **60**, 282 (1988).
44. P. D. Magill, B. Stewart, N. Smith, and D. E. Pritchard, *Phys. Rev. Lett.* **60**, 1943 (1988).
45. R. C. Forrey, N. Balakrishnan, A. Dalgarno, M. R. Haggerty, and E. J. Heller, *Phys. Rev. A* **64**, 022706 (2001).
46. R. C. Forrey, *Phys. Rev. A* **66**, 023411 (2002).
47. J. C. Flasher and R. C. Forrey, *Phys. Rev. A* **65**, 032710 (2002).
48. P. Florian, M. Hoster, and R. C. Forrey, *Phys. Rev. A* **70**, 032709 (2004).
49. A. Mack, T. K. Clark, R. C. Forrey, N. Balakrishnan, T.-G. Lee, and P. C. Stancil, *Phys. Rev. A* **74**, 052718 (2006).
50. T. Mukaiyama, J. R. Abo-Shaeer, K. Xu, J. K. Chin, and W. Ketterle, *Phys. Rev. Lett.* **92**, 180402 (2004).
51. N. Zahzam, T. Vogt, M. Mudrich, D. Comparat and P. Pillet, *Phys. Rev. Lett.* **96**, 023202 (2006).
52. F. Ferlaino, S. Knoop, M. Mark, M. Berninger, H. Schöbel, H.-C. Nägerl, and R. Grimm, *Phys. Rev. Lett.* **101**, 023201 (2008).
53. J. F. Castillo, D. E. Manolopoulos, K. Stark, and H.-J. Werner, *J. Chem. Phys.* **104**, 6531 (1996).
54. T. Takayanagi and Y. Kurosaki, *J. Chem. Phys.* **109**, 8929 (1998).
55. R. T. Skodje, D. Skouteris, D. E. Manolopoulos, S.-H. Lee, F. Dong, *Phys. Rev. Lett.* **85**, 1206 (2000).
56. N. Balakrishnan and A. Dalgarno, *J. Phys. Chem. A* **107**, 7101 (2003).
57. N. Balakrishnan, *J. Chem. Phys.* **121**, 5563 (2004).
58. E. Bodo, F. A. Gianturco, N. Balakrishnan, and A. Dalgarno, *J. Phys. B: At. Mol. Opt. Phys.* **37**, 3641 (2004).
59. P. F. Weck and N. Balakrishnan, *J. Chem. Phys.* **122**, 154309 (2005).
60. P. F. Weck and N. Balakrishnan, *J. Chem. Phys.* **122**, 234310 (2005).
61. G. Quéméner and N. Balakrishnan, *J. Chem. Phys.* **128**, 224304 (2008).
62. K. Stark and H.-J. Werner, *J. Chem. Phys.* **104**, 6515 (1996).
63. T. V. Tscherbul and R. V. Krems, *J. Chem. Phys.* **129**, 034112 (2008).
64. G. Quéméner, N. Balakrishnan, and B. K. Kendrick, *J. Chem. Phys.* **129**, 224309 (2008).
65. C. Xu, D. Xie, P. Honvault, S. Y. Lin, and H. Guo, *J. Chem. Phys.* **127**, 024304 (2007).
66. The reaction probabilities reported by Xu et al.[65] were artificially shifted by 0.00409 eV (P. Honvault, private communication).

Session VI

Undergraduate Collaborations

Deborah Watson, Jane Fox, Jean Turner, Francoise Masnou-Seeuws, Alex Dalgarno, Evelyne Roueff, Ewine van Dishoeck, Kate Kirby, Paola Cappellaro, Xi Chu

THE HFS OF H_2^+: ELECTRON-NUCLEAR COUPLING OF SPINS AND MOMENTA

JAMES F. BABB

ITAMP, Harvard-Smithsonian Center for Astrophysics,
60 Garden St., Cambridge, MA 02138
E-mail: jbabb@cfa.harvard.edu
http://www.cfa.harvard.edu/~babb

Progress in calculations on the hyperfine structure (hfs) of the hydrogen molecular ion is summarized taking a long view with emphases on Alex Dalgarno's contributions and also on some recent developments.

1. Motivation

The hyperfine transition frequency of the energy splitting of the $1S_{1/2}$ state of the hydrogen atom is 1 420 405.751 766 7(10) kHz, where the quantity in parentheses is the corresponding total error, ± 1 mHz, based on an extrapolation to a maser bulb of infinite size.[1] The corresponding precision of the sum of the theoretically calculated terms is about 1 kHz and future progress will depend on calculations of proton structure dependent terms.[2–4]

For the hydrogen molecular ion H_2^+ progress in the theory of the hyperfine structure (hfs) has only this year (2009) achieved[5] the level of precision better than 10 kHz. In this short paper, I discuss how Alex Dalgarno's attention has nurtured progress through a few well-timed studies over a time span of almost four decades.

In describing the hfs of the atom, the simple Fermi contact formula based on the interaction between the electron and the proton magnetic moments and taking account of the proton-electron reduced mass gives about 1 420.5 MHz, while inclusion of QED non-recoil terms[3] brings the theoretical value to within 45 kHz of the experimental value. Mass scaling in the Fermi contact term indicates that over 1 MHz of the H atom hfs is attributable to the reduced mass between the proton and the electron and a similar contribution can be expected for H_2^+.

Nevertheless, as of the late 70's, theoretical calculations for ortho-$H_2{}^+$ in the $N = 1$ rotational state still overestimated the Fermi contact term compared to experiment[6] by 400 kHz. This discrepancy indicated that the treatment of reduced mass effects was not complete.

2. An Introduction to the Early Oeuvre

When I started at ITAMP, Alex and I were able to build a collaboration through our shared interests in the study of the long-range interactions between an electron and an ion. I had worked with Larry Spruch at New York University on their applications to the Rydberg states of the He atom. We decided to look at the descriptions of the Rydberg states of the hydrogen molecule, which were being studied experimentally by Steve Lundeen and associates.[7] It was clear that ultimately so-called non-adiabatic effects (non-local coupling of electron and nuclear wave functions through momenta) would have to be addressed.[8] Within a week of my arrival, Alex had already generously shared his folder on $H_2{}^+$, containing papers he had accumulated in anticipation of some sort of investigation.

Our project benefited from the resources and visitors to the Institute. For example, Professor Yuri Demkov visited and shared of his findings of exact solutions for certain high nuclear charge one electron diatomic molecular ions.[9] Later, Alex would occasionally during our meetings pop into his back office and emerge with one of his reprints from the years between 1954-1960 that was relevant to the molecular ion. And so we progressed from simple variational approximations,[10] to adiabatic effects,[11] to asymptotic expressions,[12] to the hfs.[13] Actually, his student Paul Kalaghan had already combined these into one in the early 70's for his thesis working with Alex on the hfs[14,15] of $H_2{}^+$, but we decided it was worth a reinvestigation because of the discrepancy discussed in the previous section and to support on-going experiments. Eventually, the hfs results were presented in 1990, written up,[16] and refined[17] and published alongside an experimental paper.[18,19] We had improved the agreement with experimental values for the Fermi contact electron-nuclear spin coupling constants to around 50 kHz—the expected level for a complete treatment of leading order reduced mass effects (including non-adiabatic couplings).

3. Undergraduate Collaborators

During this period, through participation in an undergraduate research initiative sponsored by the Smithsonian Astrophysical Observatory, Dave

Bacon from Caltech worked with Alex and I on calculations of the spectra of H_2^+, and later Marianne (Vieira) Spurrier from Rochester worked with Alex, N. Balakrishnan, and I on calculating rovibrational deexcitation rate coefficients for H_2 colliding with He. In 1997, while still in his first undergraduate year, Jake Taylor from Harvard started research with Alex, Zong-Chao Yan, and I on calculations of the energies and electric dipole polarizability of H_2^+, which had been measured by Lundeen and associates.[20] Jake's work provided a highly accurate nonrelativistic polarizability value.[21]

4. Present Status

Vladimir Korobov visited ITAMP in 2000 for the program on wave functions and QED effects in few-electron atoms. Shortly thereafter, he completed a calculation of the polarizability of H_2^+ and found, surprisingly, that relativity accounted for only about $\frac{1}{5}$ of the difference between experiment and non-relativistic theory.[22] Yan, Zhang, and Li also calculated the nonrelativistic polarizability and found good agreement with Korobov.[23]

The nuclear spin-spin tensor interaction contributes to the hfs spectra of H_2^+ and the coupling constants were tabulated by Carrington and Brown.[24] Korobov, Hilico, and Karr[25] performed a complete calculation of the hfs by including nonadiabatic effects and by using a comprehensive Hamiltonian derived within the Breit-Pauli approximation. Their variational calculations of the hfs coupling constants are highly precise and definitive. In addition, in a parallel development Zong-Chao Yan and his collaborators[26] have also carried out calculations of the hfs of H_2^+. Their values are in harmony with Korobov *et al.*[25] In 2009, Korobov, Hilico, and Karr[5] pushed the theory of the hfs to the level of 10 kHz by including—in addition to nonadiabatic effects—QED and proton structure terms. In summary, the highest accuracy solutions for H_2^+ have resulted from the use of variational basis sets in Hylleraas coordinates through methods similar to those used with great success for few-electron atomic systems like He.

The most valuable high precision measurements of the hfs of the low-lying states of H_2^+ are those on the $v = 4$ to 8 levels carried out by Jefferts,[27] and those for the $v = 0, N = 1$ state by Fu *et al.*[19] However, additional experimental data are still desirable.

Recently, new data on H_2^+ and its isotopologues have begun to appear. Osterwalder *et al.*[28] obtained the hfs parameters of H_2^+ using a multichannel quantum defect theory analysis of measured Rydberg levels of H_2 and Cruse *et al.*[29] applied the methodology to D_2^+ using Rydberg states of D_2. Spectra of rovibrational transitions in the ground electronic state of trapped HD^+ ions (sympathetically cooled with Be^+ ions) have been obtained.[30] Rotational cooling of HD^+ by superelastic collisions with electrons has been reported by the Heidelberg TSR storage ring collaboration.[31] An experimental effort to measure two-photon transitions between the $v = 0$ and $v = 1$ levels of the ground electronic state of $H_2{}^+$ in an ion trap has begun.[32] Experimental and theoretical progress might lead to an improvement in the value of the ratio of the electron mass to the proton mass.[30,32]

There are still experimentally unexplored possibilities such as the antimatter version[33] of $H_2{}^+$, anti-$H_2{}^+$, and there have been no measurements of the hfs of the low-lying ro-vibrational states of the $D_2{}^+$ molecular ion sufficient to measure the nuclear quadrupole moment of the deuteron.

More complicated related molecular ions appear in astrophysical environments. In 1973 Alex and associates estimated the radiofrequency spectrum[34] of H_2D^+ and there are estimates of the hfs[35] though the hfs of H_2D^+ and of $HD_2{}^+$ has not yet turned out to be a useful diagnostic in radioastronomy.[36] Recent experimental work using a cryogenic multipole ion trap has provided Doppler width resolved center frequencies of the lowest transitions[37] of the molecular ions H_2D^+ and $HD_2{}^+$, though sub-Doppler measurements would be required to resolve the calculated hfs.

Through his attention to the problem around 1960, 1972, and 1992, Alex played a characteristic and unheralded role in advancing the understanding of the hydrogen molecular ion's hfs. Three particles never had it so good.

References

1. L. Essen, R. W. Donaldson, M. J. Bangham and E. G. Hope, *Nature* **229**, p. 110 (1971).
2. J. M. Jauch and F. Rohrlich, *The theory of photons and electrons: the relativistic quantum field theory of charged particles with spin one-half* (Springer-Verlag, New York, 1976).
3. J. Sapirstein and D. Yennie, Theory of hydrogenic bound states, in *Quantum Electrodynamics*, ed. T. Kinoshita (World Scientific, Singapore, 1990).
4. M. I. Eides, H. Grotch and V. A. Shelyuto, *Theory of light hydrogenic bound states* (Springer, New York, 2007).
5. V. I. Korobov, L. Hilico and J.-P. Karr, *Phys. Rev. A* **79**, p. 012501 (2009).
6. R. D. Ray and P. R. Certain, *Phys. Rev. Lett.* **38**, p. 824 (1977).
7. W. G. Sturrus, E. A. Hessels, P. Arcuni and S. R. Lundeen, *Phys. Rev. Lett.* **61**, p. 2320 (1988).
8. S. R. Lundeen, *Phys. Rev. Lett.* **62**, p. 3009 (1989); D. M. Bishop, *ibid.* **62**, p. 3008 (1989).
9. Y. N. Demkov, *Pis'ma Zh. Eksp. Teor. Fiz.* **7**, p. 101 (1968), [JETP Lett. **7**, p. 76 (1968)].

10. A. Dalgarno and G. Poots, *Proc. Phys. Soc., London, Sect. A* **67**, p. 343 (1954).
11. A. Dalgarno and R. McCarroll, *Proc. Roy Soc. London, Ser. A* **237**, p. 383 (1956).
12. A. Dalgarno and A. L. Stewart, *Proc. Roy. Soc. London, Ser. A* **238**, p. 276 (1956).
13. A. Dalgarno, T. N. L. Patterson and W. B. Somerville, *Proc. Roy. Soc. London, Ser. A* **259**, p. 100 (1960).
14. P. M. Kalaghan, PhD thesis, Harvard University, 1972.
15. P. M. Kalaghan and A. Dalgarno, *Phys. Lett.* **38A**, p. 485 (1972).
16. J. F. Babb and A. Dalgarno, *Phys. Rev. Lett.* **66**, p. 880 (1991).
17. J. F. Babb, *Phys. Rev. Lett.* **75**, p. 4377 (1995).
18. J. F. Babb and A. Dalgarno, *Phys. Rev. A* **46**, p. R5317 (1992).
19. Z. W. Fu, E. A. Hessels and S. R. Lundeen, *Phys. Rev. A* **46**, p. R5313 (1992).
20. P. L. Jacobson, D. S. Fisher, C. W. Fehrenbach, W. G. Sturrus and S. R. Lundeen, *Phys. Rev. A* **56**, p. R4361 (1997), ; *ibid.* **57** (1998), p. 4065E.
21. J. M. Taylor, A. Dalgarno, and J. F. Babb, *Phys. Rev. A* **60**, p. R2630 (1999).
22. V. I. Korobov, *Phys. Rev. A* **63**, p. 044501 (2001).
23. Z.-C. Yan, J.-Y. Zhang and Y. Li, *Phys. Rev. A* **67**, p. 062504 (2003).
24. J. M. Brown and A. Carrington, *Rotational spectroscopy of diatomic molecules* (Cambridge University Press, Cambridge, 2003), pp. 969-972, cited as A. Carrington and D. I. Gammie (1999), unpublished work.
25. V. I. Korobov, L. Hilico and J.-P. Karr, *Phys. Rev. A* **74**, p. 040502 (2006).
26. Z.-X. Zhong, Z.-C. Yan and T.-Y. Shi, *Phys. Rev. A* **79**, p. 064502 (2009).
27. K. B. Jefferts, *Phys. Rev. Lett.* **23**, p. 1476 (1969).
28. A. Osterwalder, A. Wüest, F. Merkt and C. Jungen, *J. Chem. Phys.* **121**, p. 11810 (2004).
29. H. A. Cruse, C. Jungen and F. Merkt, *Phys. Rev. A* **77**, p. 042502 (2008).
30. J. C. J. Koelemeij, B. Roth, A. Wicht, I. Ernsting and S. Schiller, *Phys. Rev. Lett.* **98**, p. 173002 (2007).
31. D. Shafir, S. Novotny, H. Buhr, S. Altevogt, A. Faure, M. Grieser, A. G. Harvey, O. Heber, J. Hoffmann, H. Kreckel, L. Lammich, I. Nevo, H. B. Pedersen, H. Rubinstein, I. F. Schneider, D. Schwalm, J. Tennyson, A. Wolf and D. Zajfman, *Phys. Rev. Lett.* **102**, p. 223202 (2009).
32. J. P. Karr, F. Bielsa, T. Valenzuela, A. Douillet, L. Hilico and V. I. Korobov, *Can. J. Phys.* **85**, p. 497 (2007).
33. H. Dehmelt, *Phys. Scr.* **T59**, p. 423 (1995).
34. A. Dalgarno, E. Herbst, S. Novick, and W. Klemperer, Ap. J. Lett. **183**, p. L131 (1973).
35. P. Jensen, I. Paidarová, V. Špirko and S. P. A. Sauer, *Molec. Phys.* **91**, p. 319 (1997).
36. T. Amano, *Philos. Trans. R. Soc. London, Ser. A* **364**, p. 2943 (2006).
37. O. Asvany, O. Ricken, H. S. P. Müller, M. C. Wiedner, T. F. Giesen and S. Schlemmer, *Phys. Rev. Lett.* **100**, p. 233004 (2008).

MOLECULAR CLOUDS AND CHEMISTRY BEYOND THE MILKY WAY

JEAN L. TURNER

Department of Physics and Astronomy, UCLA,
Los Angeles, CA 90095-1547, U.S.A.
E-mail: turner@astro.ucla.edu

The chemistry of individual molecular clouds in nearby galaxies is found to be correlated with large scale galactic structure, as revealed by principal component analysis. The high spatial resolutions of millimeter and submillimeter interferometers can resolve the complexity and confusion in other galaxies by isolating areas of common physical conditions.

Keywords: astrochemistry

1. An Undergraduate in the Dalgarno Group in the Mid-1970s

My first exposure to the field of astrochemistry was as an undergraduate astronomy concentrator in the Dalgarno group. It has been over thirty years since I wrote my junior tutorial paper and senior thesis under Alex's direction. Alex was the department chairman at this time; he was also teaching, directing theses, both graduate and undergraduate, serving as editor of the Astrophysical Journal Letters, and leading the atomic and molecular physics group. However busy he was, and he was extremely busy, he always had time for his students. My weekly meetings with Alex were sacred, not to be canceled for any reason. One time Tommy Gold, an eminent professor at Cornell University, had dropped in to say hi to Alex at the time of my weekly appointment. When I arrived, Gold was escorted out of the office, under polite but firm guidance from Valerie Sorenson. Although young and new to physics, even then I recognized that Tommy Gold was someone important and how remarkable it was that he was getting booted out of Alex's office for an astronomy concentrator in her junior year.

As a faculty member I try to follow his good example: "what would Alex do." From the other side of the teaching fence I have learned that it is not easy working with undergraduates. They don't yet know very much. They have a lot of outside distractions. Research skills are not taught in the classroom and are still to be learned. As a rule they don't get very much done, but that is okay, because it is part of the teaching process. It is remarkable to see how well Alex was able to guide his students, to motivate them, and to help them to achieve. Alex led me through my first two papers, on silicon chemistry and the quadrupole transition probabilities of H_2; his contribution to both of those papers was dominant, a true gift to a young scientist.

I went on to Berkeley for graduate school to work on the first millimeter interferometer with Jack Welch. I traded long hours feeding card decks into the CDC 6400 and reading core dumps at the CfA for long hours climbing radio antennas and soldering circuitboards. I took electrical engineering courses in addition to the numerical methods courses that Alex had recommended. The extragalactic millimeter-wave observing upon which I embarked was not amenable to the study of chemistry beyond the bright lines of CO for a long while. However, I retained my early interest in astrochemistry, and with the progress made in millimeter interferometry I was to return to it in time with my own like-minded student, David S. Meier.

2. Imaging Chemistry in Other Galaxies

Observing molecular gas in other galaxies gives us the opportunity to clearly see the gaseous structure and chemistry of galaxies, something that is difficult to do for our own Galaxy from our vantage point in the Galactic plane. A rich spectrum of molecular lines is available at millimeter and submillimeter wavelengths: a recent single dish survey in the $\lambda = 2$ mm band of the starburst Sc galaxy NGC 253 revealed lines from 25[1] of the over 35 molecular species detected to date in external galaxies. While the availability of a range of molecular lines and tracers gives great diagnostic potential for the study of gas, the challenge in extragalactic observing to date has been to interpret data at the arcminute spatial resolutions of single dish telescopes. One arcminute corresponds to a kiloparsec or more in even the closest galaxies, a sizeable fraction of the galaxy's diameter. To put this in context, the variations in chemistry that have been so successfully used to model protoplanetary disks in our own Galaxy are on scales of milliparsecs, or less. This is a reflection of the fact that most cloud chemistry takes place at visual extinctions (approximate visual optical depths) of 2-5. One might expect variations in chemistry to be limited to the edges of molecular clouds, and thus not resolvable in other galaxies. It is not surprising that single dish observations of molecular gas tracers, such as CO, HCN, H_2, and ammonia, often give conflicting results for the gas properties. Mean densities inferred over decaparsec scales range from $\sim 10^3$ cm^{-3} to $\sim 10^6$ cm^{-3} for the same regions, depending on which molecular tracer is used. Temperatures inferred from molecular lines can also vary, from $\sim$ 10–900 K.[2]

Millimeter interferometers improve spatial resolutions by an order of magnitude or more, to the arcsecond scale. This corresponds to the sizes of individual giant molecular clouds in nearby galaxies. Even at this resolution, we did not expect to see significant chemical variations from cloud to cloud; however, this is precisely what we observed in the first extragalactic imaging chemical survey of the galaxy IC 342.[4] This first study revealed that the chemistry of entire molecular clouds is closely linked to galactic structure. Below we describe a similar project for the nearby galaxy, Maffei 2.

3. Imaging Chemistry in the Spiral Galaxy, Maffei 2

Maffei 2, shown in Figure 1, is a barred spiral galaxy[3] at a distance of 3.3 Mpc, or about 10 million light years ($1''$=15 pc). An intense episode of nuclear star formation is indicated by a bright infrared and radio continuum source of luminosity $L_{IR} \sim 10^9$ $L_\odot$.[5] This starburst consists of a line of massive star clusters, or super star clusters[6] extending $\sim$ 150 pc to the north of the dynamical center. The barlike structure of the molecular gas in the nuclear region and the kinematics of this gas[7,8] suggest that the burst of star formation was prompted by the slow radial drift of gas inward along the bar. Why the starburst is concentrated to the north and is not

Fig. 1. Visible image of the central $80''$ ($\sim$1 kpc) of Maffei 2. The blue stars are foreground stars in our Galaxy. Image courtesy of Shoko Sakai and Robert Hurt.

so evident in the southern molecular peak, a feature seen in many "twin-peaked" CO galaxies, is unclear. What is it about the gas in the southern peak that has deterred star formation? What are the effects of feedback on the molecular gas near the starburst that might regulate further star formation?

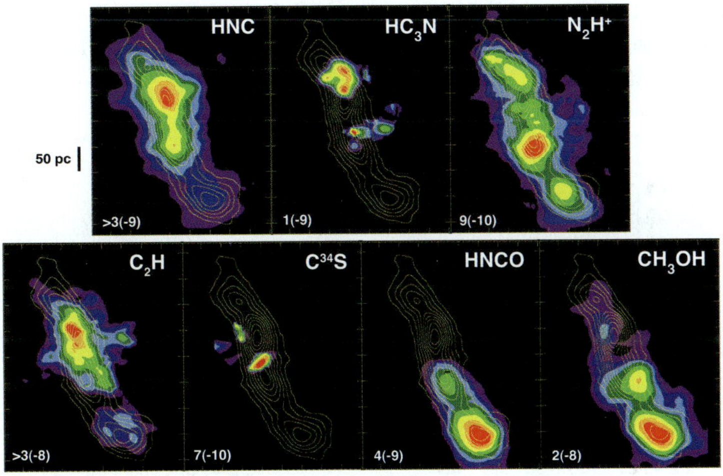

Fig. 2. Images of 3 mm transitions of molecules in the center of Maffei 2. Data from the Owens Valley Millimeter Array at a resolution of 4″. Contours of the 3 mm emission of $^{13}C^{16}O$ are shown in each panel, showing where the molecular clouds are found.

We observed Maffei 2 with the Berkeley-Illinois-Maryland Association Array (BIMA) and the Owens Valley Millimeter Array (Meier and Turner 2009, in prep.) Images of the 3 mm line emission of seven abundant molecules are shown in Figure 2. The images cover the central few hundred parsecs of the galaxy. ^{13}CO, the abundant and bright isotopologue of the optically thick $^{12}C^{16}O$, is shown in contours. ^{13}CO is a good tracer of the overall molecular gas distribution. The 5″ resolution maps of the other molecular species (in color) reveal a remarkable degree of chemical variation across the central kpc. These 3 mm lines arise from molecules with similar upper level energies and critical densities; thus the differences among the different lines appear to be due to abundance variations within the nuclear region[4] rather than excitation.

The combination of nearly a dozen different molecular lines with an image containing more than 200 independent resolution elements is a dataset sufficiently large that one can study the statistical correlations among the molecules as a function of location. Following the methods of Ungerechts *et al.* for the Orion Ridge,[9] a principal component analysis was done of the molecular line intensities. The principal component analysis extracts a set of correlations from the data by choosing independent axes that maximize variance. The axes are unbiased and statistical in nature, and do not necessarily correspond to a particular physical feature; they need to be compared to existing knowledge to be interpreted. Strong correlations are apparent in the first two principal components; a third component seems to indicate weak radial variations.

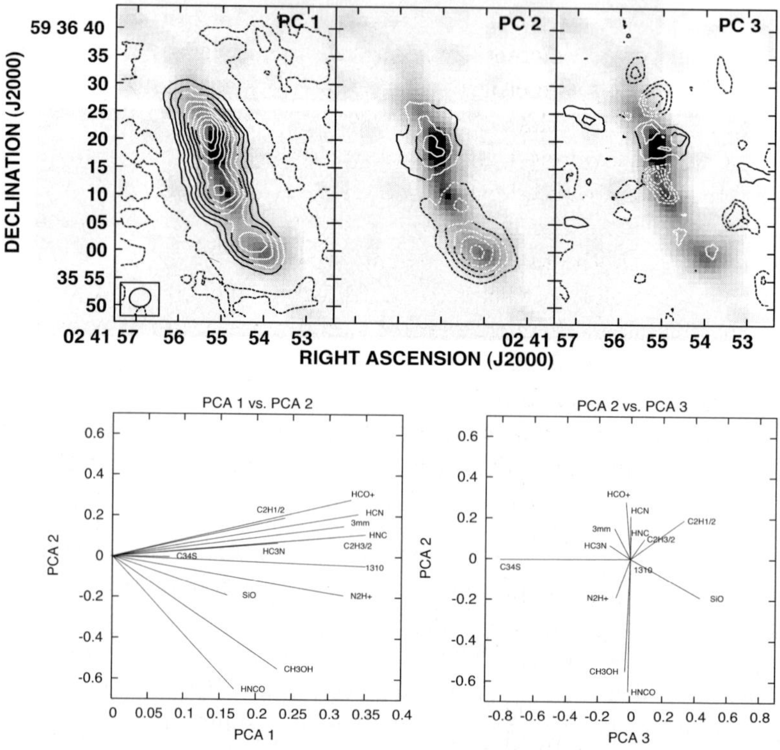

Fig. 3. Principal component maps of 3 mm lines in Maffei 2 and the projections of the molecules along the principal axes, Meier & Turner 2009, in prep.

The spatial map of principal component axis 1 is shown in Figure 3. With a resemblance to the ^{13}CO map of Figure 2, it represents the best "average" map for this set of molecules, which tend to trace a slightly denser gas component than CO. Principal component axis 1 appears to be the density-weighted mean column density. The molecules with the highest projections (i.e., best correlation with density-weighted mean column density) along PC axis 1 are ^{13}CO, N_2H^+, HNC, and HCN. In fact the PC 1 map appears to be an effective average of the ^{13}CO and HCN maps. These 3 mm lines appear to be the best general tracers of quiescent clouds and the overall denser molecular gas component characteristic of the nuclear region of a spiral galaxy.

Principal component axis 2 distinguishes two further major correlations, which are roughly anticorrelated here (modulo the requirement of PC axis 1 that molecules are found in molecular clouds). With positive projections along PC axis 2 are the emission of HNCO and CH_3OH (methanol). These molecules are found preferentially along the molecular bar arms. Observed Galactic abundances of methanol are very difficult to produce through gas-phase chemistry, so methanol is thought to be a tracer of grain chemistry. That methanol lies along the bar arm in Maffei 2 (also seen in IC 342[4]) suggests that it is produced by the processing of grain mantles in the shocks along the bar arms. The chemistry of HNCO has been less certain from Galactic studies; both gas or grain chemistry have been suggested for it, but the excellent correlation with methanol suggests that it forms in the same way as methanol, in chemical reactions on grain surfaces. These putative shock tracers are found most strongly along the southern bar arm, which also has less active star formation.

The second correlation evident in PC axis 2 of Figure 3 is that of the molecules found near the starburst, which include C_2H, HCN, HNC, HCO^+ and HC_3N, molecules often found in "photon-dominated regions," or "PDR"s. These molecules are likely to be tracers of highly irradiated molecular gas near the starburst. Axis 3 appears to represent radial variations among the molecules. $C^{34}S$ has the largest projection along this axis, and is confined to within the immediate 50–100 pc region near the dynamical center of the galaxy.

Another view of the results of the principal component analysis is the correlation matrix, a portion of which is shown in Table 1. This matrix represents the correlations between individual pairs of molecules and the 3 mm continuum. Of particular interest are the unusually low correlations and high correlations. The spatial anti-correlation of the molecular bar arm molecules and the "PDR" molecules of PC axis 2 described above is evident in unusually low values of the correlation coefficient. The correlation between HNC and HCN is very high (0.97), and perhaps this is not surprising since they are isomers. But another very high correlation, at 0.92, is between HCN and the 3 mm continuum. The 3 mm continuum in Maffei 2 is mostly free-free emission, with little contribution from dust.[8] Thus the dense gas tracers HNC and HCN are extremely well-correlated with free-free emission. This result suggests that the correlation between HCN and star-forming (infrared) luminosity found by Gao & Solomon[10] based on global HCN fluxes holds down to the scales of individual giant molecular clouds.

Table 1. Correlation matrix for 3 mm molecules in the nucleus of IC 342.

Species	^{13}CO	3 mm	C_2H	$C^{34}S$	CH_3OH	HNCO	HC_3N	HCN
^{13}CO	1.0	...	...	...	...	...	...	...
3 mm	0.88	1.0	...	...	...	...	...	...
C_2H	0.85	0.77	1.0	...	...	...	...	...
$C^{34}S$	0.20	0.24	0.17	1.0	...	...	...	...
CH_3OH	0.68	0.47	0.49	0.14	1.0	...	...	...
HNCO	0.52	0.27	0.32	0.19	0.86	1.0	...	...
HC_3N	0.54	0.53	0.41	0.15	0.32	0.16	1.0	...
HCN	0.93	0.92	0.84	0.18	0.43	0.21	0.62	1.0
HNC	0.95	0.87	0.84	0.19	0.51	0.33	0.62	0.97

While it is encouraging that the ability to resolve features such as bar arms, orbital intersections, and nuclear star clusters appears to simplify the chemical analysis, there is still more to learn about the molecular clouds in Maffei 2. The hot molecular gas chemistry remains as yet relatively unexplored.

4. The Future of Extragalactic Astrochemistry

High spatial resolution observations of millimeter-wave emission from molecules indicates that the chemical structure of nearby galaxies is variable and complex, but that it is possible to spatially isolate clouds with particular characteristics, such as clouds experiencing shocks within spiral arms, or clouds near high radiation fields. This work is at the limits of the feasibility of today's interferometers, requiring long integration times and multiple spectral tunings. The upcoming ALMA will revolutionize the study of chemistry in external galaxies due to its dramatic increase in resolution, sensitivity, and instantaneous bandwidth. Pushing out beyond the local neighborhood of galaxies, ALMA will be able to detect giant molecular clouds in the brighter lines (e.g., HNC and CH_3OH)[11] to distances of 75 Mpc in an hour. Fainter lines and isotopologues will be detectable in the

Virgo cluster. Arp 220-like systems will be detectable in species like HNC and HC_3N to redshifts of $z \sim 0.1$ in an 8 hour track. This will open up an array of galaxy types—dwarfs, ultraluminous infrared galaxies, merger systems, and potentially even ellipticals—to astrochemical scrutiny. The recent detections of HCN(3-2) and HCO^+(1-0) towards the Cloverleaf galaxy at $z = 2.56$[12,13] indicate that for the most luminous systems chemistry is within reach across the universe.

Acknowledgments

This research was funded by the National Science Foundation grant AST05-06469. I am grateful to David Meier and Mike Jura for comments on this contribution.

References

1. Martín, S., Mauersberger, R., Martín-Pintado, J., Henkel, C., García-Burillo, S., *Astrophys. J. Suppl.* **164**, 450 (2006).
2. Wiklind, T., Rydbeck, G., Hjalmarson, A., Bergman, P., *Astron. & Astrophys.* **232**, 11 (1990).
3. Hurt, R. L., Merrill, K. M., Gatley, I., & Turner, J. L., *Astron. J.* **105**, 121 (1993).
4. Meier, D. S., Turner, J. L., *Astrophys. J.*, **618**, 259 (2005).
5. Turner, J. L., & Ho, P. T. P., *Astrophys. J.*, **421**, 122 (1994).
6. Tsai, C.-W., Turner, J. L., Beck, S. C., Crosthwaite, L. P., Ho, P. T. P. & Meier, D. S., *Astron. J.* **132**, 2383 (2007).
7. Ishiguro, M. *et al.*, *Astrophys. J.* **344**, 763 (1989).
8. Meier, D. S., Turner, J. L., Hurt, R. L. 2008, *Astrophys. J.*, **675**, 281 (2008).
9. Ungerechts, H., Bergin, E., Goldsmith, P. F., Irvine, W. M., Schloeb, F. P., Snell, R. L., *Astrophys. J.*, **482**, 245 (1997).
10. Gao, Y., Solomon, P., *Astrophys. J.* **606**, 271 (2004).
11. Aalto, S., Polatidis, A. G., Hüttemeister, S., Curran, S. J.: *Astron. & Astrophys.* **381**, 783 (2002).
12. Solomon, P., Vanden Bout, P., Carilli, C., Guelin, M., *Nature* **426**, 636 (2003).
13. Riechers, D. A., Walter, F., Carilli, C. L., Weiss, A., Bertoldi, F., Menten, K. M., Knudsen, K. K., Cox, P.: *Astrophys. J.* **645**, 13 (2006).

ATOM-OPTICAL ANALOGS OF ELECTRONIC COMPONENTS AND DEVICES

R. A. PEPINO, J. COOPER, D. Z. ANDERSON and M. J. HOLLAND

JILA, National Institute of Standards and Technology and Department of Physics, The University of Colorado, Boulder, Colorado 80309-0440, USA

The research program known as *Atomtronics* aims to create one-to-one analogs of electronic components and devices with ultracold atoms trapped in periodic potentials. Such an analogy can be realized with reservoirs of neutral, ultracold atoms connected to optical lattices. Due to their highly tunable band structure, the lattices can be tailored to create resonances between certain states of the system. This allows atomic transport through specific dynamical pathways of the system, while minimizing transport through others.

In this presentation, a quantum master equation approach is outlined for treating the problem of two or more reservoirs of ultracold atoms, with arbitrary chemical potentials, connected to optical lattice configurations. The formalism is quite general and readily lends itself to the theoretical study of transport phenomena in open atomic systems.

This theoretical approach is applied to atomtronic systems, demonstrating how certain custom lattices can mimic the behavior of the electronic semiconductor diode, field effect-transistor, and bipolar junction transistor (BJT). Logic elements, such as AND and OR gates are constructed by taking two atomtronic BJTs and combining them in a manner similar to that of conventional electronics. This is promising since it is possible that we may join basic atomtronic components to construct elementary logic devices in a fully quantum mechanical system.

1. Foreword by Ron Pepino

I met Alex Dalgarno while finishing my undergraduate degree at the University of Connecticut. I was invited to work for Alex and Vasili Kharchenko for two months during the summer of 2003. As funding became available, my short tenure at ITAMP was extended over the following academic year. The main component of the project I received from Alex involved modeling the spectral emission generated by charge exchange between the highly-ionized solar wind and neutral heliospheric gas[a]. This project was perfectly

[a]This subject is discussed in Vasili Kharchenko's paper in this volume.

suited for an undergraduate pursuing a career in science: it was a tractable project, in which the successful conclusions were apparently relevant to the astrophysical community. Such a project pairing subtly speaks volumes about Alex as a mentor. Those new to the ways of research can get easily discouraged by early frustrations and failures; early research experiences can leave a lasting effect on a young student.

During my time spent with Vasili and Alex, I gained invaluable scientific experience, learning how to pose and attack physical problems, how to create computational models, and how to write scientific papers. Additionally, my positive and successful experience at ITAMP has given me the motivation and momentum to continue on in theoretical physics. Today I find myself studying open quantum systems, which is the subject of the remainder of this paper. I am incredibly grateful that I was given the opportunity to work with, and learn from these two masters. I honestly believe that if it were not for the two of them, I probably would not be in science today.

2. Introduction

'Atomtronics' focuses on establishing ultracold atom analogs of electronic circuits and devices. Foundational work has theoretically established the possibility of transistor-like device behavior from ultracold atomic systems [1,2]. Here we discuss a complete neutral atom circuits analog of the semiconductor diode as well as the first atomtronic transistor circuit to exhibit both switching behavior and linear amplification over arbitrary time scales. Afterwards, AND logic gate behavior is exhibited in an atomtronic circuit composed of two atomtronic transistors. The circuits presented here provide the rudiments of atom signal shaping and processing that may find use in atom-based sensors and possibly enable the basic elements of cold atom-based quantum logic. We also outline a quantum master equation approach for calculating the dynamics of this novel class of open quantum systems. This formalism introduces a chemical potential as a parameter into the traditional reservoir description. It also provides a quantum optics alternative for studying transport properties in optical lattices and other periodic potentials.

Our atom-optical analogy to electronic circuits begins with the definition of the 'atomtronic battery', which is composed of two reservoirs of ultracold atoms[b] having different chemical potentials (corresponding to dif-

[b]In this context, a reservoir is a large collection of ultracold atoms in thermal equilibrium, which provide a near-continuum of accessible energies to our systems.

ferent electric potentials at the terminals of a conventional battery). The 'wires' and atomtronic components are composed of optical lattices, and current refers to the number of atoms that pass a specific point in the circuit in a given amount of time. The analog of a simple circuit composed of a diode connected to a battery is illustrated in Fig. 1.

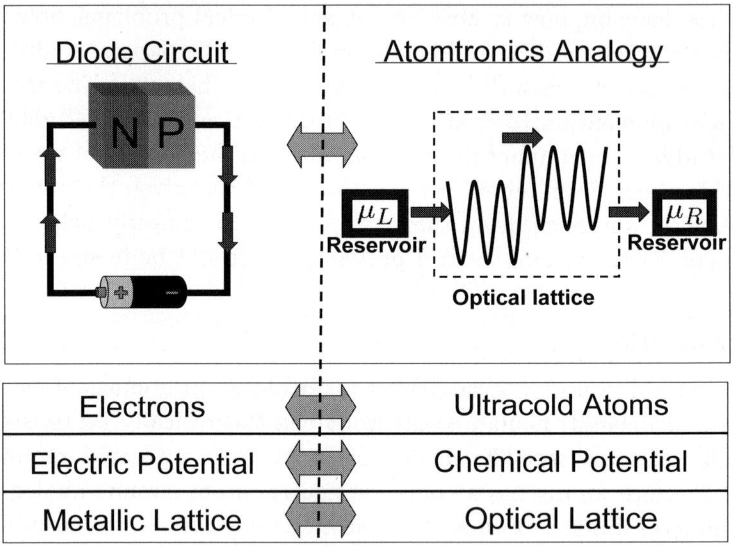

Fig. 1. Atomtronic analogy to a simple diode circuit. The atomtronic analogy of a diode formed from the joining of p-type and n-type semiconductor materials. Electrons are replaced by ultracold atoms, the battery is replaced by high and low chemical potential reservoirs, and the metallic crystal lattices (the microscopic medium that the electrons traverse) are replaced by an optical lattice. The atomtronic diode is achieved by energetically shifting one half of the optical lattice with respect to the other.

3. Atomtronic Diode

The atomtronic diode is a device that allows an atomic flux to flow across it in essentially only one direction. It is made by adding a potential step, which emulates a semiconductor junction (the boundary between p-type and n-type solid-state materials), to an energetically-flat optical lattice (Fig. 1). We find that, like the semiconductor diode, the principle behavior is a direct consequence of the junction itself. In fact, the desired current response is apparent for an atomtronic junction composed of only two lattice sites offset by a potential. For this reason, we focus here on a careful study of a two lattice site system. Despite the apparent simplification, we emphasize that our full calculations for extended systems confirm that the diode behavior is maintained irrespective of the number of lattice sites around the junction. We find that the atomtronic diode conduction asymmetry becomes optimal as the height of the potential step approaches the on-site interaction energy (the energy associated with putting two particles on the same lattice site). We refer to this as the 'resonance condition'.

Figures 2(a) and 2(b) present our calculation of the current response of a two-site optical lattice, subject to the resonance condition, as a function of the chemical potential difference. Here the external energy of the right lattice site is assumed to be equal to the external energy of the left plus the on-site interaction energy. In Fig. 2(a), the left chemical potential is set so that the left reservoir accepts all atoms on the left site, and the right chemical potential is smoothly increased to allow first zero, then one, and then two atoms on the right site. The result is a near-null current response for all values of the chemical potential difference. Conversely, in Fig. 2(b), the right chemical potential remains fixed so that the right reservoir accepts all atoms on the right site, and the left chemical potential is smoothly increased to allow first zero, then one and then two atoms on the left site. Here, there is a dramatic current increase, triggered when the left chemical potential matches the resonance condition. The knee in the curve can be likened to the exponential rise in current for a semiconductor diode in forward bias, yet unlike the electronic case, the semiconductor diode the atom current saturates.

This resonance condition leads to the desired diode-like response for a straight-forward reason, which can be understood by looking at the relative energies of the quantum states in the Fock basis. The condition sets up an exact energetic degeneracy between two of the three two-particle Fock states (either two atoms on the left site (labeled as $|20\rangle$) or an atom on each site, $|11\rangle$), and separates them energetically from the third, $|02\rangle$.

Figures 2(c) and 2(d) depict the dynamics in the Fock picture as we attempt to drive current across a two site lattice under the resonance condition, from right to left, and left to right (respectively). As seen in Fig. 2(c), if the right chemical potential (μ_R) is set to allow two atoms on the right site and the left (μ_L) is set to take all atoms from the left, the decoupling of the $|02\rangle$ state, the system evolves to a steady state in which it is most likely

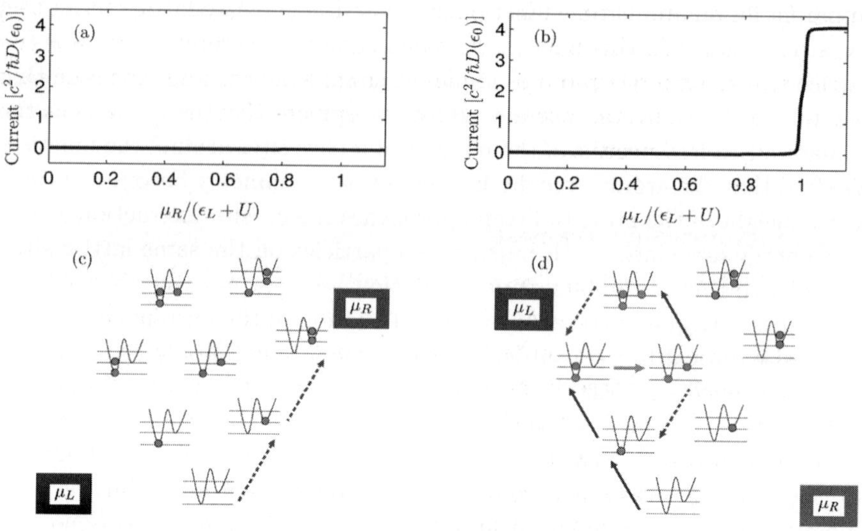

Fig. 2. Current response of the atomtronic diode. (a) Exact numerical simulation of the reverse bias characteristic of the atomtronic diode as a function of the chemical potential difference. Here the left reservoir's chemical potential is held at zero (so that the reservoir takes all atoms from the left site), and the right reservoirs chemical potential is continuously raised to the point that allows an occupancy of two atoms on the right site. The current response (here in units of the square of the coupling of the reservoir, times the average density of states of the reservoir divided by $\hbar$) is negligible. (b) Exact numerical simulation of the forward bias characteristic of the atomtronic diode as a function of the chemical potential difference. Here the right reservoir's chemical potential is held at zero, and the left reservoir's chemical potential is continuously raised to the point that allows an occupancy of two atoms on the right site. Here we see there is a significant jump in the current response when the chemical potential matches the resonance condition. (c) The Fock energy schematic of the reverse bias scenario. The chemical potentials for the left and right reservoirs (labeled as μ_L and μ_R) are set to maintain an occupancy of zero and two atoms on the left and right lattice sites (respectively). System transitions brought on by connection of the right reservoir are labeled by the dashed arrows. The combined action of the reservoirs evolves the system into the decoupled state, $|02\rangle$. (d) The Fock energy schematic of the forward bias scenario. The chemical potentials for the left and right reservoirs aim to maintain an occupancy of two and zero atoms on the left and right lattice sites (respectively). System transitions brought on by the right reservoir are labeled by dashed arrows, transitions brought on by the left reservoir are labeled by solid black arrows, and the intra-system transitions are labeled by the Gray horizontal arrow. The energetically-equal states $|20\rangle$ and $|11\rangle$ allow atoms to flow across the system.

to be found in $|02\rangle$ with very little current flow. This is the reverse-bias configuration. Interchanging the chemical potentials, so that they attempt to maintain two atoms on the left and zero atoms on the right promotes the system into the $|20\rangle$ state, we see that the resonant coupling between $|20\rangle$

and $|11\rangle$ allows one atom to hop over to the right site, putting the system in the $|11\rangle$ state. At this point, the combined action of the reservoirs causes the system to undergo two independent cycles, where the net effect of each contributes to an overall current across the system (Fig. 2(d)). This is the forward-bias configuration.

4. The Atomtronic Transistor

The desired function of an atomtronic transistor (Fig. 3) is to enable a weak atomtronic current to be amplified or to switch, either on or off, a much larger one. Transistor action requires at least three lattice sites connected to three independent reservoirs (Fig. 3(a)). The resonance condition for this device is found to be an extension of the diode case to account for the third well: the left external energy is shifted above the middle site by the on-site interaction energy and is of equal energy to that of the right site.

To investigate the device characteristics, we put a fixed bias across the transistor (arranging the left and right chemical potentials to maintain an occupancy of one atom on the left site and no atoms on the right) and monitor the system response to an increase of the middle chemical potential (Fig. 3(b)).

When the middle chemical potential is set to maintain zero atoms on the middle site, we find that the system primarily remains in the $|100\rangle$ state since it can only enter the right reservoir through the $|001\rangle$ state, which requires a second-order off-resonant process.

If the middle chemical potential is set to maintain an occupancy of one atom on the middle site, a resonance is accessed between the $|110\rangle$, $|020\rangle$ and $|011\rangle$ states. This triggers a cycle that amounts to a net current across the system. Simulations, however, reveal a competing process: atoms also leave the system through the middle reservoir since its chemical potential is set to maintain an average of one atom on the middle site, not two. Identical reservoir coupling strengths lead to an inefficient transistor.

As in the electronic transistor case, we can choose the base regime to be extremely 'thin' so that a very small middle current controls a much larger current from left to right. This can be done by coupling the middle reservoir to the system much more weakly than the other two reservoirs. Then, for example, for every ten atoms which transition from left to right, the weak coupling can allow only one atom to leave the system from the middle site. The end result is an appreciable differential gain that can be controlled.

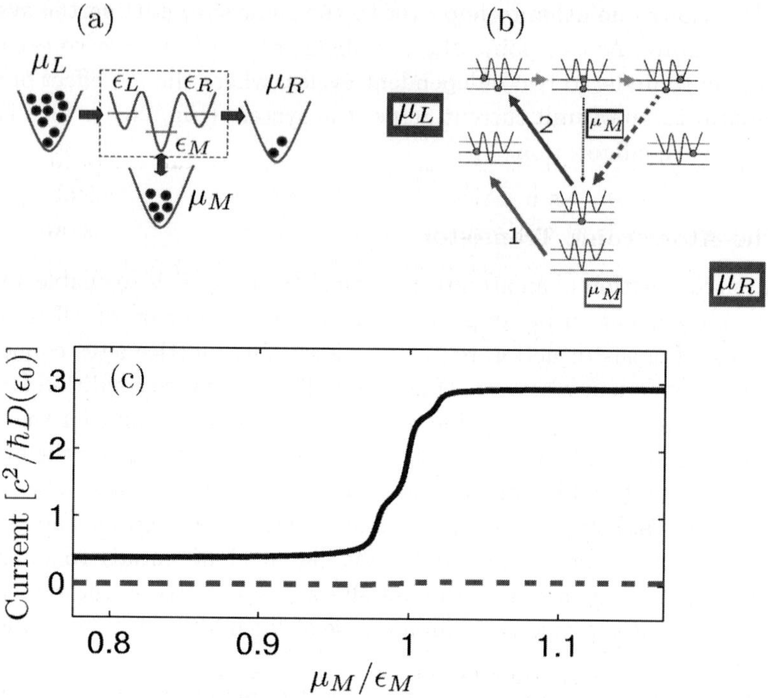

Fig. 3. Dynamics of the atomtronic transistor. (a) A cartoon of the atomtronic transistor as a three-well system, where each well is connected to its own independent reservoir. (b) An energy schematic of the relevant states of the system under the assumed resonance condition. In both illustrated cases, there is a fixed chemical potential difference across the system. In case 1, the middle chemical potential maintains an occupancy of zero particles on the middle site and most of the population remains on the left site. In case 2, the base potential is raised to put one particle on the middle site. This triggers two competing cycles that, given weak coupling of the middle reservoir, causes an avalanche of current to flow across the system. (c) An exact calculation of the current responses of the atomtronic transistor. The middle reservoir here has one-tenth the coupling strength of the left and right reservoirs. For a fixed chemical potential difference across the device, we vary the middle potential and record the response of currents leaving the system from both the right site (black, solid curve) as well as out of the middle site (Gray, dashed curve). The differential current gain for this specific system is both large and essentially linear.

Figure 3c shows the results of the numerical simulation for the case in which the middle reservoir coupling is one-tenth of the coupling of the reservoirs on either end. One can see the small increase in current via the middle leg (Gray, dashed curve) generating a substantial current across the device (black, solid curve). Furthermore the gain is approximately linear, an important practical requirement for a transistor to act as an operational amplifier.

5. Atomtronics Logic

A traditional logic element is a device with a given number of inputs and outputs that generates a series of logical responses. Such logical behavior can be expressed in a 'truth table' composed of 1s and 0s ('ons' and 'offs'). Digital logic elements are the fundamental building blocks of computing and modern electronics. In table (1a) the truth table for the AND logic gate is given as an example. The next level of complexity in emulating electronic systems is to create logic elements from the atomtronic components.

Table 1. Comparison of acoustic for frequencies for piston-cylinder problem.

AND Gate		
A	B	O
0	0	0
1	0	0
0	1	0
1	1	1

Atomtronic AND Gate Simulation		
A	B	O
9.2×10^{-3}	6.2×10^{-5}	5.0×10^{-3}
3.9×10^{-2}	6.3×10^{-4}	3.3×10^{-2}
6.5×10^{-3}	1.5×10^{-3}	1.4×10^{-1}
2.0×10^{-2}	2.0×10^{-2}	1

Note: AND Gate truth table where the two inputs are labeled by A and B, the output labeled by O, and the 1 or 0 entries represent 'on' and 'off'. (a) Truth table describing the device characteristics. (b) Numerical simulation of the Atomtronic AND Gate normalized to the maximal output current.

An AND gate is a device with two inputs (labeled here as A and B), and one output (labeled here as O) respectively. As illustrated in table 1(a), the device characteristic of the AND gate is that O remains off unless both A and B are on. In electronics, such a device can be constructed by connecting two transistors in series (as illustrated in Fig. 4(a)). By analogy, if the atomtronic BJTs are connected in the same series configuration (as illustrated in Fig. 4(b), the AND gate truth table can be generated. The data in table 1(b) has been generated in a numerical experiment of the configuration depicted in Fig. 4(b). For the transistors in this experiment the parameters, in units of the on site interaction energy U are $\epsilon_1/U = \epsilon_3/U = \epsilon_5/U = 1.3$, $\epsilon_2/U = \epsilon_4/U = 0.3$, and $J/U = 3 \times 10^{-3}$. When constructing practical logic circuits, the values of the 1s and 0s are somewhat arbitrary, as long as there is a measurable difference between the two. In table 1(b), we see that there is a discernible difference between 'on' and 'off'. Such a difference can be enhanced by increasing the on site interaction energy U.

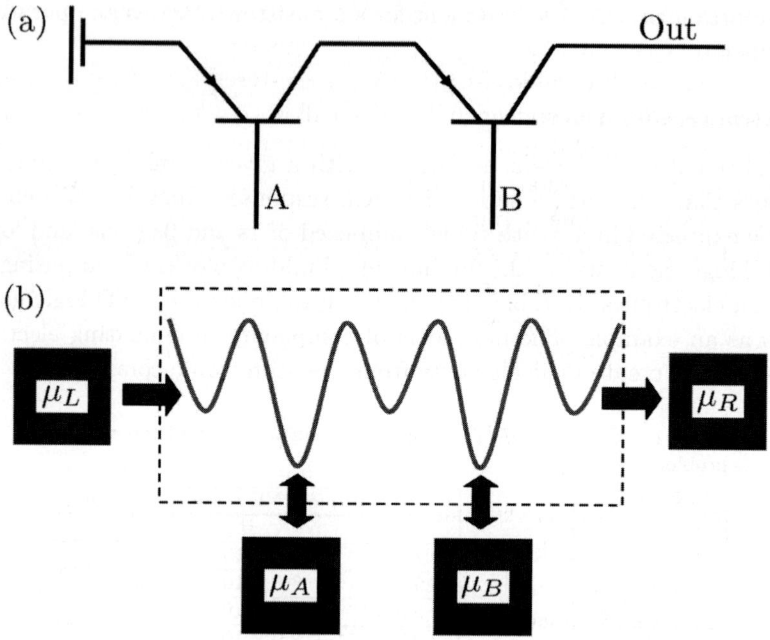

Fig. 4. (a) Electronic AND schematic. (b) Atomtronics AND schematic.

6. Methods

The atomtronic systems discussed here are open quantum systems, composed of optical lattices coupled to two or three reservoirs. The dynamics of ultracold bosons in optical lattices can be well-described by the boson Hubbard model [3–5]. For that reason, we assume the following many-body Hamiltonian for our atomtronic systems:

$$\hat{H} = \sum_{k=1}^{N} \left(\epsilon_k \hat{N}_k + \frac{1}{2} U_k \hat{N}_k \left(\hat{N}_k - 1 \right) \right) + \sum_{\langle l,m \rangle} J_{lm} \left(\hat{a}_l^\dagger \hat{a}_m + \hat{a}_l \hat{a}_m^\dagger \right) \tag{1}$$

where k is the lattice site index, N is the total number of lattice sites, ϵ_k is the external energy associated with site k, $\hat{N}_k \equiv \hat{a}_k^\dagger \hat{a}_k$ is the number operator, U_k is the two-particle on-site interaction energy on, $\hat{a}_l^\dagger$ $(\hat{a}_l)$ create (annihilate) particles on sites l, J_{lm} is the tunneling rate (or hopping term) and reflects the ease of a particle hopping between sites l and m, and the second sum is assumed to be over neighboring sites only.

We calculate observables from these system-reservoir interactions using the master equation formalism [6–9], which allows one to track the evolution of the system reduced density matrix $\hat{\sigma}_S$ for all times t such that $\tau_c \ll t \ll T_0$, where τ_c is the correlation time of the system-reservoir interaction, and T_0 is the evolution time of our system. The reservoirs are assumed to be a near-continuum of states accessible to the system, from $|0\rangle$ up to some final state $|E_\mu\rangle$, which corresponds directly to the reservoir's chemical potential. Such a fermionized boson reservoir model could be physically realized as a Mott insulator. The reservoir and interaction Hamiltonians are assumed to be of the form

$$\hat{H}_R = \sum_j \varepsilon_j \hat{R}_j^\dagger \hat{R}_j \quad \text{and} \quad H_V = \sum_j \hat{a}_q^\dagger \hat{R}_j + h.c. \tag{2}$$

respectively, where $\hat{a}_q^\dagger$ is the creation operator for an atom on the lattice site q of the system which is connected to the reservoir, and $\hat{R}_j$ is the annihilation operator for an atom in reservoir mode j. The evolution of the reduced density matrix for the system, can be written in the following approximate form [10]:

$$\frac{d\hat{\sigma}_S}{dt} \approx -i\hat{\mathscr{L}}_S\hat{\sigma}_S - \int_0^\infty e^{-\eta\tau}\mathrm{Tr}_R\left[\hat{\mathscr{L}}_V\hat{\mathscr{L}}_V(-\tau)\hat{\sigma}_R\right] d\tau\hat{\sigma}_S(t), \tag{3}$$

where $i\hat{\mathscr{L}}_S$ and $i\hat{\mathscr{L}}_V$ are the Liouville operators corresponding to the system and interaction Hamiltonians. Their action on an arbitrary operator $\hat{\Theta}$ is the following:

$$i\hat{\mathscr{L}}_X\hat{\Theta} = \frac{1}{i\hbar}[\hat{\Theta}, \hat{H}_X]. \tag{4}$$

where $X = S, V$.

The first term on the right side of Eq. 3 governs intra-system dynamics, and the second term, known as the memory kernel, dictates system evolution brought on by the reservoirs. We find that the second-order (or Born) component of the standard Born-Markov approximation is insufficient to model the dynamics of our atomtronic systems since it leads to a divergent memory kernel. Such divergences can be tamed by a more general approach to this problem [10]. Compared to the Born-Markov approximation, this approach yields a system description with improved energy conservation. Accounting for the proper behavior of the memory kernel involves explicitly including both the short-time and long-time decay of the two-time correlation function for the system-reservoir interaction [11]. In our formulation, we have confirmed that the general dynamics of the reduced density matrix remain insensitive to the exact nature of the decay in both limits, agreeing with the standard theoretical formulation of system-reservoir interactions.

7. Conclusion

We have demonstrated the characteristics of simple diode and transistor atomtronic circuits. The analogy was then expanded to a slightly more complex circuit, yielding AND logic gate behavior. Apart from the control and manipulation of atomic signals that these devices could potentially provide, it will also be intriguing to consider the connection of atomtronics with quantum information physics, since the atom devices operate in the coherent regime.

Atomtronic devices do not better their electronic counterparts in every regard. For example, the overall current flow is determined principally by the tunneling rate from one site to the next, making the atomtronic current much slower than electronic current. Compensating for this is the fact that this is a completely different physical system in which to pursue device physics: the atoms have complex internal structure and internal states, the atoms can be bosons or fermions, they are massive and affected by gravitational fields, the lattice can be dynamically varied, and many undesirable effects present in solid-state systems such as crystalline impurities, dislocations, and phonon scattering are absent.

A quantum optics approach for studying transport properties of optical lattices was also briefly discussed. This formalism is novel, not only due to the addition of the chemical potential or the reservoir as a parameter, but also the inclusion of an ansatz to the full master equation that takes the formalism beyond the standard Born-Markov approximation.

Looking forward, we aim to extend this analogy further by attempting to develop pulse-driven devices and possibly extend the theory to larger systems via density matrix renormalization group or time-evolving block decimation techniques.

Support for this project is acknowledged from the U.S. Department of Energy, the National Science Foundation and the Air Force Office of Scientific Research FA9550-07-1-0083.

References

1. A. Micheli, A. J. Daley, D. Jaksch and P. Zoller, *Phys. Rev. Lett.* **93**, p. 140408 (2004).

2. J. A. Stickney, D. Z. Anderson and A. A. Zozulya, *Phys. Rev. A* **75**, p. 013608 (2007).
3. M. Greiner, O. Mandel, T. Esslinger, T. T.W. Hänsch and I. Bloch, *Nature* **415**, 39 (2002).
4. D. Jaksch, C. Bruder, J. Cirac, C. Gardiner and P. Zoller, *Phys. Rev. Lett.* **81**, 3108 (1998).
5. J.H. Denschlag *et al.*, *J Phys B* **35**, 3095 (2002).
6. P. Meystre and M. Sargent, *Elements of Quantum Optics* (Springer, 1999).
7. D. F. Walls and G. J. Milburn, *Quantum optics* (Springer, 1994).
8. D. N. Zubarev, V. Morozov and G. Röpke, *Statistical mechanics of nonequilibrium processes. 2. Relaxation and hydrodynamic processes* (Akad.-Verl., 1997).
9. C. Cohen-Tannoudji, J. Duport-Roc and G. Arynberg, *Atom-Photon interactions* (Wiley, 1992).
10. R. A. Pepino, J. Cooper, D. Meiser, D. Z. Anderson and M. J. Holland, *Preprint* (2009).
11. S. G. Bhongale, R. Walser, and M. J. Holland, *Phys. Rev. A* **66**, p. 043618(Oct 2002).

QUANTUM METROLOGY AND SIMULATION

J. M. TAYLOR[1,2]

[1] *Massachusetts Institute of Technology, Cambridge, MA 02139 USA*
[2] *National Institute of Standards and Technology, Gaithersburg, MD 20877 USA*
E-mail: jacob.taylor@nist.gov

Experimental approaches to implementing ideas from quantum information science focus on developing well-understood physical systems that have only a few quantum bits. I will discuss potential applications of these quantum systems, including high-resolution magnetic field sensing and simulating quantum mechanical models from condensed matter physics. By relying upon proven technologies, we can take full advantage of the tremendous experimental advances in preparing, controlling, and measuring isolated mesoscopic and atomic systems made over the past decade.

Keywords: quantum computation; magnetometry; three-body loss

1. Introduction

Theoretical atomic and molecular physics is the detailed study of quantum mechanics, an idea introduced to this author by Alex Dalgarno. The richness of phenomena exhibited in nature, and in mathematical approaches to understanding those phenomena, have a natural culmination in the field of quantum information science. Some problems can be solved using the robustness of underlying ideas, such as the variational principle, which through Alex's guidance becomes applicable to a wide array of calculations. This provided a fulfilling entre to the world of quantum mechanics, while cleverly avoiding problems whose computational complexity exceed the capacity of classical computers to solve.

Perhaps not surprisingly, such "hard" quantum mechanics problems–defined by the difficulty of computing solutions using any reliable approximation techniques–correspond to computational difficult classical problems. Inverting that concept, one realizes that quantum mechanical systems can exhibit the potential for solving certain classical computation problems. However, implementing these ideas requires going beyond probing what nature provides, and requires quantum control techniques to manipulate the system. Early work showed the potential for secure cryptographic key distribution,[1] for exponentially fast simulation of Hamiltonian systems,[2] and for algorithms with exponentially better performance than their classical computation counterparts.[3] Motivated by these applications, experimental progress in isolating and controlling quantum systems has been rapid over the past decade.

A key element for any such program of research is identifying and analyzing single quantum bits (qubits) to use as a fundamental unit for computation. We focus on two promising systems here: a fluorescent center in diamond: the Nitrogen-Vacancy (NV) center[4–6] and ultra-cold atomic gases in optical lattices. The NV center is a single-qubit system, which has a high sensivity to magnetic fields. In contrast, our exploration of cold atomic gases does not have individual quantum bits *per se* but rather uses basic atomic physics, such as three-body loss, to simulate various condensed matter systems. Both of these avenues of investigation have been reported elsewhere.[8–10]

2. Magnetometry with Nitrogen-Vacancy Centers in Diamond

NV centers have remarkable spin coherence and exhibit optically-detectable electron spin resonance. Early theoretical work in quantum information focused on using such centers as a minimal, two-qubit device for building a quantum communication system.[5] While the long-term goal of such research is to isolate systems with sufficiently many quantum bits to allow for high fidelity entanglement generation–which in turn allows large-scale, distributed quantum computation[11]–we focus on an alternative application of these quantum bits: exploiting the extreme sensitivity of quantum devices to external perturbations to create a high-resolution magnetic field sensor.[8,9,12,13]

In many respects the NV system behaves as a small diatomic molecule with a restricted C_{3v} symmetry about the N-V axis. The ground state of the NV^- system, considered here, is a spin 1 triplet with a zero-field splitting of the $m_s = \pm 1$ states from the $m_s = 0$ state of 2.88 GHz (Fig. 1). Application of a microwave magnetic field near 2.88 GHz allows for high fidelity electron spin resonance and good quantum control over the ground state triplet. Additional quantum control techniques (described below) allow for the polarization, manipulation, and measurement of individual centers.

Single spins, such as the NV center, are primarily sensitive to the local magnetic field. For example, a standard Ramsey interferometry pulse sequence–a $\pi/2$ rotation of the spin to create a spin coherence, followed by Larmor precession for a time t, then another $\pi/2$ rotation and readout of the spin–yields a probability of ending in a "1" state which responds linearly to small changes in the Larmor precession frequency due, e.g., to an unknown magnetic field. This observation forms the basis for atomic magnetometers and for the approach for magnetometry we now describe. The details and experimental observation are published elsewhere[8,14]–here we briefly review the underlying concepts.

In addition, the NV center exhibits spin-dependent fluorescence under excitation above the zero-phonon line at 638 nm both at cryogenic and room temperatures. While the specifics of the mechanism are beyond the scope of this paper (see, e.g., Ref. 15), the fluorescence from a given NV center is reduced for the $m_s = \pm 1$ states. After several microseconds under moderate excitation power, the NV center is pumped into the $m_s = 0$ state and all spin-dependent fluorescence ends. Thus, the initial transient signal provides a means of measuring an individual NV center's spin state, while the steady-state value leads to optical pumping of the NV center into the $m_s = 0$ state. This allows for good readout and preparation of the electron spin degrees of freedom, and full quantum control of single NV centers has been demonstrated by several groups.[4,16,17]

2.1. *ESR Near Zero Magnetic Field*

In single crystalline diamond, there are four classes of NV centers, one with the N–V axis aligned with our chosen z axis (along 111), and three others according to tetragonal symmetry $(11\bar{1}, 1\bar{1}1, \bar{1}11)$. Let us assume we have an oscillating magnetic field (ESR field), controlled externally, which we apply in the $x - y$ plane: $\vec{B}_{ESR} \propto \cos(\omega t + \upsilon)[\cos\phi, \sin\phi, 0]$. The overall Hamiltonian for the 111 NV centers (those aligned with z) is

$$H_{(111)} = \hbar\Delta\hat{S}_z^2 - \hbar\Omega(t)\cos(\omega t + \upsilon)(\cos\phi\hat{S}_x + \sin\phi\hat{S}_y) \tag{1}$$

In the rotating frame and rotating wave approximation (with $\omega \sim \Delta = 2.87$ GHz and $\delta = \omega - \Delta$),

$$H_{(111)} = \hbar\begin{pmatrix} 0 & 0 & 0 \\ 0 & \delta & 0 \\ 0 & 0 & 0 \end{pmatrix} - \hbar/\sqrt{2}\Omega\begin{pmatrix} 0 & e^{i(\phi-\upsilon)} & 0 \\ e^{-i(\phi-\upsilon)} & 0 & e^{i(\phi+\upsilon)} \\ 0 & e^{-i(\phi+\upsilon)} & 0 \end{pmatrix} \tag{2}$$

This indicates that υ (the phase of the ESR field) does not influence the ±1 subspace, while the angle of the field ϕ does.

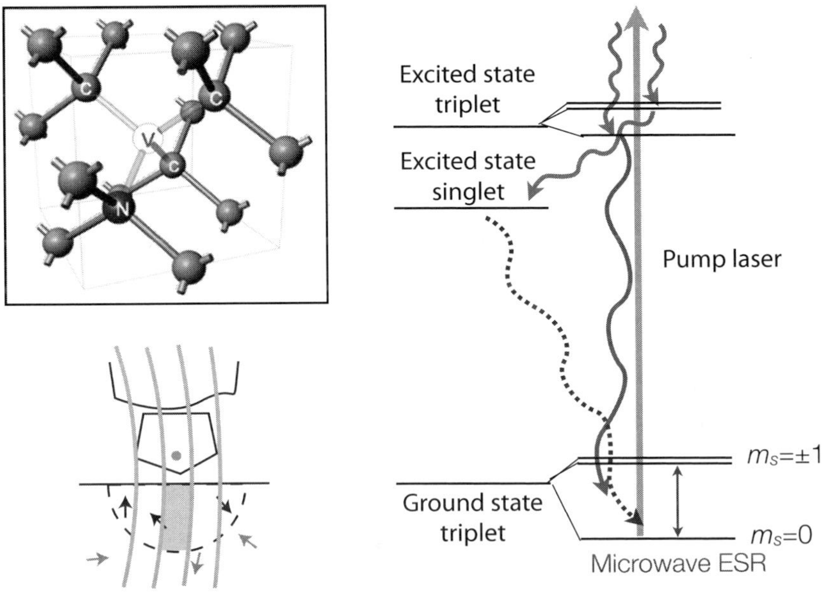

Fig. 1. (a) A Nitrogen-Vacancy impurity in the diamond lattice. The N-V axis, along 111, is used to define z. (b) An ESR microwave field couples between the $m_s = 0$ and $m_s = \pm 1$ states of the ground state triplet, with a zero-magnetic field splitting of 2.88 GHz due to the crystal field. Pumping with green (510-530 nm) laser light leads to spin dependent fluorescence and eventual optical pumping into $m_s = 0$. (c) Schematic of a scanning-tip NV-based magnetometer, with a single NV center in a diamond nanocrystal at the end of scanning tip. Local magnetic fluctuations due to Larmor precession of paramagnetic or nuclear spin impurities in the substrate provide a local field signal on the order of μT to nT, while an external field gradient isolates a small region of the substrate to be at the chosen detection frequency of the NV magnetometer.

For the zero-field case we will use the manifold $|1\rangle, |-1\rangle$ for all of our effective manipulations. We define a bright and dark state with respect to the microwave field:

$$|B(\phi)\rangle = \frac{1}{\sqrt{2}}\left[e^{-i\phi}|1\rangle + e^{i\phi}|-1\rangle\right] \tag{3}$$

$$|D(\phi)\rangle = \frac{1}{\sqrt{2}}\left[e^{-i\phi}|1\rangle - e^{i\phi}|-1\rangle\right] \tag{4}$$

Rewriting the Hamiltonian in the basis $|0\rangle, |B(\phi)\rangle, |D(\phi)\rangle$:

$$H_{(111)} = -\hbar \begin{pmatrix} -\delta & \Omega e^{-i\Phi} & 0 \\ \Omega e^{i\Phi} & 0 & 0 \\ 0 & 0 & 0 \end{pmatrix} \tag{5}$$

In this same basis, an external magnetic field *not* oscillating near Δ, with the interaction $V = -\hbar\gamma\vec{B}\cdot\vec{S}$, becomes

$$V = -\hbar\gamma B_z \begin{pmatrix} 0 & 0 & 0 \\ 0 & 0 & 1 \\ 0 & 1 & 0 \end{pmatrix} \tag{6}$$

We remark that application of a small, static external magnetic field perpendicular to the z axis leads to a large effective "detuning" of the other three classes of NV centers, removing them from the magnetometry response, while the effect on the 111-oriented centers is reduced by the 2.88 GHz zero-field splitting, and can be safely neglected.

2.2. *Pseudo-Spin 1/2 for the* $|m_s| = 1$ *Manifold*

As the sequence for magnetometry we now describe only uses the $|\pm1\rangle$ states of the NV center, it is convenient to define a set of pseudo-spin operators:

$$\sigma_x = |D(0)\rangle\langle B(0)| + |B(0)\rangle\langle D(0)| \tag{7}$$

$$\sigma_y = |B(\pi/2)\rangle\langle B(\pi/2)| - |D(\pi/2)\rangle\langle D(\pi/2)| \tag{8}$$

$$\sigma_z = |B(0)\rangle\langle B(0)| - |D(0)\rangle\langle D(0)| \tag{9}$$

For example, starting in the state $|0\rangle$, a π-pulse with $x-y$ angle 0 produces the state $|B(0)\rangle$ (and similarly maps $|B(0)\rangle$ to $|0\rangle$ for measurement). This allows for creation and measurement of σ_z eigenstates. A 2π-pulse with $x-y$ angle ϕ produces an overall minus sign on the $\{|0\rangle, |B(\phi)\rangle\}$ subspace. This is equivalent to the operation $-[\cos(\phi)\sigma_z + \sin(\phi)\sigma_y]$ on the pseudo-spin space. Finally, the interaction V is given by $V = -\hbar\gamma B_z\sigma_x$ in the pseudo-spin space.

These are the sufficient operations to perform a spin echo sequence. Specifically, we

(1) Start in $|0\rangle$ and apply an ESR π-pulse with an $x-y$ angle $\phi = 0$.
(2) Wait a time τ. An overall phase $\Phi = \int_0^\tau \gamma B_z(t)dt$ accumulates via V.
(3) Apply a 2π-pulse with $x-y$ angle $\phi = \pi/4$.
(4) Wait a time τ. A countering phase $\bar{\Phi} = \int_\tau^{2\tau} \gamma B_z(t)dt$ is accumulated.
(5) A final π-pulse with $x-y$ angle $\phi = 0$ maps $|B(0)\rangle \to |0\rangle$, allowing for an optical $|m_s|$ measurement to distinguish the final two states.
(6) The probability of ending in $|0\rangle$ is (see Eqn. 13)

$$P_{\phi=\pi/4} = \frac{1}{2}\left[1 - \sin\left(2\langle\bar{\Phi} - \Phi\rangle\right) e^{-(\tau/T_2)^3}\right]$$

Experimental investigations of the NV center have demonstrated that the main dephasing mechanism for the electron spin degree of freedom is coupling to Carbon-13 nuclei, with natural abundance of 1.1%, in the diamond lattice.[16] This leads to a inhomogeneous broadening of ~ 1 MHz. However, the long correlation time of this nuclear spin bath allows for spin-echo techniques to extend the coherence time several orders of magnitude, with decay of the echo on a characteristic timescale $T_2 \sim 250\ \mu$s.[14] In a sense, dynamic noise in the unknown field B_z due to local environmental effects will lead to a fundamental limit to the sensitivity of this system to magnetic fields.

2.3. *AC Magnetometry*

We now investigate the specifics of detecting an oscillating magnetic field using this NV center-based magnetometer. Consider a signal $B_z(t) = B_0 \cos(\nu t + \upsilon) + \tilde{a}(t)$, where $\tilde{a}(t)$ is a Gaussian noise term with $\langle \tilde{a}(t) \rangle = 0$ and $\langle \tilde{a}(t')\tilde{a}(t) \rangle = \int S(\omega) e^{i\omega(t-t')} d\omega$. The first moment of $\Delta\Phi$ is

$$\langle \Delta\Phi \rangle = \frac{4B_0}{\nu} \left[\sin^2(\nu\tau/2) \sin(-\nu\tau - \upsilon) \right] \tag{10}$$

This is maximal for $\upsilon = \pi/2$ and $\tau = \pi/\nu$, giving $\langle \Delta\Phi \rangle = \frac{B_0 \tau}{\pi}$.

The next moment includes the errors due to the Gaussian noise term, a. We write

$$\begin{aligned} \langle \cos^2(\Delta\Phi + \phi) \rangle &= \frac{1}{4} \left[2 + \langle e^{2i(\Delta\Phi+\phi)} \rangle + \langle e^{-2i(\Delta\Phi+\phi)} \rangle \right] \\ &= \frac{1}{2} \left[1 + \cos(2\langle \Delta\Phi \rangle + 2\phi) \exp\left(\langle -(\Delta\Phi - \langle \Delta\Phi \rangle)^2 \rangle \right) \right] \end{aligned} \tag{11}$$

The usual decay of an echo signal is due to the last term; for dipolar-interactions of the nuclear spin bath,

$$\exp\left(\langle -(\Delta\Phi - \langle \Delta\Phi \rangle)^2 \rangle \right) \sim \exp[-(\tau/T_2)^3] \tag{12}$$

The signal at our optimal echo pulse angle $\phi = \pi/4$ is (including noise):

$$P_{\pi/4} = 1/2 \left[1 - \sin\left(2\langle \Delta\Phi \rangle \right) e^{-(\tau/T_2)^3} \right] \tag{13}$$

At $\tau = \pi/\nu$, the optimal measurement point described above,

$$S = 1/2 - P_{\pi/4} = \frac{1}{2} \sin\left(\frac{2B_0\tau}{\pi} \right) e^{-(\tau/T_2)^3} \tag{14}$$

For example, for $\nu T_2 = \pi(T_2/\tau) \ll 1$, the signal goes to zero exponentially fast. This sets a lower bound on the measurable frequencies, i.e., we require $\nu > 1/T_2$.

2.4. *Measurement Bandwidth, Bandcenter, and Sensitivity*

We now consider other possible signals near a bandcenter frequency ν_0 and with fixed $\tau = \pi/\nu_0$. We write $B_z(t) = b(t) + \tilde{a}(t)$, where the noise term $(\tilde{a}(t))$ is the same as above. However, now $b(t)$ varies over a range of frequencies:

$$b(t) = \frac{1}{2\pi}\int_{-\infty}^{\infty} e^{i\omega t} b(\omega)\, d\omega \tag{15}$$

Working with the Fourier-transformed form of b, the echo phase is

$$\begin{aligned}\langle\Delta\Phi\rangle &= \frac{1}{2\pi i}\int_{-\infty}^{\infty} b(\omega)\frac{(1-e^{i\omega\tau})^2}{\omega} d\omega \\ &= \tau\frac{i}{\pi}\int_{-\infty}^{\infty} b(\omega)e^{i\omega\tau}\frac{\sin^2(\omega\tau/2)}{(\omega\tau/2)} d\omega\end{aligned} \tag{16}$$

The features of the windowing function (the usual sinc function, above), shown in Fig. 2a, determine the effective bandwidth possible for the system. It has extrema at $\omega = \tan(\omega\tau/2)/\tau = \pm 2.33\ldots/\tau$ and minimal FWHM $= \pi/\tau$.

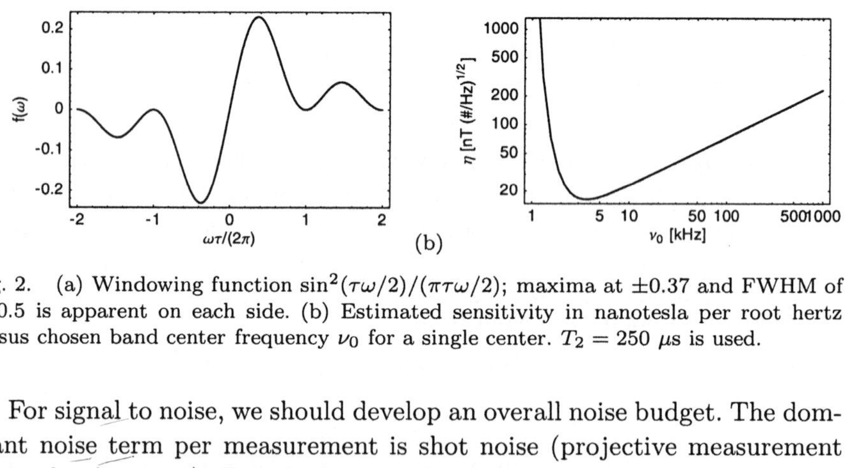

Fig. 2. (a) Windowing function $\sin^2(\tau\omega/2)/(\pi\tau\omega/2)$; maxima at ± 0.37 and FWHM of ~ 0.5 is apparent on each side. (b) Estimated sensitivity in nanotesla per root hertz versus chosen band center frequency ν_0 for a single center. $T_2 = 250\ \mu$s is used.

For signal to noise, we should develop an overall noise budget. The dominant noise term per measurement is shot noise (projective measurement of two-level system). For N measurements taking a time $T_{\text{tot}} \approx 2\tau N$, the shot noise $\approx 1/\sqrt{N}$, while the signal near the optimal frequency $\nu_0 = \pi/\tau$ is $\frac{B_0}{\pi\tau}e^{-(\tau/T_2)^3}$ for $B_0\tau < 1$. The final signal to noise for a single center is

$$S/N = \frac{B_0\tau\sqrt{T_{\text{tot}}/(2\tau)}}{\pi}e^{-(\tau/T_2)^3} = B_0\sqrt{\frac{\tau T_{\text{tot}}}{2\pi^2}}e^{-(\tau/T_2)^3} \tag{17}$$

In a similar way, the sensitivity in frequency per root hertz is

$$\eta = \pi\sqrt{2/\tau}e^{(\tau)^3/(T_2)^3} \tag{18}$$

The approximate sensitivity for a wide range of band centers ν_0 and an estimated electron spin echo dephasing time $T_2 = 250$ μs is plotted in Fig. 2b. We use $g\mu_B/\hbar$ for conversion from frequency to field values (i.e., bare electron g-factor). As a useful side remark, the high sensitivity for bandcenters $\nu_0 \approx 1/T_2$ can be replicated at higher frequencies, in contrast to the apparent trend in Fig. 2b. In particular, a series of n echo pulses with measurement only at the end, such that the total interrogation time $2n\tau \approx T_2$, yields similar sensitivity as the ideal scenario but for oscillating fields near a bandcenter $\nu_0 \sim n/T_2$.

3. Using Three-Body Loss for Elastic Interactions

We now change topics to consider a system with many particles: ultracold, trapped atoms in a deep optical lattice. Cold atomic gases provide a remarkable test-bed for understanding and realizing ideas and models for condensed matter and quantum information systems.[18] Of particular interest are simulating systems with 3- or 4-body interactions, which might allow for the observation of quantum states with non-abelian statistics sufficient for quantum computation.[19] Such interactions may be generated from an elastic two-body interaction, but are usually weak[20] as they rely on perturbation theory.[21,22] In contrast, the natural 3-body interaction in cold atomic systems—3-body recombination—can be large even as the scattering length goes to zero, such as in recent experiments with Cesium.[23] For fast dissipation, losses can be suppressed by the effect of impedance mismatch, whereby the quantum mechanical tunneling process necessary to form the precursor state is suppressed by the rapid loss of the precursor. This is supported by recent observations of long-lived quantum states in the presence of fast 2-body losses in an optical lattice.[24]

We consider a scenario in which atoms, confined to the lowest Bloch band of an optical lattice, develop a strong on-site 3-body loss term γ_3. The tight confinement of the optical lattice leads to a high density of atoms and fast recombination when three or more atoms are on a lattice site, due to the small size of the Wannier wavefunction. Curiously, we find that strong on-site loss leads to a dramatic suppression of the quantum tunneling of atoms when such a process would lead to decay. Thus, by incorporating the intrinsic 3-body process directly into the model of the system, we find regimes where the lifetime of states within the lattice is enhanced as the recombination rate increases. The metastable region allows the simulation of so-called 3-hard core bosons in a Bose-Hubbard hamiltonian.[25] With the addition of a magnetic term, via rotating lattices or time-dependent lattice potentials,[26,27] the 3-hard core repulsion simulates an effective (elastic) short-range 3-body interaction in the dilute limit which has as its ground state at unit magnetic filling ($\nu = 1$) the Moore-Read state,[20,28,29] and is known to have non-abelian anyons with sufficient complexity to allow for quantum computation.[19]

To investigate this regime, we examine the zero temperature phase diagram in the limit of $\gamma_3 \to \infty$. One new phase of interest occurs when the on-site interaction is moderate but attractive. The restriction to 2 atoms per site makes the system quasi-stable, and we predict a superfluid paired-boson phase should be observable.

3.1. *Three Body Loss in a Tightly-Confined Region*

We start by considering three-body loss processes in a cold atomic gas using an effective field theory with an imaginary term. We expect from dimensional analysis an interaction of the form $-i\hbar^2(\rho_3)^4/(2m) \times \hat{\psi}^{\dagger 3}\hat{\psi}^3$, where ρ_3 is the recombination length.[10,30] In the limit of large elastic scattering length a, a universal regime of recombination is predicted from Efimov resonances,[31–33] in which case $\rho_3 \sim \kappa^{1/4}a$, where κ varies from 0 to 22. However, at small scattering lengths, a background 3-body loss process is observed,[23] consistent with the effects of deeply bound states, which are largely independent of the first bound state that dominates the low-energy elastic scattering. This recombination length has been observed to be 500-1000 a_0 near a zero of the scattering length a in Cesium. This suggests that a regime with large three-body recombination and weak two-body interaction is accessible.

Considering the effective interaction in a deep optical lattice, we start by working in a basis defined by the Wannier function $w_0(r)$ of the lowest Bloch band,[18] with a characteristic size d. This yields a master equation whose unitary evolution is governed by the Bose-Hubbard Hamiltonian

$$H = -J \sum_{<ij>} b_i^\dagger b_j + U/2 \sum_i n_i(n_i - 1) \tag{19}$$

and the dissipative component has a Linblad-type structure:

$$\dot{\rho} = i[H, \rho] - \frac{\gamma_3}{12} \sum_{i=1,2} \left[2b_i^3 \rho (b_i^\dagger)^3 - \{\rho, (b_i^\dagger)^3 (b_i)^3\}\right] \tag{20}$$

We can estimate the on-site two-body interaction from the elastic scattering, U, and for the on-site 3-body recombination rate, γ_3:

$$U = \frac{4\pi\hbar^2 a}{m}\int d^3r|w_0(r)|^4 \approx 4\pi\frac{\hbar^2}{ma^2}\left(\frac{a}{\sqrt{2\pi}d}\right)^3 \tag{21}$$

$$\hbar\gamma_3 = \frac{\hbar^2\rho_3^4}{m}\int d^3r|w_0(r)|^6 \approx \frac{\hbar^2}{m\rho_3^2}\left(\frac{\rho_3}{\sqrt{3^{1/2}\pi}d}\right)^6 \tag{22}$$

With this simplified model (developed in more detail below), it should be possible to explore the Bose-Hubbard hamiltonian with a tunneling rate J for the lowest Bloch band in a regime where $\gamma_3 \gg |U|, J$, and where U can be made positive, negative, or zero.

3.2. *Suppression of Three-Body Loss in a Lattice*

We now explore a toy model with just two lattice sites, to illustrate the main effects of three-body loss for atoms in a lattice. This two-site system is shown diagrammatically in Fig. 3. For an occupation with a total of 2 or fewer bosons in the system, no decay will occur. The first non-trivial situation is 3 bosons in the system at $t = 0$. There are four states for this system: $|3,0\rangle, |2,1\rangle, |1,2\rangle, |0,3\rangle$, where $|n,m\rangle$ has n bosons in site 1 and m bosons in site 2. Breaking this into a symmetric and anti-symmetric subspace with $|3,0\rangle \pm |0,3\rangle$ and $|1,2\rangle \pm |2,1\rangle$, and labeling these states "3" and "2", the density matrix equations of motion are

$$\dot{\rho_{33}} = -\gamma_3\rho_{33} + \sqrt{3}Ji(\rho_{23} - \rho_{23}^*) \tag{23}$$

$$\dot{\rho_{22}} = -\sqrt{3}Ji(\rho_{23} - \rho_{23}^*) \tag{24}$$

$$\dot{\rho_{23}} = -[\gamma_3/2 + 2i(U \pm J)]\rho_{23} - \sqrt{3}Ji(\rho_{22} - \rho_{33}) \tag{25}$$

When $\gamma_3 \gg J$ or $|U| \gg J$ we can adiabatically eliminate "3" by setting $\dot{\rho_{23}} = 0$ and $\rho_{33} = 0$, and find

$$\rho_{23} = \frac{-i\sqrt{3}J\rho_{22}}{\gamma_3/2 + 2i(U \pm J)} \tag{26}$$

$$\dot{\rho_{22}} = -\frac{3J^2\gamma_3}{4(U \pm J)^2 + \gamma_3^2/4}\rho_{22} \tag{27}$$

Examination of this result indicates that the states $|2,1\rangle \pm |1,2\rangle$ decay with an effective rate

$$\Gamma \approx \frac{12J^2}{\gamma_3} \tag{28}$$

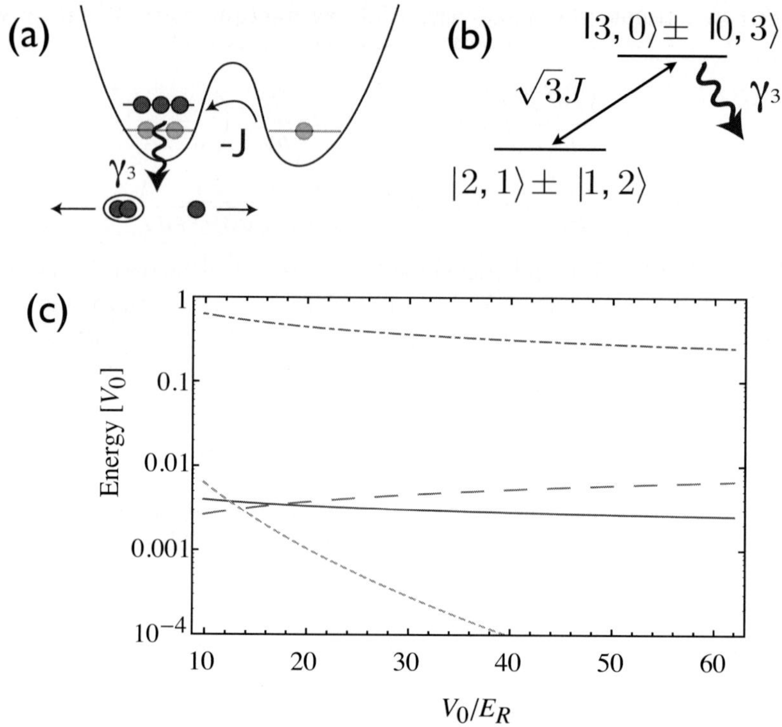

Fig. 3. (a,b) "Toy" model, where the on-site three body decay into a deeply bound molecular state and an untrapped free particle occurs with rate $\gamma_3 \gg J$, which in turn suppresses the tunneling from right to left. (c) The parameters ω_B (dash-dotted), U (solid), γ_3 (dashed), and J (dotted) of the Bose-Hubbard hamiltonian in the deep lattice limit for ^{133}Cs near $B = 17$ G and 760 nm lattice light. In this region the recombination length $\rho_3 = 800 - 1000a_0$ while the scattering length switches sign continuously from -200 to 200 a_0. $a = 5a_0$ is shown.

which *decreases* as the 3-body recombination rate increases. This is a well known phenomenon: the rapid dephasing of the 3-particles-on-a-site state suppresses the coherent tunneling to that state, which in turn prevents the decay process from occurring, analogous to reflection of light from a metal (due to impedance mismatch) or a weak excitation to a broad resonance in quantum optics. As a side remark, we note that the real component of ρ_{23} indicates an energy shift of the state "2" induced by virtual transitions to "3", of order $U\frac{J^2}{\gamma_3^2}$, which we neglect.

To extend this result to a lattice with many sites, we move to an equivalent stochastic Schrödinger equation description of the loss process.[34] Then we can develop a low energy Hamiltonian for the system, and find new, reduced loss terms associated with the low energy space. We start with the full Bose-Hubbard Hamiltonian for the lowest energy band, and add an imaginary energy term corresponding to the loss operator $b_i^3/\sqrt{6}$:

$$H = -J \sum_{<ij>} b_i^\dagger b_j + U/2 \sum_i (b_i^\dagger)^2 b_i^2 - i\gamma_3/2 \sum_i \frac{(b_i^\dagger)^3 (b_i)^3}{6} \tag{29}$$

We can now make a low energy expansion assuming a large, imaginary, gap, $i\gamma_3$. Formally, we introduce the projector P which projects into states with at most 2 atoms per site, and define $Q = 1 - P$. We then write the energy-dependent Hamiltonian for the subspace defined by P as

$$H_P = PHP + PHQ \frac{1}{E - QHQ} QHP \tag{30}$$

For our system,

$$QHP = \frac{-J}{2} \sum_i \sum_{j \in nn_i} (b_i^\dagger)^3 (b_i)^2 p_{01,j} b_j \tag{31}$$

where nn_i is set of the nearest neighbors to site i (in toy case, for $i = 1$, $j = 2$) and $p_{01,j} = 1 - \frac{n_j(n_j-1)}{2}$ is the projector onto a site with 0 or 1 atom. This corresponds to destroying three atoms, two in one site and one in an adjacent site, to create three atoms in the single site.

At this point, we assume E is small and approximate $E - QHQ$ with $-i\gamma/3/2$. This is appropriate so long as $\gamma_3 \gg J, |U|$, and reflects the assumption that the energy difference between the state before tunneling (with 2 atoms per site at most) and after tunneling (with one site with three atoms) is dominated by the value $i\gamma_3/2$.[a]

We can now define "loss" operators associated with this virtual decay channel as

$$c_j = \frac{b_j^2}{2} \sum_{k \in nn_j} p_{01,k} b_k \tag{32}$$

and a rate $\Gamma = 12J^2/\gamma_3$. The corresponding stochastic Schrödinger equation is

$$H_P = PHP - \frac{i\Gamma}{2} \sum_j c_j^\dagger c_j \tag{33}$$

We remark that this reproduces, for the two-site toy model, the estimated decay found above. In addition, this effective low-energy Hamiltonian still

[a]In practice, we will be able to relax the assumption on U provided $|U| \gg J$.

has an appropriate structure for solution with the stochastic Schrödinger equation, using the new loss operators c_j.

3.3. *Phase Diagram*

One intriguing aspect of the low energy Hamiltonian H_P is the effect of projection: a strong 3-body loss process prevents 3 bosons from occupying the same site. This provides a robust realization of 3-hard core bosons, as first investigated by Ref. 25. To understand the effects of 3-hard core repulsion, we now develop a mean field phase diagram, appropriate for describing the system in high dimensional lattices.

We use a Gutzweiler ansatz for the wavefunction: $\hat{\psi} = \prod_i (c_0 + c_1 b_i^\dagger + c_2 b_i^{\dagger 2}/\sqrt{2})$. The mean field phase diagram is plotted as a function of interaction U and the average number of particles per site n for a fixed tunneling rate J and number of nearest neighbors z. This diagram is of immediate use when looking at the time dynamics of the system, where the background BEC which provides the chemical bath is removed, and changes in the number of atoms only occur due to 3-body loss events. The phase diagrams are calculated in the infinite γ_3 limit, where the loss from the low energy space $\Gamma \to 0$.

In addition to the expected phases—a Mott insulator at large, positive interaction U and superfluid phase for smaller U—the regime of attractive interaction U supports a superfluid phase and a confinement-induced dimer superfluid phase (DSF). This contrasts with the unprojected Hamiltonian, where the ground state for attractive interactions is all of the atoms on a single lattice site. Thus, the effect of 3-body recombination is to stabilize the system against this attractive point, allowing for the potential observation of superfluidity for attractive interactions and even a superfluid phase of paired bosons.

4. Conclusion

We have discussed how a molecular-like system, trapped in diamond, can be used as a sensitive nano-scale magnetometer using ideas and techniques from atomic physics and quantum information. The resulting device has unprecedented sensitivity within a $\sim 10^{-18}$ cm^3 volume, opening up new regimes of nanometer-scale magnetic field detection. We have also investigated the novel implications of the well-studied three-body loss problem in cold atomic gases for atoms trapped in a deep optical lattice. The resulting dynamics can be largely elastic, effectively simulating large, elastic

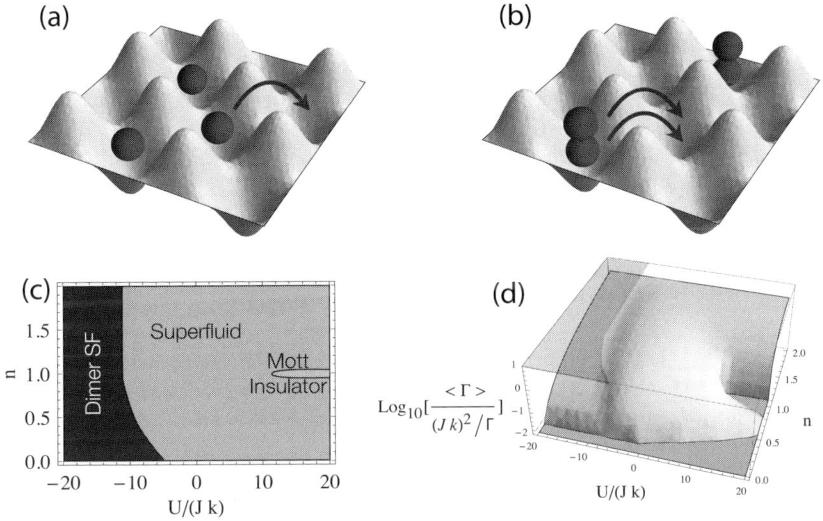

Fig. 4. (a) Atoms hopping independently on an optical lattice; (b) Attractively paired atoms hopping jointly on the lattice, where three-body loss prevents pair collisions; (c) Mean-field phase diagram as a function of two-body interaction U and density n for a lattice with k nearest neighbors per site; (d) Effective loss rate due to the imaginary term $ic_j^\dagger c_j$ in the renormalized effective Hamiltonian, evaluated for the mean-field solution.

three-body interactions, allowing for the exploration of complex models from condensed matter physics using cold atoms to simulate the quantum mechanical systems in question.

References

1. C. H. Bennett and G. Brassard, *Proceedings of IEEE International Conference on Computers Systems and Signal Processing* , p. 175 (1984).
2. R. P. Feynman, *Int. J. Theor. Phys.* **21**, p. 467 (1982).
3. A. Ekert and R. Jozsa, *Rev. Mod. Phys.* **68**, p. 733 (1996).
4. F. Jelezko, T. Gaebel, I. Popa, M. Domhan, A. Gruber and J. Wrachtrup, *Phys. Rev. Lett.* **93**, p. 130501 (2004).
5. L. Childress, J. M. Taylor, A. S. Sorensen and M. D. Lukin, *Phys. Rev. A* **72**, p. 052330 (2005).
6. P. Hemmer, S. Prawer, E. Trajkov, J. Wrachtrup, F. Jelezko, N. Manson and M. Sellars, Vlsi quantum computer in diamond *Advanced Optical and Quantum Memories and Computing III* **6130** (SPIE, 2006).
7. W. Hofstetter, *Phil. Mag.* **86**, p. 1891 (2006).
8. J. M. Taylor, P. Cappellaro, L. Childress, L. Jiang, D. Budker, P. R. Hemmer, A. Yacoby, R. Walsworth and M. D. Lukin, *Nat. Phys.* **4**, p. 810 (2008).

9. J. R. Maze, P. L. Stanwix, J. S. Hodges, S. Hong, J. M. Taylor, P. Cappellaro, L. Jiang, M. V. G. Dutt, E. Togan, A. S. Zibrov, A. Yacoby, R. L. Walsworth and M. D. Lukin, *Nature* **455**, p. 644 (2008).
10. D. Jaksch, C. Bruder, J. I. Cirac, C. W. Gardiner and P. Zoller, *Phys. Rev. Lett.* **81**, p. 3108 (1998).
11. L. Jiang, J. M. Taylor, A. Sørensen and M. D. Lukin, *Phys. Rev. A* **76**, p. 062323 (2007).
12. C. L. Degen, *Applied Physics Letters* **92**, p. 243111 (2008).
13. G. Balasubramanian, I. Y. Chan, R. Kolesov, M. Al-Hmoud, J. Tisler, C. Shin, C. Kim, A. Wojcik, P. R. Hemmer, A. Krueger, T. Hanke, A. Leitenstorfer, R. Bratschitsch, F. Jelezko and J. Wrachtrup, *Nature* **455**, p. 648 (2008).
14. J. R. Maze, J. M. Taylor and M. D. Lukin, *Phys. Rev. B* **78**, p. 094303 (2008).
15. E. van Oort, N. B. Manson and M. Glasbeek, *J. Phys. C* **21**, p. 4385 (1988).
16. L. Childress, M. V. G. Dutt, J. M. Taylor, A. S. Zibrov, F. Jelezko, J. Wrachtrup, P. R. Hemmer and M. D. Lukin, *Science* **314**, p. 281 (2006).
17. R. J. Epstein, F. M. Mendoza, Y. K. Kato and D. D. Awschalom, *Nat. Phys.* **1**, p. 94 (2005).
18. O. Morsch and M. Oberthaler, *Reviews of Modern Physics* **78**, p. 179 (2006).
19. C. Nayak, S. H. Simon, A. Stern, M. Freedman and S. Das Sarma, *e-print: 0707.1889* (2008).
20. N. R. Cooper, *Physical Review Letters* **92**, p. 220405 (2004).
21. H. P. Büchler, A. Micheli and P. Zoller, *Nature Physics* **3**, p. 726 (2007).
22. S. P. Jordan and E. Farhi, *Phys. Rev. A* **77**, p. 062329 (2008).
23. T. Kraemer, M. Mark, P. Waldburger, J. G. Danzl, C. Chin, B. Engeser, A. D. Lange, K. Pilch, A. Jaakkola, H. C. Nagerl and R. Grimm, *Nature* **440**, p. 315 (2006).
24. N. Syassen, D. M. Bauer, M. Lettner, T. Volz, D. Dietze, J. J. Garcia-Ripoll, J. I. Cirac, G. Rempe and S. Dürr, *Science* **320**, p. 1329 (2008).
25. B. Paredes, T. Keilmann and J. I. Cirac, *Phys. Rev. A* **75**, p. 053611 (2007).
26. S. Tung, V. Schweikhard and E. A. Cornell, *Phys. Rev. Lett.* **97**, p. 240402 (2006).
27. M. Hafezi, A. Sørensen, E. Demler and M. D. Lukin, *Phys. Rev. A* **76**, p. 023613 (2007).
28. G. Moore and N. Read, *Nucl. Phys. B* **360**, p. 362 (1991).
29. N. Read and E. H. Rezayi, *Phys. Rev. B* **59**, p. 8084 (1999).
30. L. Pitaevskii and S. Stringari, *Bose-Einstein Condensation*, International Series of Monographs on Physics (Oxford, 2003).
31. E. Braaten and H. W. Hammer, *Phys. Rep.* **428**, p. 259 (2006).
32. E. Braaten and H. W. Hammer, *Ann. of Phys.* **322**, p. 120 (2007).
33. P. Massignan and H. T. C. Stoff, *e-print: cond-mat/0702462* (2007).
34. C. W. Gardiner and P. Zoller, *Quantum Noise: A Handbook of Markovian and Non-Markovian Quantum Stochastic Methods with Applications to Quantum Optics*, 3rd edn. (Springer-Verlag, 2004).

Session VII

ITAMP History and Highlights

Lisa Bastille, James Babb, Hossein Sadeghpour, Alex Dalgarno, Kate Kirby, Mikhail Lukin

THE BEGINNINGS OF ITAMP — PROPOSALS FROM THE AMO COMMUNITY AND THE NSF RESPONSE

R. H. PRATT
Department of Physics and Astronomy, University of Pittsburgh
Pittsburgh, PA 15260

1. Introduction

The Dalgarno Celebratory Symposium included a session September 11, 2008 on the history of ITAMP (now the Institute for Theoretical Atomic, Molecular and Optical Physics), for which Alex Dalgarno is largely responsible. ITAMP history has two parts, (1) 1984-88, leading to its founding, i.e. pre-history, and (2) after 1987, during its existence. Kate Kirby and I described the first period in our talks, while she and the other speakers discussed what happened afterwards. I spoke because I was an NSF Program Director for Theoretical Physics, responsible for Atomic Theory, as a rotator (i.e. temporary, on loan from the University of Pittsburgh) in 1987-89, during the period when ITAMP was funded and founded. The pages that follow correspond to the text of my talk.

The pre-history of ITAMP has three aspects: (1) activities in the community, which can be related to the founding of TAMOC in 1984 and to NRC reports in 1986 and 1987, (2) the preparation of the Harvard/Smithsonian proposal, and (3) the response of NSF to the community and then to the ITAMP proposal. Here I will sketch the community activities and the subsequent NSF responses, while Kirby will describe the ITAMP proposal and its preparation.

2. The TAMOC Community and the NRC Reports, 1984-1987

The informal organization TAMOC (Theoretical Atomic, Molecular and Optical Physics Community), begun in 1984 and still in existence, meeting once a year at DAMOP meetings, was important in focusing attention on the possibility of an atomic theory institute. For a history of the early years of TAMOC see [1].

TAMOC first met at the DEAP (later DAMOP) in Storrs, CT May 19, 1984, attended by 32 people. It was organized by Lloyd Armstrong, Alex Dalgarno and Neal Lane, in response to concerns about the health of AMO theory, including the scarcity of AMO theory faculty positions at top universities. A theory center was proposed, to host workshops and provide high profile postdoctoral positions, as well as to set up a visitors program for informal scientific interactions.

At the 1985 TAMOC meeting five preliminary proposals for an AMO theory institute were described (U. Tennessee, Texas A & M, USC, LSU, and Nebraska/Kansas State).

In 1986 it was reported that three institutions were applying for funds to establish an AMO Theory Center (Texas A&M, Nebraska/Kansas State, Harvard/Smithsonian).

In 1987 it was stated that four universities were developing proposals for an AMO theory institute (Texas A&M, Kansas State/Nebraska, Harvard/ Smithsonian, and Cal Tech/USC).

In 1988 it was stated that there were now six proposals, with new proposals coming from LSU and JILA.

In 1989 (after the funding of ITAMP) two proposals for theory centers were described (Texas A&M and JILA).

Thus in all seven proposals were mentioned at various times by TAMOC. Not all of them were in fact submitted to NSF Physics, and most of those that were submitted were submitted only once. (Some related proposals, none of them funded, were submitted to the NSF Science and Technology Centers (STC) program.) NSF funded ITAMP in 1988 and JILA, on a lesser scale, in 1989.

In the early 1980s there was a survey of the status of physics in the US, conducted through the Board on Physics and Astronomy of the National Research Council for the National Academy of Sciences. The Physics Survey Committee, chaired by William F. Brinkman, produced a report "Physics through the 1990s", in 1986, including separate reports on the various subfields. The report of the Panel on Atomic, Molecular, and Optical Physics was chaired by Daniel Kleppner [2]. That panel activity also triggered a separate NRC report "the State of Theoretical Atomic, Molecular, and Optical Sciences in the United States" from a panel chaired by Lloyd Armstrong, published in 1987; see [3].

The 1986 NRC report of the Panel on AMO Physics (chaired by Kleppner) did very briefly mention the problems of AMO theory. The theorists on the Panel were Alex Dalgarno, Neal Lane and Joe Macek. Lloyd Armstrong was particularly thanked, singled out, by the panel for his assistance. The key statement regarding theory was:

> *A major problem is the dispersed nature of the theoretical community in the United States.... There is a critical need to focus efforts in the country, to strengthen the field, and to attract students. The panel recommends that the agencies invite and support proposals addressing this issue, for example by creating centers, workshops or summer schools ....*

The more extensive 1987 NRC report of the Panel on Theoretical AMO Science in the US (chaired by Lloyd Armstrong) in fact attracted more attention. (Alex Dalgarno and Eric Heller were among its members.) The report noted that (unlike in chemistry) the field in physics no longer had a strong representation in the major university physics departments. It noted as a major initiative under discussion in the community the formation of an Institute of Theoretical AMO Science:

> *Such an institute could serve to focus theoretical research in AMO science, and to increase the visibility of the field. We make no recommendation for or against the formation of such an institute, because we believe the appropriateness and viability of an institute will depend crucially on the specific details of its operation.*

The report however then goes on at some length to list the criteria such an institute should satisfy (broad approach, interactions with experiment and among workers from different subfields, research at a level of novelty not possible for an isolated individual researcher, associated with and involving personnel from a host university). Finally:

> *Obviously, formation of such an institute would require that some agency, or agencies, provide considerably increased funding to this field. Such an institute would be quite expensive by the present standards of funding in the field, probably costing from \$0.5M to \$1.5M per year to operate at a meaningful level. As the current funding levels in the field are already inadequate, a significant portion of these operating costs could not be obtained by reallocating present funding without causing severe damage to the field.*

3. Some Relevant NSF Activities Prior to September 1987

Previous related Institutes that had been funded by NSF Physics included ITP in Santa Barbara and JILA in Boulder.

At the 1985 TAMOC meeting Norm Bardsley, then the rotator at NSF handling atomic theory, "reported on the funding possibilities at NSF for an AMO theory institute" [1].

Two proposals to NSF Physics for atomic theory institutes (I am not allowed to say which) were submitted for consideration during 1986-87, while Peter Mohr was the NSF rotator responsible for atomic theory. One was very large in its concept, the other more modest. They did not review especially well, and both were declined. (I do not know whether funds could have been found for a stronger proposal.) Two additional proposals had been broached, and would subsequently be submitted.

As of when I came to NSF, toward the end of September 1987, the cast of characters at the agency had largely changed. Erich Bloch remained Director of the National Science Foundation. But Rich Nicholson was the new Associate Director for Mathematical and Physical Sciences. Marcel Bardon, the Director of Physics, was away on leave at NATO, and he was replaced for the year by Gary Crawley, from Michigan State. I replaced Peter Mohr, who went to NIST. (Also Boris Kayser, the only permanent person in the Theoretical Physics Program - handling high energy theory - was away for the year, replaced by Vic Teplitz.)

As I shall describe, both Eric Bloch, Director of NSF, and Gary Crawley, Acting Director of Physics, played a more personal role in the decision to fund ITAMP than might have been expected, and they deserve credit.

4. NSF September 1987-October 1988, A Chronology

The Harvard/Smithsonian proposal for an Institute for Theoretical Atomic and Molecular Physics was initially received at NSF February 8, 1988. It was approved and funded effective November 1, 1988 [4].

Some chronology of what happened during that year is available in weekly personal letters I wrote to my family, who had stayed in Pittsburgh while I commuted weekly between Pittsburgh and Washington. (In general I spent four days and three nights a week in Washington, three days and four nights a week in Pittsburgh.) NSF was then located at 18th and G Streets, close to the White House.

I give below excerpts from those letters [5]. Dates are dates of the letters.

4.1. ***Chronology***

September 25, 1987
I was sworn in and spent the next three days with my predecessor Peter Mohr learning about the job. ... Met this year's Physics Director, Gary Crawley of MSU.

October 11, 1987
The annual physics allocations meetings, also attended by Nicholson, new Directorate head ...Considerable discussion of whether theory is under-funded in comparison to experiment

November 1, 1987
The twice-yearly meeting of the Physics Advisory Committee... recommending a change in the theory/experimental balance... I succeeded in fairly well clearing my desk, also proceeding with discussions for new initiatives (like an institute) in atomic theory - this may be my major issue for the year.

December 28, 1987
I left Maine. ...Alex Dalgarno of Harvard Smithsonian met me and we drove to Harvard. I first met with Gerry Gabrielse, recently moved from U Washington, and then met with the Smithsonian theory group (they had brought in sandwiches for lunch) to discuss their proposal for a theory .institute. Some touring, meeting with Shapiro, the Director. (I also visited the other group in the process of submitting an institute proposal)

January 19, 1988
Met Bob Schrieffer (Nobel laureate of BCS superconductivity fame), Director of the ITP, to which he was returning after a Los Alamos visit, discussing potential relations and conflicts between ITP and the new theory center proposals.

January 25, 1988
Our theory group met with Gary Crawley to discuss allocation of 600K "extra" he was giving us: I will get 250K toward an atomic theory institute, Bill McDonald will get 250K for new nuclear theory funding, and Vic Teplitz will get 100K toward the new cosmology initiative. We still don't know fully the funding which will be available.

[There were issues in how theory funding was allocated among the components of the theory program.]

[The Smithsonian proposal was submitted Feb 8.]

March 6, 1988
Review of the theory program by our outside panel. (I discussed during the week that the institute proposal submitted through the Smithsonian is going to involve us in all sorts of policy problems of one government agency funding another, even though that is not the reality.)

April 11, 1988
To Denver ... to the JILA tower, seeing Gordon Dunn (the Chair) and Dave Norcross (main theorist), as they prepared for the next day's building dedication and 25th anniversary celebration. It turns out they had thought there was no point (in terms of money) to propose a theory institute this year, were surprised I may be able to fund one, but thought I should go ahead without them.

[The referee reports on the two proposals were mainly received mid March to mid April.]

May 4, 1988
The Associate Director of NSF, Rich Nicholson, came by to learn why Livermore was concerned about atomic theory - Bloch was giving a speech there Wednesday and LLNL was having him meet Teller to discuss these issues. .... Crawley and I met Nicholson to discuss problems in dealing with the Smithsonian on an atomic theory institute proposal.

May 16, 1988
A major upset developed when Bloch was called to Capitol Hill by (XX) Senators & Congressmen to complain about reviews of last year's (XX) center proposal which said (4 out of 8) that (XX) was not a geographically suitable location. Crawley and I met with Nicholson, and we may need to meet with Bloch. (I am not allowed to indicate which state and which proposal was involved.)

May 22, 1988
I arrived at NSF to discover that Rich Nicholson and I had an early afternoon meeting with Erich Bloch, Director of the Foundation, to discuss the Congressional upset over last year's treatment of the (XX) Atomic Theory Institute proposal. (Bloch was going out of town immediately after our appointment, and didn't want to wait until after Gary Crawley got back.) Bloch decided that the process had been flawed, he rejected the idea of a new decision on the proposal and told us to stop any action (on Harvard) this year, and send out a formal notice requesting proposals for next year. My colleagues set to work to propose how to spend the $250,000 I had been holding.

Tuesday afternoon Crawley and I met with Nicholson. Crawley decided to ask Bloch to reconsider his decision, and I outlined a message for Crawley to send before leaving for the airport to fly to California ... I talked with Art Toor & Norm Bardsley [at Livermore] about atomic theory institutes, including results of the meeting between Edward Teller and Erich Bloch.

May 31, 1988
I found that Bloch had agreed to a compromise, where (XX) has a month to submit a new proposal, which could lead to funding an Institute by early fall.

[The budget level was not to exceed $650K. The new proposal came in by the end of June, and I had all reviews before the end of July.]

August 8, 1988
The week began with our site visit to Boston. ... Gary Crawley, coming from Washington, met me. ... there was time enough to accomplish our objectives After lunch we walked over to the main campus from Garden Street, talking to the Dean, the Physics Chair and faculty...During the week I continued to work toward a possible Institute award; today I may get a go-ahead to write a funding recommendation.

August 15, 1988
This week I finally got a go-ahead on the Harvard theory institute project. Monday was the first day back from vacation for both Nicholson and Bloch. Our meeting with Nicholson was postponed to Tuesday morning, and he said "fine". He and Crawley briefed Bloch that afternoon in the

course of a meeting on an unrelated subject (building a gravity wave detector). I stayed up until 5 am Wednesday night writing the analysis and recommendations for the project. The Harvard people were to Federal Express in their revised documents Friday. If no further problems develop I hope to have everything on Crawley's desk before I leave Thursday for Maine and Romania, he will forward them to Grants & Contracts before he leaves NSF to return to Michigan State a week later, and the award could be made by the end of September. (I have also gotten agreement for my plan for consideration of further proposals.)

August 22, 1988
The theory institute action got through the physics office & to Crawley's desk; he reviewed it with Marcel Bardon (the returning director, who dropped by Thursday), and presumably it is on its way to Grants & Contracts. This will be Crawley's last week, and then he returns to Michigan State.

September 19, 1988
Marcel Bardon, the returning physics director, came by as soon as I arrived from the airport last Monday to report that we are supposed to make a report on the Atomic Theory Institute to the National Science Board in October - they are concerned about the establishment of new (permanent) activities like "institutes".

October 16, 1988
Thursday morning I presented the atomic theory institute to the program committee of the Science Board as an informational item. We learned that Grants & Contracts signed the award Wednesday, in anticipation of the meeting. The Board talked to me 30 minutes rather than the intended 15. Generally friendly, though one member expressed doubt it would be useful for industry and (I think) would have liked an opportunity to vote on it. (They only saw it at all because of the use of the word "Institute".)

October 31, 1988
Occupied with the Physics Advisory Committee. I made a presentation on our new atomic theory institute, which surprised them as they did not feel properly informed. .. Thursday I wrote declination letters & analysis for the other atomic theory institute proposals.

November 20, 1988
Discussions with the public affairs office of a press release on the atomic theory institute

[The press release was issued Dec 2.]

5. Further Comments

The ITAMP proposal was signed by me August 16, by Crawley August 26, but then also by Bardon September 2, after his official return. The official award letter was dated October 12, with a start date of November 1. A press release was issued December 2.

On August 11 Bloch had agreed (1) to proceed, (2) to the two grant approach (one to Harvard and one to Smithsonian) as consistent with PAM, and (3) to the strategy of considering further proposals later in the year. He said to deal with the declinations after the new award had been made.

The initial award for the start-up year was $300K, intended to increase to $600K in the next year, and to continue to scale up to $750K by the fifth year. The awards beyond the third year were contingent on progress in filling the full professor position in the Harvard Physics Department, for which the Smithsonian was paying.

The proposal had initially been submitted only through Smithsonian, February 8, 1988. Although content was unchanged, budgets were split and resubmitted August 12, 1988, to achieve funding consistent with rules on funding to federal-related institutions (PAM), while also taking advantage of Smithsonian overhead policies for visitors.

The 1987-88 proposals were reviewed by the same referees as the 1986-87 proposals had been, except for one referee who was out of country and not available, plus three new referees, for a total of ten. The ITAMP proposal was the clear first choice, receiving mainly "excellent" ratings, although a couple referees were not convinced as to the need for an institute. One of the other proposals of that year was also well regarded, and a number of the referees encouraged trying to fund it also. The commitment of a full professorship at Harvard was an important factor for a number of the referees in their preference for ITAMP.

I mentioned the 1986 NRC report and quoted extensively from the 1987 NRC report in justifying my recommendation to fund ITAMP.

It is interesting to note, both in terms of the original NRC reports and the later success, that the original ITAMP proposal did not mention workshops.

It should be noted that Gary Crawley played an important role, both in providing most of the initial year funding (and committing the future), and in successfully appealing Bloch's initial May decision to delay any funding a year in view of the flaw in the process of the preceding year.

I have no idea as to the importance of Erich Bloch's discussion of atomic theory with Edward Teller in the successful outcome. I do not know why the discussion came about, and I do not know its nature, even though I apparently heard about it when I was out at Livermore consulting a couple weeks later.

References

1. http://www.aps.org/units/damop/resources/tamoc/history, A (Brief) Early History of TAMOC (1984-1994), prepared by Deborah Watson (private communication),
2. Atomic, Molecular, and Optical Sciences (1986), National Academies Press, Panel on Atomic, Molecular, and Optical Physics, Daniel Kleppner, Chairman, Committee on Atomic and Molecular Science, National Research Council.
3. The State of Theoretical Atomic, Molecular, and Optical Science in the United States, National Academy Press (1987), Panel on Theoretical Atomic, and Optical Sciences, Lloyd Armstrong, Jr., Chairman, Committee on Atomic and Molecular Science, Board on Physics and Astronomy, National Research Council.
4. Original NSF proposal "jackets" for awards 8812592 and 8821409, retrieved from U.S. Government Archives by B. Schneider (private communication).
5. R. Pratt, personal letters (unpublished). Available in University of Pittsburgh Archives.

HISTORY OF ITAMP: PART II – THE INSTITUTE STRIKES BACK

BARRY I. SCHNEIDER
Physics Division, National Science Foundation
Arlington, VA 22230

1. Introduction

I want to thank the organizers for the opportunity to attend this scientific meeting in honor of Alexander Dalgarno on this, the 80th year of his birth. Alex has been a wonderful colleague, a truly extraordinary physicist, and a real friend to many of us in this room. I sincerely hope that Alex will continue to produce the science we have all seen from him, for many years to come. The organizers asked me to speak about ITAMP—in particular the years following its birth, which were so well described by Dick Pratt in the previous talk. I took over the responsibility of running the Theoretical Atomic Physics Program at the NSF from Dick in FY90, coming as a rotator from my home institution, Los Alamos Scientific Laboratory. I went back to Los Alamos after that first year and John Cooper from NIST manned the program for FY91. I returned to the NSF as a permanent program director in October of 1991 and have been in the Physics Division since then.

2. The First Five Years

After ITAMP was founded, significant accomplishment in itself, Alex and ITAMP were faced with many challenges. The award stipulated that a new Professor of Physics be recruited into the Harvard Physics Department in the area of theoretical atomic and molecular physics and that the individual succeed Alex as the next director of ITAMP. In addition, there were space problems at 60 Garden Street and in the physics building at Harvard. The separation of the Smithsonian from the Harvard physics building was another difficulty facing the fledgling institute. Given that some members of the AMO community were skeptical of the need for such an institute, there was obviously a lot of work to be done to ensure that the Institute would be the success that Alex and the NSF envisioned. Running a viable visitor program and recruiting talented graduate students and postdocs is hard enough. Coupled with issues such as finding suitable candidates for the new professorship and dealing with lack of space only exacerbated the situation.

For one reason or another, the recruitment process took a very long time. Whether because of the inherent inertia of such procedures, or because the Harvard administration was still not fully committed to the conditions of the award, is unclear to me. I do know it took a long time, and on several site visits there were discussions about the need to move forward. Clearly Alex and the other ITAMP people were doing their best to jump start the process. Finally a set of candidates were identified, a finalist chosen and an offer made. Unfortunately, the candidate rejected the offér and, according to Harvard rules, the process had to be started all over again.

One good thing to emerge from the initial search was that faculty members in the Chemistry Department, such as Bill Klemperer and Dudley Herschbach., became interested in ITAMP. This should have been the case from the outset as there is a close connection between theoretical atomic and molecular physics and many aspects of theoretical chemistry. This had been pointed out in the 1987 NRC report on the State of Theoretical Atomic and Molecular Physics, which discussed the potential impact of an institute of TAMOP on the health and vibrancy of the field.

The widening of the search to include candidates from theoretical chemical physics was, in my opinion, a positive development. Eventually Harvard made an offer to Rick Heller, which he accepted. Rick joined Harvard sometime in 1993 and was appointed director of ITAMP soon after. In retrospect, it is interesting that Rick had come full circle, beginning as a graduate student at Harvard under the supervision of Bill Reinhardt in the Chemistry department and returning twenty years later as professor of physics and director of ITAMP. In fact our paths had intersected on a number of other occasions, first when I tried to convince him to come to Los Alamos as a postdoc, and later when he and I were both staff members in T-12, the Theoretical Chemistry Group in T-Division at LANL. So now ITAMP had a new director and at least one of the early stipulations of the award had been satisfied.

Before ending this part of the talk, it is essential to recognize the important contributions made by Kate Kirby to ITAMP. She was there on a day-to-day basis to make sure that the visitor program functioned smoothly, that workshops were properly organized and that graduate students and postdocs were given the support they needed to be productive. Kate's involvement only increased over the years and it is beyond doubt that much of the Institute's success can be attributed to her contributions.

3. The Second Five Years

During the next five years, ITAMP moved into full gear. The long- and short-term visitor program was flourishing, more workshops were organized, the postdoctoral program began to achieve some real successes and ITAMP played a key role in catalyzing theoretical AMO science here in the US, as well as providing a place where AMP scientists from abroad could spend productive time. ITAMP also became the fulcrum for atomic and molecular science in the Cambridge area, organizing activities such as the Cambridge Area Open House, where people from neighboring institutions could get together to discuss their work in an informal setting. Regular seminars in AMO physics were also organized at Harvard and MIT. ITAMP had clearly established itself as an important community resource and with the exception of the continuing issues of lack of space and the separation of Harvard-Smithsonian Center for Astrophysics from the Harvard Physics Department, things were going pretty well.

Returning for a moment to Alex—although he was no longer director of ITAMP, he stayed involved, particularly on the scientific end, mentoring postdocs and graduate students and providing his usual source of knowledge and ideas for anyone wishing to benefit from his expertise.

A few words about the NSF perspective on ITAMP are in order here. Looking back at the written record, it is clear that PHY was very supportive of ITAMP during this formative period. While the reviews of ITAMP were always solid, there were skeptics. And since ITAMP comprised 20 percent of the Theoretical Atomic and Molecular Physics Program budget, it was always a target for budget cuts in tough times. However, based on the reviews and the panel discussions, the community was getting a good return on its investment. There was one period in which ITAMP did suffer a substantial reduction in funding due to a general NSF funding cut. But those funds were eventually restored, and I believe that in the long run the cuts had no negative impact on the Institute.

4. The Third Five Years

By the end of the first ten years of the establishment of ITAMP, Rick Heller decided to step down as director of ITAMP. He wanted more time for other pursuits and hoped to establish a closer connection with the Chemistry Department. While he continued his involvement with ITAMP, he began to devote more of his time to research, as well as supervising students in the two departments with which he was connected.

Rick has an artistic side. At Los Alamos I remember him as a serious photographer, using an 8 by 10 view camera to take pictures of New Mexico. This led to his blending of science and art in a fascinating way. A number of his works grace the hallways of the NSF, as well as other more reputable artistic venues.

The challenge facing ITAMP was to find a new director, and a search was initiated. I have some impressions concerning the search, which may or may not be accurate. It was unclear whether or not the Harvard Physics Department felt an obligation to create another full professorship position in AMO theory. After all, Heller was still there and a member in good standing of the department. They had fulfilled their obligation. On the other hand, the Smithsonian appeared to be supportive of ITAMP and willing to continue funding the directorship of the Institute. The search identified several promising candidates, one of whom had to withdraw due to serious health issues. Another leading AMO theorist was selected but ultimately could not be convinced to leave his present institution. Eventually the decision was made to end the search for the present and Kate Kirby was appointed as Director in 2001. Kate took over, and as history shows, provided and continues to provide outstanding leadership for ITAMP.

During this period, ITAMP underwent another review, and while the community was still supportive of the Institute, there were some lingering issues. It was clear that the postdoctoral program was a resounding success. The ITAMP postdocs were selected from among the best graduate students in the country. ITAMP not only gave these talented people the flexibility to work in a host of different areas of AMO theory, it also exposed them to numerous visitors with diverse interests and a top-notch local faculty at Harvard and Smithsonian, not only in AMO physics but other areas of physics as well. The training they received at ITAMP produced the most attractive candidates for academic and laboratory positions in the US, and it became evident that a high percentage of ITAMP "graduates" were being offered these positions. Clearly, a big success.

On the other hand, the graduate student program appeared to be much less productive and its funding was cut during one of the renewal reviews. The visitors program also faced challenges. ITAMP found it more and more difficult to attract really outstanding long-term visitors, particularly US scientists. Two-career couples are unwilling to separate, and families face the additional difficulty of uprooting their children to a new environment for long periods of time. ITAMP is focusing on these changing demographics as they try to come up with creative solutions for these difficult problems.

5. Present

Misha Lukin, who first came to ITAMP as a postdoc, was appointed Assistant Professor of Physics at Harvard. His expertise in quantum optics and other aspects of cold atom physics and quantum information theory made him an extremely attractive candidate. His outstanding research enabled him to rise quickly to a full professorship and to enlarge his scientific repertoire to include experiment as well as theory. A few years ago Misha agreed to become co-director of ITAMP with Kate Kirby.

It is now almost twenty years since ITAMP was founded. It continues to be an important element of TAMOP in the US, and a resource and model for other countries. Many people, including some in this room, were involved in laying out the rationale for the creation of ITAMP. Others, such as Dick Pratt, and Gerry Crawley, worked hard to get the NSF to recognize the importance of such a resource for the AMO community, fought for additional funds to make it happen and dealt with some of the bureaucratic obstacles that could have quickly killed the idea. Others, like myself, have played a role in stewarding ITAMP, as well as being productively involved in workshops here at the Smithsonian. However, without the vision and dedication of Alex Dalgarno, who persisted under less than ideal circumstances, none of this would have happened. Happy birthday Alex, and may the force be with you!

ITAMP HISTORY: PART III

KATE KIRBY

ITAMP, Harvard-Smithsonian Center for Astrophysics,
60 Garden St., Cambridge, MA 02138

At this 3-day symposium, we have all come together to honor the scientific contributions of a remarkable scientist, Professor Alex Dalgarno. Because Alex's scientific reach has been so broad, many of us may know only a small slice of the science that Alex has impacted. For instance, if you are an astrophysicist, you may be aware of the important work Alex has done in molecular astrophysics, early Universe chemistry, or on the origin of cometary x-rays, but you may not be aware of his seminal work in atomic and molecular scattering theory, or his work on long-range forces, or his exceptional contributions to the understanding of atomic and molecular processes in the terrestrial ionosphere and thermosphere.

However, what should not be forgotten is Alex's extraordinary legacy benefitting the entire theoretical Atomic, Molecular and Optical Physics Community—the creation of the Institute for Theoretical Atomic, Molecular and Optical Physics (ITAMP) at Harvard University and Smithsonian Astrophysical Observatory.

Before ITAMP there was TAMOC (Theoretical Atomic, Molecular and Optical Community), as has been noted previously. The first meeting, held at the University of Connecticut (Storrs) took place during the American Physical Society's DEAP (Division of Electron and Atomic Physics of the APS, later to become DAMOP) meeting in 1984. Two things about that DEAP meeting stand out in my memory:

- torrential rains throughout the meeting
- my considerable trepidation, as I was taking over as secretary-treasurer of the Division from Dr. Pat Dehmer, and I was realizing that it would be a huge amount of work.

The second meeting of TAMOC took place in 1985 at the DAMOP meeting in Norman, Oklahoma, and two things stand out in my memory of that meeting:

- exceptional heat (high 90's Farenheit)
- I had just given birth to my 4th child and I brought him with me to the meeting—so I was very busy.

At both TAMOC meetings, discussion focused on the issue of establishing a center devoted to fostering the field of AMO Theory.

As various proposals for theoretical institutes at other institutions materialized over the next several years in response to concerns about the welfare and future of atomic and molecular theory, I remember urging Alex most enthusiastically to submit a proposal. When he did, the proposal formalized what already existed in our theoretical atomic and molecular physics group here at Harvard-Smithsonian Center for Astrophysics, in that we always had a number of postdocs and graduate students, as well as a constant stream of stimulating visitors (especially in the summer). Many visitors were former students and postdocs of Alex, as well as on-going collaborators.

The original proposal was submitted through the Smithsonian Astrophysical Observatory (SAO) because its overhead at that time was considerably less than that of Harvard University. There was probably some interest as well on the part of CfA director, Dr. Irwin Shapiro, in trying to establish a precedent for NSF funding SAO science. However, as we have heard, this ultimately caused a problem, and the final decision was to propose approximately half the program on the Harvard side and half on the SAO side (with the SAO side charging no overhead for the Visitor program and the postdoctoral fellowship stipends). We still continue to maintain this joint institutional proposal—always trying to maximize programmatic funds by minimizing overhead.

The proposal was funded starting in November 1988, and so this is also ITAMP's 20th Birthday Year. Alex served as Director for the critical first five years of ITAMP's existence.

A prominent goal in the first proposal was to be the appointment of a senior Atomic and Molecular theorist to an SAO federal position with a professorial appointment in the Harvard Physics Department. Alex had the unenviable task of organizing and chairing the Search Committee of Harvard physics faculty. There were very few AMO physicists in the Harvard Physics Department at the time—Roy Glauber, Jerry Gabrielse, Frank Pipkin and Tim Chupp. Norman Ramsey had retired in 1986. The process did not go smoothly or rapidly. After a lengthy process the first offer was declined. But Alex and his committee persisted and the search process was restarted. Ultimately we were very fortunate to be able to hire Rick Heller from the University of Washington. Rick took over (as planned) as ITAMP's second director in the fall of 1993, and immediately attracted a large number of physics graduate students to his group.

The evolution of ITAMP over the past two decades has been quite remarkable, and many of its most positive aspects bear the thumb-print of its creator—Alex Dalgarno.

For example, reflecting the breadth of research interests that has always been a hallmark of Dalgarno's work, ITAMP prides itself on featuring a broad spectrum of research in its workshops. ITAMP attempts to bring together diverse communities within and outside of AMO Physics to focus on specific topics. Earlier this year we organized a workshop on Atomic and Molecular Physics of the Early Universe. The various fields of physics beyond AMO include: Astrophysics, condensed matter physics, quantum information science, medical physics, nuclear physics, theoretical chemistry, and mesoscopic physics.

Again, taking the lead from Dalgarno, ITAMP has always recognized that its people are its most important aspect. We have had a succession of exceptional ITAMP-funded and ITAMP-associated postdocs, each contributing to the intellectual vitality of the Institute, starting with the first ITAMP postdoc, Dr. Jim Babb, who is now a research physicist at SAO and ITAMP scientist as well. Later, Dr. Mikhail Lukin, an ITAMP postdoc from 1998 to 2001, was appointed to an assistant professorship in the Harvard Physics Department in 2001, and then 3 years later to tenured full professorship at Harvard. Since 2006 Misha has served as co-Director of ITAMP.

Almost all of our other postdocs have gone on to take research and teaching positions at outstanding academic and research institutions around the world: ETH Zurich, Georgia Institute of Technology, University of Copenhagen, LPTMS (CNRS, Orsay, France), Max Planck Institute PKS (Dresden), University of British Columbia, Ben-Gurion University of the Negev (Israel), Los Alamos National Laboratory, Argonne National Laboratory, University of Connecticut, Temple University, University of Montana, Kansas State University, JILA/University of Colorado, Penn State, University of Nevada - Las Vegas, University of Nevada - Reno, and so on. And the wonderful thing about ITAMP postdocs is that they come back and visit, and organize workshops, and send their students and postdocs to visit. They all have become part of a vital network of researchers in the AMO Theory community which Alex helped to establish.

Finally, in our list of ITAMP programs, I come to the Cambridge-area AMO Physics Openhouses. ITAMP started these in 1995 as a way to support and stimulate a community of AMO physicists in the Cambridge area by featuring an afternoon of talks by postdocs from different research groups at Harvard and MIT, and then followed by a social hour/pizza dinner. A number of these were held in partnership not only with the Harvard-MIT Center for Ultracold Atoms, but also with the UConn physics department. Then in April of this year, the Cambridge-area AMO Physics Openhouse took place on a Friday, jointly sponsored by ITAMP, CUA, UConn, Yale, and Boston University. Over 80 posters were featured and close to 250 students, postdocs, researchers, and professors participated. Remarkably, this attendance is about 10–15% of that at the annual Division of AMO Physics (DAMOP) meeting.

I have focused on just a few highlights in ITAMP's evolutionary history. Alex Dalgarno's original proposal launched an exceptional enterprise which still continues to flourish. And fortunately Alex continues to flourish and to contribute to the intellectual vitality of ITAMP programs. Happy Birthday, Alex.

Neal Lane, Alex Dalgarno

ALEX DALGARNO SYMPOSIUM DINNER REMARKS

NEAL LANE

James A. Baker III Institute for Public Policy and
Department of Physics and Astronomy, Rice University, Houston, TX 77025

It is an honor to be invited to speak at this special symposium to honor a truly extraordinary scientist, mentor, colleague and human being, Alex Dalgarno.

Belfast and Queen's

I want to begin by providing some personal background. My connection with Alex really started in Belfast. In the late summer of 1964, with a fresh physics PhD in hand from the University of Oklahoma and a one-year NSF postdoctoral fellowship, I left Norman Oklahoma for Belfast to spend a post-doc year with Alex in the Applied Mathematics Department of Queen's University of Belfast. I was accompanied by my young wife, Joni, and two-year-old daughter, Christy. While I knew who Alex Dalgarno was, I was not sure where Belfast was. But we found it and settled in before the short days and long nights of Ulster winter arrived.

Belfast in 1964 was something of a shock to a couple of kids from Oklahoma. Most winter days were cold and damp, with morning and evening thick fog that had me hugging the curb—dodging parked cars—while more seasoned Ulster drivers whizzed by. Food in 60's Belfast was variable. On occasion, several of us would lunch at the Stepaside Restaurant, just down the street from the Department. There, for two shillings (28 cents US in those days), one could purchase "beans on toast", should anyone want "beans on toast". But you couldn't get a cheeseburger for any amount of money. My fellow physicists were sports minded. I was invited by my officemate, Ray Flannery, to try a game of squash. He didn't tell me that in walking from the showers to the lockers, one had to walk past open windows through which the bracing breeze from the Irish sea flowed freely. Dress was an issue. Oklahoma was a casual place, even on campus. But in Belfast, even the gardeners wore coats and ties—and I wanted to be at least as well dressed as the gardeners. On Sunday, my wife and I took our daughter to the park, only to find that the swings were chained up. We had not realized that swinging, at least on Sunday, was sinful. Belfast was well patrolled by police—large humorless red-faced men who were members of the Royal Ulster Constabulary. This was not Norman Oklahoma.

But, all that said, Joni and I had no problems in Belfast. Our son, John, was born in Belfast the following January—family planning! And, Queen's University was an intellectually exciting and friendly place, especially the Department of Applied Mathematics, occupying two unassuming row houses just off Botanic Avenue. I had the very good fortune not only to learn a great deal of physics but to make new friends and colleagues, including the Department's Chairman, David (later Sir David) Bates, Ray Flannery, Ron Henry, Ken Bell, Arthur Kingston, Mike Rudge, as well as some fellow Americans, Jim Brown, George Victor and Bob Mapleton.

I felt immediately welcomed by everyone in the Department; and Joni and I enjoyed the warm hospitality of Alex and his family, David and Barbara Bates, and everyone else who made our time there so pleasant. Ray Flannery, my office mate, showed great kindness and patience dealing with this kid from Oklahoma, teaching me squash, and helping me learn the Ulster version of English. In contrast to the "Stepaside" restaurant, the food served in the homes of the Dalgarnos, Bates, Jim and Gail Brown, George and Sally Victor and other new friends at Queens was excellent—not cheeseburgers, but excellent.

Queen's was also where I think I first learned of the surprising two-body potential barrier—not due to a curve crossing—in the interaction potential between a metastable helium atom and a ground state helium atom. It was Alex's PhD thesis, done on a hand calculator. I think he actually did the calculation twice. At Rice, my students and I used it—rather the physics underlying it—to explain the observed spectra of excited He in liquid helium. Tom Hartquist, who later studied with Alex and is a presenter at this symposium, worked on that project as a Rice University undergraduate.

My wife, Joni, and I often talk about our Belfast days. It was a very special time in our lives; and we have only the most positive recollections of Northern Ireland, Belfast—fog, chained swings, and all—and of Queen's University and the many friends we made there. Indeed, those are among our most treasured memories.

But the purpose of this symposium is to acknowledge Alex's extraordinary contributions to science—an incredible record of scientific achievement. And we are here to honor Alex, the person, as a colleague, mentor and friend.

Alex's Influence on My Career

The organizers of this symposium suggested that I comment on how my early experiences working with Alex influenced my career. I can certainly imagine that Alex might not want to accept responsibility for my going into policy, although it was nice of him to mention my diversion in his paper, *A Serendipitous Journey.* But, in fact, I did learn some things working with Alex that came in handy in the policy world. I'll come back to those.

Alex's paper *A Serendipitous Journey,* which I commend to anyone who has not read it, should remind all of us that our life's journey is largely unplanned, rather a sequence of 'forks in the road' including opportunities made possible, often, by what our teachers and mentors think of us and say or write about us, at various stages of our careers.

Joni and I grew up in Oklahoma City and were High School sweethearts. We both attended O.U., Neal in physics and Joni in mathematics. I was very fortunate that a young Chinese physicist, Chun Lin, was starting his career at the University of Oklahoma, where I was in school, and for whatever reason, took me under his wing. Chun Lin was like Alex in many ways: A brilliant scientist, gifted teacher and caring mentor. Lin had laboratories as well as a theoretical group. After a few mishaps in his lab, he advised me to take up theory and handed me a copy of "Mott and Massey" (*Theory of Atomic Collisions,* by N. F. Mott and H. S. W. Massey). With his thoughtful guidance, I received my PhD in the summer of 1964.

Chun Lin knew Alex Dalgarno and recommended me to Alex for postdoctoral work at Queen's University of Belfast, after helping me to obtain an NSF postdoctoral fellowship to support my family for the year. My work at Queen's and Alex's recommendation got me to JILA, where I was appointed a Visiting Fellow. My wife, Joni, wrote many of the programs that I and many of my colleagues used in those early years. (Indeed, on one occasion when Alex was introducing me, he said "Neal is famous for a program that his wife wrote!") Those experiences—at Queen's and JILA—along with Alex's recommendations got me a job at Rice University, as Assistant Professor, making $1,000 per month! At Rice I was very fortunate to have the support of other mentors, particularly King Walters, who helped me with initial research funding, and outstanding students, who did all the work. Over the years, we published papers, gave presentations, and obtained federal research grants. But, funding for the field of theoretical atomic and molecular physics was not good; and over the years several of us began to lobby the federal agencies to do something about it. Alex, Lloyd Armstrong and I conspired to form an organization, TAMOC (Theoretical Atomic, Molecular and Optical Community), which met regularly at the time of the annual APS DAMOP (formerly DEAP) meetings. I joined the NSF Physics Advisory Committee, where my continued concerns about funding apparently earned me an appointment as NSF Division Director of Physics for the year 1979–80. I learned a lot in one year about NSF, Washington, and how difficult it is to make funding tradeoffs among different fields of physics. I had a taste of university administration as Chancellor of the University of Colorado at Colorado Springs, then Provost of Rice University. In 1993, newly elected President Bill Clinton asked me to become Director of NSF, a position I held for five years until he moved me over to the White House, as Assistant to the President for Science and Technology (the President's "Science Advisor") and Director of the Office of Science and Technology. I left Washington at the end of the Clinton Administration and am now, happily back at Rice University, with appointments in Physics and Astronomy and the James A. Baker III Institute for Public Policy. I still have my original copy of "Mott and Massey" on my shelf. Looking back, it is hard to see any anticipated pattern of career moves.

My point, in giving this personal biography, is that we are all on serendipitous journeys. One thing leads to another, opportunities present themselves, and choices are made. Most importantly, special people around us, like Alex, are there at the various forks in the road and make a huge difference in determining our future.

But in addition to getting a helping hand from Alex and my other mentors throughout my academic career, I learned some lessons that, earlier, I said came in handy, even in the policy jobs. So, let me comment on some "lessons" I think I learned from Alex.

I'll call these "lessons" Dalgarno's Laws.

"Dalgarno's Laws"

Law Number 1. Motivation matters: Choose problems that are important or at least interesting—let curiosity be a guide and be open to new directions.

Alex's own description of his life's journey from mathematics to atomic and molecular physics to aeronomy and astronomy and astrophysics illustrates the point. It is full of examples where his curiosity and ability to see beyond the problem at hand to an even more interesting one led him to open new windows of discovery.

Law Number 2. Confidence matters: Start from the assumption that nothing is impossible and go from there!

There are many examples in Alex's career where he chose to tackle a tough calculation, one not attempted before, and to devise a new mathematical technique or adapt one from another application to answer an important question about how atoms and molecules behave, both in small numbers and, collectively, in the atmosphere or regions of interstellar space.

Alex is also well known not only for squeezing the most out of his own calculation but also pointing out to other researchers what they have missed in their own work by failing to see a connection to an important unanswered question. I have had that experience myself.

Later I came to learn that this particular law—at least the version "nothing is impossible!"—doesn't apply equally to everyone. What is *not impossible* to Alex may well be impossible for some of the rest of us!

Law Number 3. Skill matters: Try a new idea and use all the tools available—or invent new ones.

All of us who have worked with Alex know his skill with mathematics and theoretical techniques.

But skill extends beyond technical matters. People skill (social skill) is also important. And an important element of that is humor—even if very dry!

I suspect you have all seen Alex use humor for effect. I recall an occasion when he was giving an invited paper to a large audience in a very long hall, shaped rather like a large "bowling alley". After his insightful talk, someone in the very back of the dimly lit hall rose to criticize Alex for his slides, saying they were entirely too small to be seen. As was his way, Alex kindly acknowledged the comment—but added something like "I'm sorry that you are sitting so far away that I can't make out who is speaking!"

Law Number 4. Patience and tenacity matter: Never give up.

Again, you will have your own stories. I recall that when we were working on some calculation—I think it was the rate constant for free-free absorption of radiation from several gases—I had gone through the derivation and had what I thought was the final equation, which I proudly took to Alex. He put it in his briefcase, and sometime in the middle of the night, looked it over. Next morning he said, with his kindly smile, that it was looking pretty good, we were only 10 to the power 20 or so apart! I strongly suspect that the error was my side; but with steadfast commitment we closed the gap. Alex gave his students plenty of rope, but he did not let them flounder long.

Alex also is legendary for his summations of conferences, where he does not brush over silliness or outright stupidity, but he makes his points with grace and dignity. Once, in a conference session Alex was chairing, during a heated exchange between a speaker and member of the audience (which promised to be endless), Alex wandered over to the blackboard and wrote down a few equations and numbers, which settled matters. I don't know, to this day, whether what he wrote actually answered the question or simply placed both the speaker and questioner in a vulnerable position of not quite knowing, themselves, and standing in front of a large audience. But, we were able to move on.

Law Number 5. People matter: Value them, treat them with respect, mentor them, and celebrate their success.

Words so often associated with Alex are accessibility, warmth and kindness. Each of us will have examples. I have already recounted how Joni and I were warmly welcomed to Belfast and Queen's by the Dalgarno's; and Alex's door was always open to his students and colleagues. Even when he was very busy, he would take the time to listen, sympathetically, about the hole you had gotten yourself into and suggest a way out of it.

But Alex would also spend social time with students and colleagues. At a conference (I think it was at Palo Alto) I recall climbing over rocks and hills with Alex while others were preparing their talks. And his prowess with sports is well known—soccer in his early days, then squash and tennis, which I experienced first-hand in Belfast. Alex's Belfast schedule was always something of a mystery to me. As best I can remember, it went something like this. He seemed to come in around 10am—we were watching his door, so we could talk with him about what had or had not worked the previous evening. Around 11am, it was time to go to the faculty Common Room, a few doors down the street, for coffee—sometimes called 'elevenses' by people in the U.K. Lunch was served about 1pm, usually back in the Common Room, where the entrée varied between Russian fish flan and American fish flan—probably the same fish, but on alternate days—followed by more coffee and conversation. Alex's working day seemed to begin in earnest around 2 or 2:30pm. But at 4pm, it was time for squash—or tennis, weather permitting. And on many evenings there were dinner parties, either at Alex's or other faculty homes. The following morning, Alex might hand you a handwritten draft of the paper that he had written sometime before sunrise (every choice of words had been carefully considered, although he often skipped many of the letters in the words!). Even with this schedule, Alex was out-publishing most of his colleagues around the world.

Law Number 6. Integrity matters: Be open and honest with everyone. This one requires no documentation. On the morality side, you know that you can always count on a straight story from Alex. And in the professional context, Alex consistently demands the highest professional standards in his work and that of others. The care he puts into a calculation, the interpretation of the results, and how it is presented to the community all support the integrity of science. His over-two decades of service as editor of Astrophysical Journal (ApJ) Letters have extended those high standards to the community.

So, this is my list of lessons or "Dalgarno's Laws":

- Motivation matters - choose the right problems
- Confidence matters - assume nothing is impossible
- Skill matters - use all the tools
- Patience and tenacity matter - never give up
- People matter - human relationships are special (also mentoring young people assures the future of the science)
- Integrity matters - honesty and high standards.

These are some of the things I think I learned from my early days of science, especially my days with Alex. Indeed, throughout my career, I have tried to use the Dalgarno wisdom in the science and policy world, where data is often sparse and there aren't any equations, let alone, solutions.

There are hundreds of examples I think I could come up with from my NSF and White House days—putting together Clinton's National Nanotechnology Initiative for one—but I will spare you those. It is my opinion—

and my experience—that when applied with care, Dalgarno's laws are as valid in policy making as in science. Certainly, I am grateful to have had them at hand.

The Future of U.S. Science?

I'll close with a word about the future of U.S. science. We have an important election coming up. And it is fair to ask a question: Why should the new President and Congress—and those that will follow them—pay more attention to science than their predecessors have done?

Many of you knew Rick Smalley, long-time professor of chemistry at Rice, who shared the Nobel Prize with Rice's Bob Curl and the UK's Harry Kroto, for the discovery of buckyballs (carbon 60 or buckminsterfullerene), and who sadly succumbed to cancer in 2005. Rick often used a slide called "Humanity's Top Ten Problems for the Next 50 Years". His list, roughly in order of seriousness, went as follows: Energy, water, food, environment, poverty, terrorism and war, disease, education, democracy and population. I think if Rick were here, today, he might reorder some of these (some of us might put population higher). But I recall three points that Rick made about the list. First, all of these world challenges are connected in one way or another. Second, energy is at the top, because unless the world meets its energy needs, we will not be able to deal with the others. And third, without major (even revolutionary) advances in science and technology, we will not succeed in dealing with energy needs—or any of the others. And Rick felt that nanotechnology offered that positive outcome. Whether or not nanotechnology will turn out to be the holy grail, it is clear that science will be critical to dealing with all of these problems.

But perhaps the most important reason science needs to be on the minds of our nations' leaders is that science reminds us that truth has value and evidence is important. Moreover, science offers humanity hope—perhaps, the only hope—for the future.

Our new President, members of Congress, and the people the President brings to the White House and to head up the agencies need to understand that. Both candidates sound like they understand. But after the election, the new President will have a lot on his plate—and much of it will require an immediate response. That's always the way in Washington.

I might just add that our space program and NASA are in shambles. It will take early action by the new Administration to rescue the U.S. space program—the science and human exploration.

As a society, we are not going to make progress in removing human misery and replacing that with a decent quality of life for most of the world's population without major progress in science, which requires significant investments in research.

But having been at those policy tables, I can tell you 'it's a tough sell!'

Policy is based on smooth extrapolations—and special interests. It is assumed that nothing new will be discovered or invented. Most of the time is taken up with crises and political fighting. And, when it comes to Congress, everybody has to get something out of it. There is an old saying that "There are two things you don't want to see being made—sausage and legislation"– former Chancellor of Germany, Otto von Bismark (1815–1898).

The notion of evidence-based decision making and the reality of a long history of science-spawned technological revolutions have not made their way into policy discussions. That is why the gap between science and public policy (hence, between scientists and policy makers and politicians) is so large. We need to close that gap. And—like it or not—many more of us are going to have to get involved in policy and politics, and become what former Congressman from California, George Brown, called 'citizen scientists' or 'civic scientists' a term many of us use today.

Our children and grandchildren and those around the planet are facing a world very different than the one we lived in. We owe it to them to give them a helping hand, while we still can. However you may wish to define the 'civic scientist' that's what the notion is all about.

Alex as the President's Science Advisor

As I was thinking about these matters, I began to wonder how Alex would deal with many of the challenges a President's Science Advisor faces. What would Alex Dalgarno, President's Science Advisor do?

First, Alex loves physics, astronomy and mathematics—and he knows a lot about the universe and how it works—both the theory and what is observed. He likely would awe Presidents, Vice Presidents, White House staff and members of Congress with his passion and deep understanding of the science but also how important science and technology are to people's lives. The politicians might just respond by cranking up all the research budgets to more rational levels.

Second, Alex cares about humanity—especially people who are the captives of fear, ignorance, poverty, disease and other tragic circumstances life has dealt them—whether here in the US, UK, or somewhere else on the planet. He would undoubtedly form partnerships in Washington, and internationally, to couple the power of science and technology to sensible domestic and foreign policy actions for the US. Because he doesn't give up, he could well succeed where others have failed.

Third (but the list could be much longer), I know that Alex would not tolerate political interference with the scientific record, uses of misinformation in the name of science to score political points, or the muzzling of government scientists to prevent their efforts to inform the public. The integrity of science—the public's trust in the truth and value of science and the honesty of scientists is too important to risk.

Alex, as the Presidents Science Advisor, could generate that combination of a renewed respect for the integrity of science—the power of evidence and logic and the value of truth and knowledge—and a commitment to the well-being of the world's humanity and the planet might cast a much more favorable light on this country and its image, which has greatly suffered in recent years.

So, I think it's time to send Alex Dalgarno to the White House!

Closing

As a final note, I want to personally thank Alex for coming to Rice to attend a two-day science policy conference focused on "bridging the gap between science and society." I know it was a very different kind of meeting from the many he had attended over the years. But toward the end, he commented that he found many aspects of the conference program to be "interesting". I was greatly relieved to hear that. Maybe it means Alex might consider that White House move after all.

Thank you Alex—for all you have done for me and all of us here, and many more who could not be with us, but also for what you have done for science and humanity. That contribution will be a legacy that will stand very tall, long after all of us are gone.

So, please join me in a toast to Alex. I'll use a very short Irish one: "To Alex – May you live to be a hundred years, with one extra year to repent."